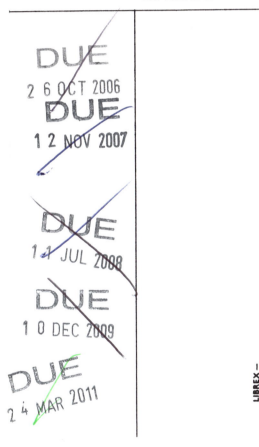

H. Mollet, A. Grubenmann

Formulation Technology

Hans Mollet, Arnold Grubenmann

Formulation Technology

Emulsions, Suspensions, Solid Forms

Translated by H. R. Payne

Weinheim · New York · Chichester · Brisbane · Singapore · Toronto

Dr. Hans Mollet (†)

Dr. Arnold Grubenmann
Chemin des Cossettes 1
CH-1723 Marly
Switzerland

Library of Congress Card No. applied for.

British Library Cataloguing-in-Publication Data:
A catalogue record for this book is available from the British Library.

Die Deutsche Bibliothek – CIP Cataloguing-in-Publication-Data:
A catalogue record for this publication is available from Die Deutsche Bibliothek.

1st Edition 2001
    1st Reprint 2004

ISBN 3-527-30201-8

© WILEY-VCH Verlag GmbH. D-69469 Weinheim (Federal Republic of Germany), 2001
Printed on acid-free paper.

Printing: Strauss Offsetdruck, D-69509 Mörlenbach
Bookbinding: Osswald & Co., D-67433 Neustadt (Weinstraße)

Printed in the Federal Republic of Germany.

Preface

What do we mean by the phrase *"formulation technology"*? The word *"formulation"* has long been established as a synonym for *"recipe"*. For many people, this term suggests something of a black art rather than an exact scientific discipline. The source of the oldest formulations is probably pharmacy, in which the skills associated with the development and execution of recipes has grown into an independent discipline, galenics. In other fields of chemistry, particularly industrial chemistry, formulations are amongst a company's most closely guarded trade secrets on account of their often considerable economic value. With a few exceptions, such as for pigments, foodstuffs, cosmetics, and agrochemicals, there is no general work covering the whole area of formulation technology; formulation chemists are forced to rely on widely scattered, although admittedly numerous, references in the most varied journals.

A large proportion of chemical substances, whether inorganic or organic, natural or synthetic, must be refined and formulated before they can be used in medicine, industry, agriculture, foods, cosmetics, and so on. Often this is merely a question of grinding and mixing; pure dyes and pharmaceuticals must also be combined with suitable auxiliary substances simply to permit reasonable dosage. However, a recipe alone is not usually sufficient; in addition, knowledge of the necessary raw materials, of their preparation, and of the application is required. Most important of all is the processing of the formulation into its optimal form for trade and for use. Here we should mention freeflowing, dust-free powders of optimal particle size, agglomerates and granulates, stable concentrated solutions and suspensions, emulsions, microemulsions, instant products, slow-release preparations, microcapsules, liposomes, and so on.

It has long been recognized that the application properties of a substance to be formulated can be improved by suitable measures, such as an increase in solubility, solubilization, division of solids into a colloidal form, agglomeration of the substance to be formulated, above all the use of efficient tensides – all these create numerous effects, improvements, and new possibilities for use in the field of formulation. Often the competitiveness in the marketplace of a synthetic product that is excellent of itself is determined by its commercial formulation, as has been demonstrated by vitamins formulated to flow freely or dust-free dyestuffs.

The art of formulation is thus a scientific discipline, with a pronounced inter-disciplinary character centered around physics, physical chemistry, colloid and interface chemistry, analysis, and not least process technology. Modern commercial forms and forms for application rely on many methods from process technology and on advanced modern analytical techniques. Thus the discipline of formulation has developed into formulation technology, which rests on solid scientific supports, and in which empiricism is increasingly being replaced by scientific criteria. This is not to say that creativity and inventiveness should lose their importance in the solution of problems and the creation of new or better commercial formulations.

A huge store of empirical knowledge is available to formulation chemists; this is useful, but not sufficient. The ability to diagnose current problems and, on the basis of accumulated knowledge, to relate them to solutions found earlier does indeed result in progress, but is not enough for a rapid and certain solution of problems of formulation. A more advanced method is the empirical deduction of relationships between the composition of a formulation and its properties and the expression of these in equations that correlate with the experimental data. Various computer-aided techniques common for correlation analysis are used for this purpose. Basically, these involve empirical trial-and-error schemes and regression methods. This methodology is very efficient if all the components of the formulation have already been selected by experiment or from practical requirements.

However, the soundest scientific approach is the understanding of the relationships between the components of a formulation and its properties (such as the stability of an emulsion or a suspension) in terms of molecular theory. Nowadays this is possible in simple cases, but not for complex systems. Simplifying assumptions have to be made, thus weakening the connection to the theory. So, for the time being, we cannot get by without empiricism. Nevertheless, awareness of the theoretical basis of colloid and surface chemistry, such as DLVO theory in the case of the stability of dispersions, may protect us from attempting solutions forbidden by theory. Whether formulation chemists are accustomed to approaching problems from a purely empirical angle or to seeking correlations between the components of a formulation and its application properties with the assistance of statistical computing methods, a knowledge of the physicochemical and technical basics relevant to formulation technology will be useful to them in making progress.

The present monograph is intended to fill the publication gap concerning the manufacture of optimized formulations, commercial forms, and forms for application. The aim of the book is a holistic treatment of the separate disciplines that play a role in the formulation of an active ingredient into its commercial form, in particular of colloid and surface chemistry and process technology, and the establishment of a coherent, interdisciplinary theory of formulation technology.

This general treatment of the subject, independent of individual products and of substance-specific formulation problems, makes up the heart of the book. Alongside it, the practical aspects of selected individual topics are summarized in order to provide an overview of the state of the art and the problems existing in these areas, such as pharmaceutical technology, dyestuffs and pigments, and cosmetics.

Finally, we wish to thank the many colleagues who have aided us in the realization of this project. First and foremost amongst these is Professor H. F. Eicke of Basel University, to whom we are indebted for his knowledgeable assistance and many suggestions and corrections. Amongst the many experts from industry who helped us with valuable contributions, we would like to single out Dr. U. Glor (Novartis), Dr. R. Jeanneret, Dr. E. Neuenschwander, and Dr. U. Strahm (Ciba SC), and Mr. A. Schrenk (Nestlé).

Thanks are also due to the publishers and authors who have allowed us to reproduce figures and tables. The references name the relevant sources and can be found in the literature sections of the individual chapters.

Hans Mollet, Arnold Grubenmann

Contents

1 Colloids, Phases, Interfaces

1.1 General Remarks

Colloid chemistry deals with systems that contain either large molecules or very small particles. With respect to particle size, they lie between solutions and coarse particulate matter; the size range is roughly 1–1000 nm, that is, 10 Å to 1 μm. This corresponds to 10^3–10^9 atoms per molecule or particle.

Interface chemistry is concerned with the phenomena and processes of heterogeneous systems, in which surface phenomena play a major role. Examples include adsorption and desorption, precipitation, crystallization, dispersion, flocculation, coagulation, wetting, formation and disruption of emulsions and foams, cleaning, lubrication, and corrosion. The specific characteristics of the interfaces which are important in such phenomena are controlled by electrochemical properties (charges) or by the use of certain organic compounds called tensides (also known as detergents or surfactants), which contain both polar and nonpolar groups in each molecule.

Table 1.1. Examples of colloidal states (s = solid, l = liquid, g = gas).

s/s	s/l	s/g
solid pharmaceutical preparations	dispersions	aerosols
	suspensions, lime slurries	smoke
reinforced plastics	latex	
magnetic tape		
l/s	**l/l**	**l/g**
gels	emulsions	aerosols
gel permeation chromato-	creams	fog
graphy separating gels	milk	spray
g/s	**g/l**	**g/g**
foam products	foam	–
aerogels	foam rubber	
foamed concrete, meerschaum mineral	whipped cream	

Examples of colloidal states are given in Table 1.1. This area of chemistry contrasts with the field of homogeneous-phase chemistry, which comprises the major part of synthetic chemistry. The usual training of a chemist concentrates on the homogeneous phase, and chemistry in the heterogeneous phase, interface chemistry, does not receive the attention that its great technical and biological importance merits. Exceptions that attract notice at the level of an undergraduate course in chemistry are, for example, adsorption processes and surface tension. It is already 90 years since W. Ostwald described colloids as a "world of neglected dimensions". Very little has changed since then; colloid and interface chemistry remain neglected disciplines in education. We intend to reduce the level of such neglect with this book; we hope that the "art" of formulation, as it once was, can thus be developed into *formulation technology*, with a scientific basis and a pronounced interdisciplinary character, the emphasis being on chemistry, physical chemistry, especially interface chemistry, and process technology.

A leading interface chemist, Pradip K. Mookerjee, once wrote:
"No school teaches about mixing things together so that they do what you want and don't react with each other."

1.2 Physical Behavior of Atoms and Molecules inside Phases and at Interfaces and Surfaces

The three phases – gas, liquid, and solid – are depicted schematically in Figure 1.1.

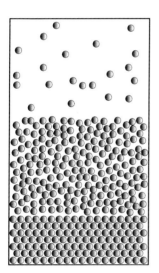

Gas: The molecules are distant from one another; little or no attraction. High mobility results in elastic collisions.

Liquid: Molecules in constant motion. Cohesive forces between the molecules influence their motion. Only in special cases are these forces sufficient to form areas of local order.

Solid: Strong forces hold the molecules in a regular arrangement.

Figure 1.1. The three phases: gas, liquid, and solid.

For a solid or a liquid to hold together, there must be strong attractive forces between its atoms. An atom inside a phase is completely surrounded by other atoms and is in a state of dynamic equilibrium (Figure 1.2).

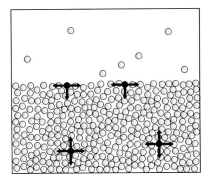

Figure 1.2. Surface forces and internal forces in a liquid.

The atoms at the surface are in a very different situation. Because of the equilibrating forces in the outer sphere, they are in a state of *surface tension*. Molecules at the surface have fewer neighbors, that is, fewer intermolecular interactions compared with the molecules in the bulk of the liquid. This leads to an attractive force normal to the surface acting to pull the surface molecules into the liquid. The surface tension γ is defined as the force necessary to counteract exactly this inward force, measured in mN (milliNewton) acting on a line of 1 m length parallel to the surface (formerly dyn acting on a line of 1 cm), and thus has the units mN/m.

$$\left[\frac{mN}{m}\right] = \left[\frac{dyn}{cm}\right] \qquad (1.1)$$

The *free surface energy* of a liquid is defined as the work necessary to increase the surface by 1 cm^2; units milliJoule/m^2.

$$\left[\frac{mJ}{m^2}\right] = \left[\frac{erg}{cm^2}\right] \quad \text{(dimensionally equivalent to:} \left[\frac{mN}{m}\right] = \left[\frac{dyn}{cm}\right] \text{)} \qquad (1.2)$$

(Note that m can stand for milli and for meter.)

The units of surface tension and free surface energy are therefore dimensionally equivalent! The surface energy is equal to the work that is required to bring atoms or molecules out of the interior of a liquid to the surface. Accordingly, the surface tends to contract; thus droplets form (smallest possible surface).

When two immiscible liquids are in contact, the forces of attraction acting on a molecule at the interface will be somewhat different than at a surface. There are interactions between the differing molecules at the interface (van der Waals forces; see below). Often, the interfacial tension γ_{L1L2} is somewhere between the surface tensions γ_{L1} and γ_{L2} of the two individual liquids (in contrast, the dispersion fraction γ_d and the polar fraction γ_p are always between the two individual contributions; see Section 1.7.2).

Example: The interfacial tension between hexane and water (γ_{L1L2}) lies between the surface tension of hexane (γ_{L2} = 18.43 mN/m) and that of water (γ_{L1} = 72.79 mN/m): γ_{L1L2} = 51.10 mN/m.

Table 1.2. Surface tension and interfacial tension with water of liquids at 20°C [mN/m]; water: γ_{L1} (from [1]).

Liquid	γ_{L2}	γ_{L1L2}	Liquid	γ_{L2}	γ_{L1L2}
water	72.75 (γ_{L1})	–	ethanol	22.3	–
benzene	28.88	5.0	*n*-octanol	27.5	8.5
acetone	27.6	–	*n*-hexane	18.4	51.1
acetic acid	23.7	–	*n*-octane	21.8	50.8
CCl$_4$	26.8	45.1	mercury	485	375

The attractive force between molecules in the bulk of the liquid is known as the *internal pressure* or the *cohesion energy* ΔE_v. The cohesion energy density $\Delta E_v / V$ is an important quantity; it is defined in Equation 1.3:

$$\frac{\Delta E_v}{V} = \frac{\Delta H_v - RT}{V} = \delta^2 \tag{1.3}$$

ΔH_v: enthalpy of vaporization [J·mol^{-1}]
V: molar volume [m^3·mol^{-1}]
R: gas constant = 8.314 J·K^{-1}·mol^{-1}
T: absolute temperature [K]

A quantity of great practical use is the solubility parameter $\delta = \sqrt{\delta^2}$ (cf. Chapter 8, Solubility Parameters). The quantities necessary for its calculation, ΔH and V, are easily obtained from reference books. The mutual solubility of two components can be determined from δ; the closer their δ values, the greater their mutual solubility.

Example: phenanthrene δ = 20.0 MPa$^{1/2}$ [1 MPa$^{1/2}$ = (10^6 N·m^{-2})$^{1/2}$]
                  carbon disulfide δ = 20.5 MPa$^{1/2}$
                  n-hexane δ = 14.9 MPa$^{1/2}$

Phenanthrene is therefore more soluble in carbon disulfide than in *n*-hexane. This rule applies only to nonpolar substances; for polar substances, see for example references [2] and [3].

1.2.1 Disperse Systems

Simple *colloidal dispersions* are two-phase systems consisting of a disperse phase (for example a powder) finely distributed in a dispersion medium. *Sols* and *emulsions* are the most important types of colloidal dispersions. The fine distributions of solids in a liquid formerly known as sols (the expression sol was used to distinguish colloidal from macroscopic suspensions) are now called suspensions or simply dispersions. Unlike these, emulsions consist of liquid droplets distributed in an immiscible liquid dispersion medium.

The classification of dispersions created by W. Ostwald over 80 years ago is still valid today (Figure 1.3). In principal there is always an inner, disperse or discontinuous phase that is immiscible with an outer, continuous or homogeneous phase.

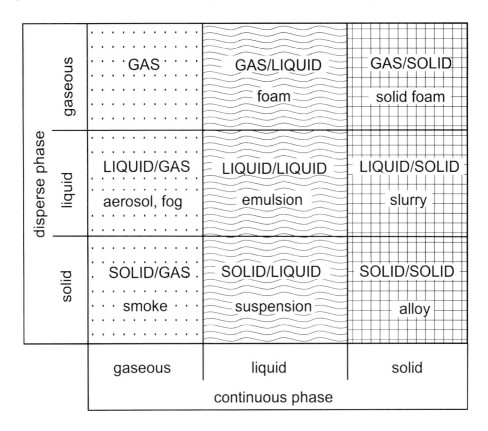

Figure 1.3. Classification of disperse systems according to W. Ostwald.

Table 1.3, which should be read as supplementary to Figure 1.3, lists a selection of typical colloidal systems.

Table 1.3. Some typical colloidal systems (from reference [4]).

Examples	Class	Disperse phase	Continuous phase
Disperse systems			
fog, spray, vapor, tobacco smoke, aerosol sprays, flue gases	liquid or solid aerosols	liquid or solid	gas
milk, butter, mayonnaise, asphalt, cosmetic creams	emulsions	liquid	liquid
inorganic colloids (gold, silver iodide, sulfur, metallic hydroxides)	sols or colloidal suspensions	solid	liquid
clay, mud, toothpaste	slurry	solid	liquid
opal, pearls, colored glass, pigmented plastics	solid dispersions	solid	solid
foam	liquid foams	gas	liquid
meerschaum mineral, foamed plastics	solid foams	gas	solid
Macromolecular colloids			
jelly, glue	gel	macromolecules	solvent
Association colloids			
soap/water, detergent/water	–	micelles	solvent
Biocolloids			
blood	–	cells	serum
Triphasic colloidal systems			
oil-bearing rocks	porous stone	oil	water/stone
mineral flotation	mineral	water	air
double emulsions	–	aqueous phase	water

Table 1.4 distinguishes between particle sizes of the colloidal dispersion state on the one hand and those of smaller molecules and of coarse heterogeneous systems on the other. These size ranges are only guidelines; in some special cases, such as suspensions and emulsions, particles of diameter greater than 1 μm are generally present. The threshold at which colloidal behavior becomes the behavior of a molecular solution lies at about 1 nm.

Table 1.4. Distinction between size ranges of the colloidal state, smaller molecules, and coarse discontinuous states.

area of defined size	heterogeneous systems; coarse discontinuities	colloids 1–1000 nm	homogeneous systems small molecules; ions
examples:	macroemulsions; dispersions	metal sols; biocolloids; macromolecules; micelles; microemulsions	water; dodecane Ca^{2+}
range of optical resolution:	magnifying glass $\rightarrow$ microscope $\rightarrow$ ultra- microscope	$\rightarrow$ electron microscope	
	1 mm 100 µm 10 µm 1 µm	100 nm 10 nm 1 nm 1 Å	

It is not necessary for all three dimensions of a colloid to be smaller than 1 µm. Colloidal behavior can also be observed for fibers, only two of whose three dimensions fall within the colloidal region; in the case of films, only one dimension does.

A cube can be divided into colloidal systems of various types (Figure 1.4): laminar, fibrillar, and corpuscular.

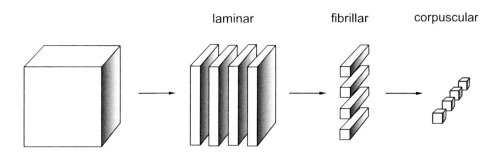

Figure 1.4. Colloidal systems (from reference [5]).

The increase in surface energy that results from this division explains the unique properties of the colloidal state.

Laminar: 1 cm^3 stretched into a film of 10 nm $\rightarrow$ total surface area 2×10^6 cm^2.
Fibrillar: 1 cm^3 divided into fibers of 10 nm $\rightarrow$ total surface area 4×10^6 cm^2.
Corpuscular: 1 cm^3 split into cubes of 10 nm $\rightarrow$ total surface area 6×10^6 cm^2.

1.2.2 The Importance of Surfaces and Interfaces

When a solid is crushed or ground, its surface area increases considerably, in relation to the degree of division. The technical importance of the process of division lies in the possibility of the manufacture of fine powders and dispersions with a very large surface area. The increase in surface area is measured in terms of the *specific surface area* S_w, units:

$$S_W = \left[\frac{cm^2}{g} \right] \quad or \quad \left[\frac{m^2}{g} \right] \qquad \text{with respect to mass} \qquad (1.4)$$

$$S_V = \left[\frac{cm^2}{cm^3} \right] \quad or \quad \left[\frac{m^2}{cm^3} \right] \qquad \text{with respect to volume} \qquad (1.5)$$

Conversion from the diameter of spherical particles:

$$S_W = \frac{6}{\rho \cdot d} \qquad (1.6)$$

ρ: true density
d: diameter [μm]

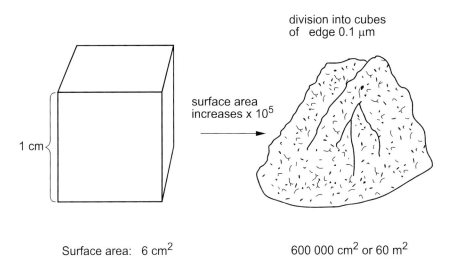

division into cubes
of edge 0.1 μm

surface area
increases x 10^5

1 cm

Surface area: 6 cm^2 600 000 cm^2 or 60 m^2

Figure 1.5. Increase in surface area on division of a cube. Powder dispersed in 6 cm^3 water. An interface of 8.6 m^2 per cm^3 of the dispersion is available (60 m^2/7 cm^3) [6].

The increase in surface area is shown by the following example (Figure 1.5): a cube of edge 1 cm is divided into cubes of edge 0.1 μm; this yields a surface area of 60 m^2 from an original area of 6 cm^2, that is, an increase by a factor of 10^5. If this powder is dispersed in 6 cm^3 of water, the result is an available interface of 8.6 m^2 per cm^3. *The more finely the solid is divided, the greater the interface between liquid and solid phases, and the more the properties of the interface determine the behavior of the resulting suspension.*

The more finely a material is divided and its surface area increases, the greater the proportion of atoms/molecules found at the surface rather than in the bulk of the material. For a cube of edge 1 cm, only 2–3 molecules out of every ten million are found at the surface. When the cube has an edge of 1 μm, there is one molecule at the surface for every 450 molecules, and at 10 nm every fourth molecule is positioned at the surface; see Table 1.5. *At dimensions less than 10 nm it is no longer possible to differentiate between surface and bulk molecules.*

Table 1.5. Relationship of surface and bulk molecules.

Specific surface area:			
S_v = surface area/volume (with respect to volume)			
S_w = surface area/mass = S_v/ρ (with respect to mass);		ρ = density	
spheres:	surface area $\pi{\cdot}d^2$	volume $\pi{\cdot}d^3/6$	$\rightarrow$ $S_v = 6/d$
cubes:	surface area $6d^2$	volume d^3	$\rightarrow$ $S_v = 6/d$

Example:	AgBr crystal	
	molecular volume	0.05 nm^3
	distance between layers	0.37 nm

In Figure 1.6, the variation in the number of surface molecules with the particle size is shown as a graph for this example. It is clear that when AgBr is divided into particles of 10 nm diameter, about 20 % of the ion pairs are located in the surface layer, whereas for particles of 0.1 μm, this is the case for only 2 % of ion pairs.

The chemical composition of the surface often differs from that of the bulk. Likewise, the arrangement of the atoms and the electronic structure at the surface are not the same as those inside the solid. The great importance of surfaces was recognized as a result of miniaturization in microelectronics. It has even been suggested that *the surface region of a finely dispersed solid should be regarded as a new phase of matter.*

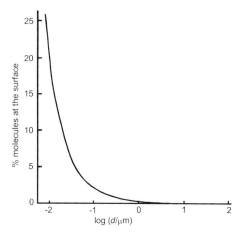

Figure 1.6. AgBr crystal: proportion of molecules at the surface as a function of particle size [4].

1.3 Some Essential Concepts in Colloid Chemistry

Monodisperse or isodisperse systems: Systems in which all particles are approximately the same size.

Polydisperse systems: Systems containing particles of varying sizes.

Lyophobic or hydrophobic colloids: The particles are incompatible with the dispersion medium, which is organic in the case of lyophobic colloids, aqueous in the case of hydrophobic. Such systems demand special methods of manufacture, in particular dispersion accompanied by a reduction in size of the particles. The thermodynamic instability of such systems is visible in their tendency to clump, to aggregate, agglomerate, and flocculate.

Lyophilic or hydrophilic colloids: The particles are compatible with the medium. They interact with the dispersion medium. Unlike the lyophobes (hydrophobes), they form spontaneously and are *thermodynamically stable*. Examples: macromolecules, polyelectrolytes, association colloids.

Amphiphilic colloids or association colloids: The molecules have an affinity for both polar and nonpolar solvents. These form the large class of surface-active and related substances that includes micelles. They are thermodynamically stable.

1.3.1 Structure and Nomenclature of Particles

For a long time there was no uniformity in the nomenclature of particles; the DIN 53 206 standard, which has attained international recognition, has now remedied this unsatisfactory state of affairs (Figure 1.7).

Primary particle or single particle	Can be recognized as individual particles by suitable physical means (e.g. light or electron microscopy). Note: As a special case, a crystalline primary particle may be a single crystal or may consist of several coherently scattering lattice domains (crystallites) which may be distinguished with appropriate radiation (e.g. X-rays).
	blocks spheres rods irregular coherently scattering lattice domains (crystallites) primary particles
Aggregate	Primary particles assembled face-to-face; their surface area is smaller than the sum of the surface area of the primary particles.  aggregates
Agglomerate	Primary particles and/or aggregates not permanently joined together but attached e.g. at edges and corners; their surface area does not differ markedly from the sum of that of the individual particles. 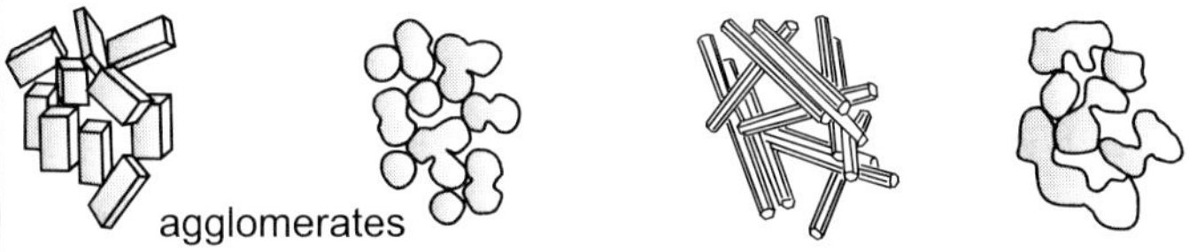 agglomerates
Flocculate	Agglomerate found in suspensions (e.g. in pigment–binder systems); can easily be separated by small shearing forces. 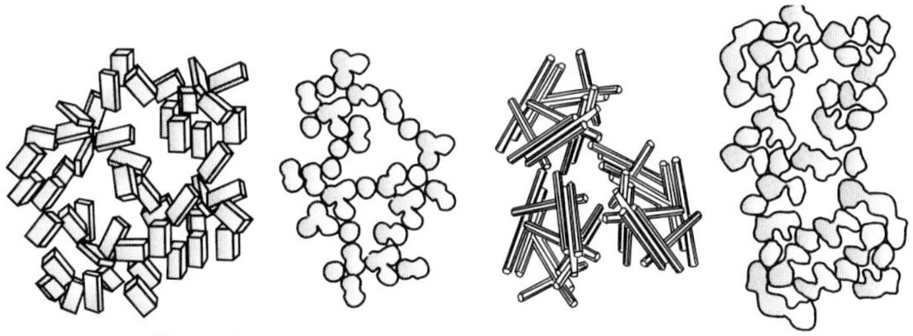 flocculates

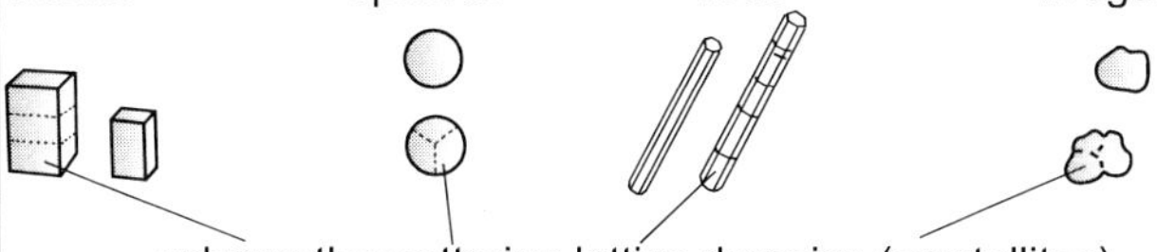

Figure 1.7. Structure and nomenclature of particles according to DIN 53 206, with the addition of flocculate.

1.3.2 Analysis of Particle Size

Since colloidal systems occur over a large range of particle size, a number of different measuring techniques are required for size analysis. For particle sizes in the range 0.001 to 100 μm the methods shown in Figure 1.8 are suitable. The most important of these are represented schematically in Figure 1.9. The choice of a valid method of measurement for a collection of particles is of the utmost importance if one wishes to avoid completely false results. Figures 1.10 and 1.11 depict schematically some additional methods for the measurement of particle sizes.

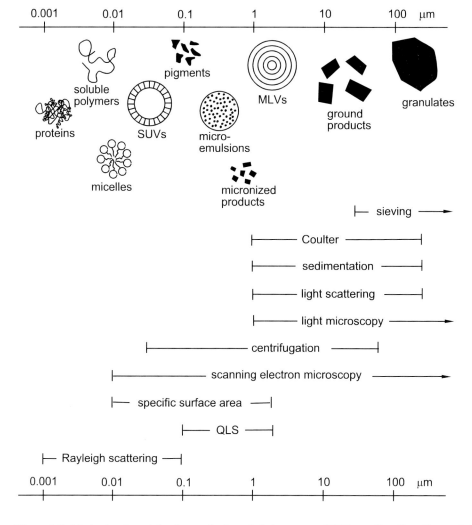

Figure 1.8. Methods of particle size analysis and their ranges. (SUV: small unilamellar vesicle; MLV: multilamellar vesicle.)

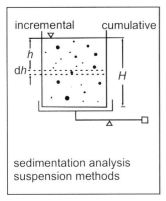

sedimentation analysis
suspension methods

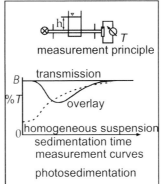

photosedimentation

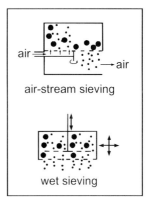

wet sieving

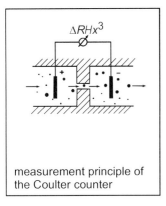

measurement principle of
the Coulter counter

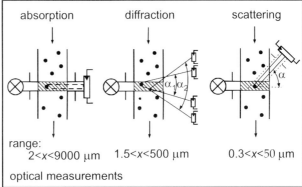

optical measurements

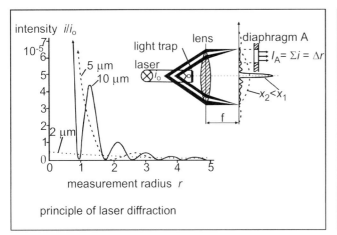

principle of laser diffraction

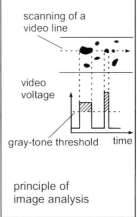

principle of
image analysis

Figure 1.9. Analysis of particle size; methods from reference [7].

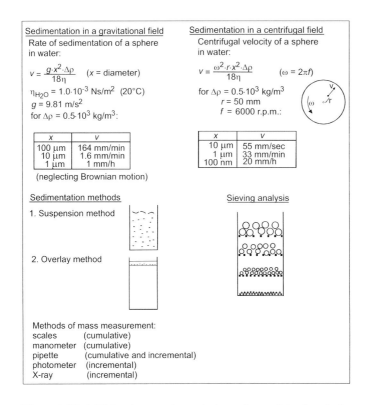

Figure 1.10. Additional measuring techniques for particle sizes (reference [8]).

Optical methods should not be used unquestioningly for the analysis of particle size. Particle size and form and complex refractive indices, of the material of the particles and of the dispersion medium, must all be taken into consideration before a decision is made. Figure 1.11 classifies light scattering according to particle size.

The following optical methods exemplify those suitable for measurements on groups of particles:

- Methods:
  - diffraction of light (Fraunhofer diffraction)
  - quasi-elastic light scattering (QLS)
  - photocorrelation spectroscopy (PCS) or other designations.

- Advantages
  - online measurement possible
  - measurement possible at high concentrations (up to approx. 1 %)
  - suitable for broad distributions
  - wide choice of dispersion media possible
  - user-friendly apparatus available

- Disadvantages
  - limited resolution
  - working back to the values of the physical parameters: size distribution by mathematically ill-defined problems
  - suitability for polydisperse systems problematic (in QLS)

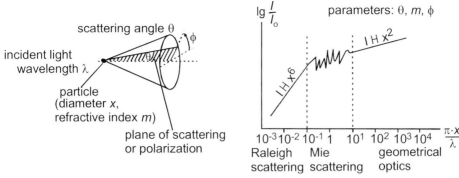

Figure 1.11. Light scattering [8].

1.3.3 Important Principles and Foundations of Interface Physics for Formulation Chemists

In order to describe the stability, formation, and decay of a colloidal system (a dispersion, an emulsion, ...), we need to know something about bond energies, surface and interface energies, and kinetic processes, amongst other things.

Two of the most important questions about colloidal dispersions and emulsions are:
1. Under what conditions is the disperse state stable?
2. Under what conditions does it flocculate or coagulate?

A fundamental principle of thermodynamics states that, at constant temperature, a system tends to alter spontaneously in such a way as to reduce its free energy.

The simple example of a weight under the influence of gravity will serve as a mechanical analogy: Figure 1.12.

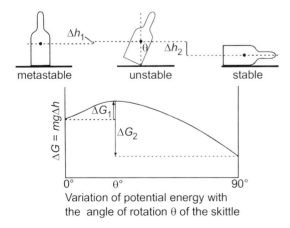

Figure 1.12. Free energy and stability [4].

The energy of a gas, liquid, or solid that is available for performing useful work is known as the *free enthalpy G* of the system:

$$G = H - TS \tag{1.7}$$

G: free enthalpy (Gibbs free energy)
H: enthalpy
S: entropy
T: temperature

On account of the "disorder" of the molecules in the system, it is not possible to convert all the thermal energy it contains into mechanical energy. All systems tend to the greatest possible freedom of movement, to increased randomization or disorder, or, expressed in physical terms, to an increase in entropy. This is the second law of thermodynamics.

For a system to reach a stable state, its free surface energy tends to diminish. An increase in free surface energy ΔG in disperse systems is achieved by division of the particles and therefore by the increase ΔA in the total surface area A.

$$\Delta G = \gamma_{SL} \cdot \Delta A \tag{1.8}$$

γ_{SL}: Interfacial tension between the liquid medium and the particles.

As ΔG of the system diminishes, the system becomes more stable; equilibrium is reached when $\Delta G = 0$. This can be achieved either by a reduction in γ_{SL} or by a decrease in the area of the interface. The latter occurs through flocculation or coagulation. The interfacial tension can be reduced by addition of a tenside, but not to the extent that $\gamma_{SL} = 0$. Thus, the system remains unstable. It can be stabilized by the introduction of a repulsive force or a potential threshold, as described in Chapter 5 (for example, by the use of emulsifiers in emulsions). In brief: at constant temperature, the particles of a dispersion always tend to become coarser, to flocculate or to coagulate, if the system is not somehow stabilized. *The more finely dispersed the particles, the less stable the system is unless it is protected, that is, unless there is a sufficiently high energy barrier resulting from an electrical double layer or steric protecting layer.* This is the basic requirement for the manufacture of stable dispersions and emulsions. However, we use the word "stable" not in the thermodynamic sense, but in the practical sense to refer to behavior over a useful time period. Nevertheless, thermodynamically stable dispersions, the microemulsions, do exist (see Chapter 3).

1.4 Intermolecular Binding Forces

For the molecules in gases, liquids, and solids to form aggregates, they must be held together by intermolecular forces. These are manifested in *the cohesion of similar molecules and the adhesion of dissimilar molecules.* A knowledge of these forces is necessary for the understanding not only of the properties of gases, liquids, and solids, but also of interface phenomena such as the stabilization of emulsions, flocculation in suspensions, the removal of dirt from surfaces, and so on.

1.4.1 Repulsive and Attractive Forces

When molecules meet, both attractive and repulsive forces come into play (Figure 1.13). Attraction occurs if, when two molecules approach one another, those sites bearing opposite charges coincide more nearly than those with the same charge. Molecules repel each other when their electron clouds interpenetrate. This force increases exponentially as the distance between the molecules decreases. At an equilibrium distance of 3–4 Å (r_e) the attractive and repulsive forces balance. At this point the potential energy of the two molecules is at its minimum; they are in a stable state. This holds true not only for atoms and molecules but also for larger entities such as colloidal particles and droplets in dispersions and emulsions.

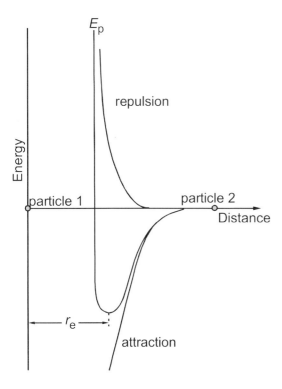

Figure 1.13. Repulsion and attraction energies and their sum, potential energy (E_p), as a function of the distance between two particles. The energy minimum occurs at the equilibrium distance r_e.

1.4.2 Van der Waals Forces

There is always an attraction between two atoms, molecules, colloidal particles, or macroscopic particles, the van der Waals force. This force, however, only makes itself felt when the atoms or molecules are very close together, for *it increases in inverse proportion to the sixth power of the distance r* (Equation 1.9). In contrast, the potential energy of repulsion changes faster with respect to distance, as shown in Figure 1.13, so that there is an energy minimum at the equilibrium distance r_e.

Van der Waals Attraction:
The energy of attraction V between two interacting molecules or atoms is given by Equation 1.9 (λ_{12}: London constant, r: distance between the particles):

$$V = -\frac{\lambda_{12}}{r^6} \tag{1.9}$$

The van der Waals energy experienced by a pair of atoms or molecules is additive. The energy of attraction V_{att} between two particles of volume V_1 and V_2 is:

$$V_{att} = -\iint \frac{q_1 \cdot q_2 \cdot \lambda_{12}}{r^6} dV_1 dV_2 \tag{1.10}$$

It follows that, for two spheres of radius a at a distance H (A: Hamaker constant $\sim 10^{-20}$ J) [9]:

$$V_{att} \cong -\frac{A \cdot a}{12H} \tag{1.11}$$

As an approximation:
−   Attraction between two atoms at the distance of their radii: $V \sim kT$.
−   Attraction between two colloidal particles with a radius of 50 nm at a distance of 50 nm: $V \sim kT$.
−   Attraction between two spheres of 1 cm radius at a distance of 1 cm: $V \sim kT$.

The most important facts about van der Waals forces:
1. They are long-range forces.
2. Particles always attract.
3. V is greater than kT at distances less than the radii of the particles.
4. The force decreases as the distance to the particle increases.

Van der Waals interactions are closely connected with the condensation of gases, the formation of metal complexes, the solubility of solids, and so on, but above all with the stability of colloidal systems. Van der Waals forces result from the fact that dipolar molecules line up such that the positive pole of one points at the negative pole of its neighbor. In this way, very large groups of molecules can associate by means of weak attractions. Permanent dipoles can also induce dipoles in molecules that are themselves not polar but that are easily polarizable. But even between nonpolar molecules, dipoles can be mutually induced by movement of the electrons. The resultant forces are known as *dispersion* or *London forces*.

The attraction between two atoms is on the order of the thermal energy kT when the distance between the atoms is on the order of their radii.

kT is the unit for the thermal energy. This recurs frequently in the theory of the stability of dispersions, and so it is explained in more detail here:

k: Boltzmann constant = $1.38 \cdot 10^{-16}$ erg per kelvin and per molecule,
T: absolute temperature in kelvins.

At a temperature of 300 K:
$$1\ kT = 1.38 \cdot 10^{-16} \cdot 10^{-7} \cdot 300 \cdot 6 \cdot 10^{23} = 2.5 \cdot 10^{3}\ \text{J/mol} \cong 580\ \text{cal/mol}$$

Van der Waals forces between atoms or molecules are additive and independent of scale. Thus, whether for colloidal particles (radius ~ 50 nm) 50 nm apart or for pebbles of radius 1 cm at a distance of 1 cm, the van der Waals attraction is ~ kT.

Equation 1.10 can be integrated for various colloidal particle shapes; for the spherical particles that interest us most, integration yields Equation 1.11.

In the colloidal microcosm, van der Waals adhesive forces are much larger than particle weights. The reason: weight decreases as the cube of particle size, but attraction decreases approximately linearly. At a particle size of 1 μm, the adhesive force can be 10^6 times greater than the force due to gravity. If van der Waals forces did not exist, the surface of the earth would be a cloud of dust!

The essential facts to remember about the fundamentally important topic of van der Waals forces:

– they increase as the distance between surfaces diminishes;
– they are greater than kT at distances slightly less than the particle size;
– in the presence of a liquid medium they decrease;
– they act over long distances, that is, they are long-range forces;
– they are always present as the attractive force between colloidal particles. The force can be many times more than the particle weight.

1.4.3 Hydrogen Bonding

For the formulation chemist, hydrogen bonds are of great importance for their involvement in the adsorption of molecules at interfaces. Figure 1.14 depicts some structures containing hydrogen bonds.

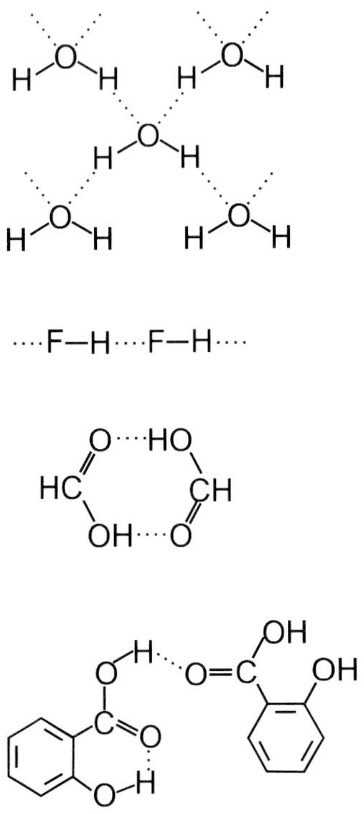

Figure 1.14. Hydrogen-bonded structures.

In the water molecule, the electrons from the hydrogen atoms are attracted to the oxygen atom and form covalent bonds. The protons thus exposed can easily attach themselves to any electron-rich structure, such as an oxygen with a completed octet, or other electronegative atoms: fluorine, nitrogen, and so on. In water, such bonds exist between alcohols, carboxylic acids, aldehydes, esters, and polypeptides. As an example of their effects, we cite the increase in expected boiling point of the substance. In the cases of formic and acetic acid, they cause dimerization. Such dimers can even exist in the vapor state. Hydrogen fluoride exists in the gaseous phase as a polymer held together by hydrogen bonds (F–H ...).

1.4.4 Survey of Intermolecular Forces and Valence Bonds

The strength of bonds may be judged from their bond energies, such as those given in Table 1.6.

Table 1.6. Intermolecular forces and valence bonds (from reference [10]).

Bond type		Bond energy [kcal/mol]
a) Van der Waals forces and other intermolecular attraction		
dipole–dipole interactions (Keesom force)		
dipole-induced dipole interactions (Debye force)		
dispersion (London) force		1–10
ion-induced dipole force		
hydrogen bonds	OH ········· O	6
	CH ········· O	2–3
	OH ········· N	4–7
	NH ········· O	2–3
	FH ········· F	7
b) ionic or electrovalent bond, heteropolar bond		100–200
c) atomic or covalent bond, homopolar bond		50–150

The strongest bonds are ionic bonds, at 100–200 kcal/mol, and covalent bonds, at 50–150 kcal/mol. Hydrogen bonds, with a bond energy of 2–8 kcal/mol, are relatively weak. The van der Waals forces create bonds with an energy of 1–10 kcal/mol. The bond energies in "weakly condensed matter" of the type found in flocs have only recently been examined; they lie in the region of kT, the thermal energy. Flocs are extremely unstable structures.

1.5 The Liquid–Gas and Liquid–Liquid Interface

1.5.1 Surface and Interfacial Tension

The terms surface and interfacial tension have already been introduced in a general form in Section 1.1.

An experiment along the lines of that depicted in Figure 1.15 demonstrates that surface tension is a force per unit length (mN/m or dyn/cm). A wire frame is dipped into a soap solution, so that a thin soap film can form over the area *ABCD*. This film can be stretched by the application of force to the sliding crosspiece *AB* of length *L* (for example by attachment of a weight); the force acts in the direction opposite to that of the surface tension of the soap film. The force necessary to disrupt the film is measured. *Surface tension can thus be defined as the change in free surface energy per increase in surface area.*

Or: The surface tension is equal to the work in mJ necessary to create 1 cm^2 of new surface.

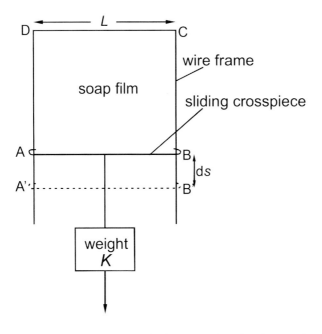

Figure 1.15. Apparatus for the demonstration of surface tension.

From Figure 1.15, the surface tension is given by Equation 1.12:

$$\gamma = \frac{K}{2L} \tag{1.12}$$

From this we can derive the connection between surface tension and the work done to increase the surface area of the film, dW:

$$dW = K \cdot ds = \gamma \cdot 2L \cdot ds = \gamma \cdot dA \tag{1.13}$$

$$W = \gamma \cdot \Delta A \tag{1.14}$$

The surface tension γ can therefore be considered as the change in surface energy per unit area created, with units:

$$\gamma: \quad \left[\frac{erg}{cm^2} \right] \quad or \quad \left[\frac{J}{m^2} \right] \tag{1.15}$$

(γ: surface tension; K: force necessary to disrupt film; dA: increase in surface area; W: work necessary to form film = increase in free surface energy)

Table 1.2 is supplemented by the information contained in Tables 1.7 and 1.8, which list further surface and interfacial tensions measured with water.

Table 1.7. Examples of surface tensions.

Substance	Surface tension at 20°C [mN/m]
water	72.8
oleic acid	32.5
benzene	28.9
chloroform	27.1
carbon tetrachloride	26.7
castor oil	39.0
liquid paraffin	33.1
mercury	486
silver (liquid)	920 (1000 °C)
copper (liquid)	1270 (1120 °C)
copper (solid)	1430 (1080 °C)
iron (solid)	2300 (1450 °C)

Table 1.8. Examples of interfacial tension with water.

Substance	Interfacial tension at 20°C [mN/m]
carbon tetrachloride	45.0
benzene	35.0
chloroform	32.8
n-hexane	51.1
n-octane	50.8
n-octanol	8.5
olive oil	22.9

1.5.2 Surface Activity

Substances such as short-chain fatty acids and alcohols are soluble in both water and oil; they are amphiphilic. The HC (hydrocarbon) part of the molecule is responsible for its oil solubility, while the polar carboxy or hydroxy group has sufficient affinity for water to carry a small HC chain with it into aqueous solution. Figures 1.16 and 1.17 depict two typical amphiphiles, stearic acid and a phospholipid. When these molecules are present at an o/w (oil/water) or w/a (water/air) interface, the hydrophilic head group will be buried in the water phase and the lipophilic HC chain will be stretched out into the oil phase or the air; see Figure 1.18.

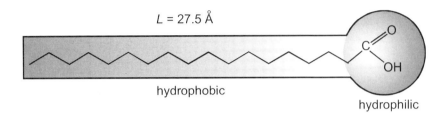

Figure 1.16. A typical amphiphilic molecule, stearic acid ($C_{17}H_{35}COOH$).

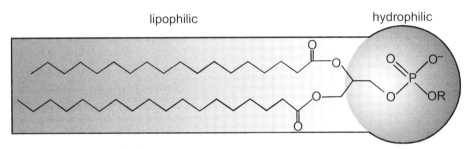

Figure 1.17. A phospholipid.

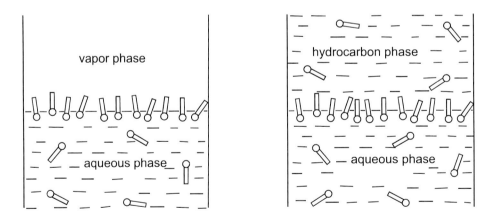

Figure 1.18. Adsorption of tenside molecules at air/water and oil/water interfaces.

These amphiphilic molecules, also known as tensides or surfactants, line up per-pendicular to the interface. This alignment is associated with a decrease in free energy.

As a consequence of the accumulation of tensides at the interface, the interface tends to spread because of the action of surface or interfacial pressure. This tendency is counteracted by the surface tension γ_0, since work must be done to increase the surface area ($W = \gamma dA$). Thus, the addition of a surfactant decreases the surface tension γ.

$$\gamma = \gamma_0 - \pi \qquad \left[\frac{mN}{m}\right] \qquad\qquad (1.16)$$

γ_0: Surface tension in the absence of tenside
γ : Surface tension in the presence of tenside
π : Surface pressure of the tenside, i.e. film pressure

The film pressure or surface pressure is therefore the difference between the surface tensions of the solvent without (γ_0) and with (γ) the film (Equation 1.17).

$$\pi = \gamma_0 - \gamma \qquad\qquad (1.17)$$

If $\pi \geq \gamma_0$, spontaneous mixing or emulsification occurs, a phenomenon of interest in emulsification technology.

When present in sufficient quantity, tensides form a monomolecular layer on top of the liquid (Figure 1.19). This can be compressed in a Langmuir trough, Figure 1.20.

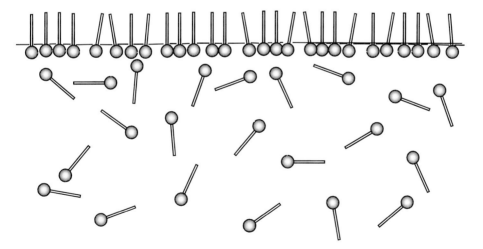

Figure 1.19. A monomolecular layer (monolayer) of tenside molecules.

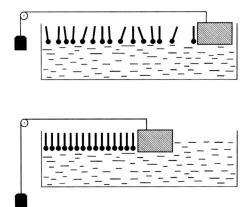

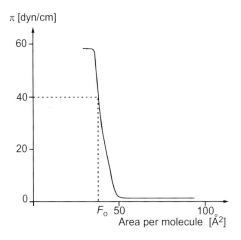

Figure 1.20. Schematic depiction of a Langmuir film balance.

This method of measurement provides information about the film thickness and film pressure, the mean geometrical cross-section of the adsorbed molecules, the general rheological and elastic properties of the film, and, qualitatively, about the lateral inter-actions of the adsorbed molecules. In addition, it is possible to investigate the interac-tions between tensides and cotensides (see Chapter 2 on the stability of emulsions).

The variation of π with the area of the surface film A can be represented by a π–A graph (Figure 1.21); A is expressed in units of $\mathring{A}^2$/molecule. Such graphs can be recorded from a Langmuir film balance. In particular, we want to know about interfacial films of emulsifiers and mixtures of emulsifiers and tensides, which are important in the stabili-zation of emulsions.

π [dyn/cm]

```
60 ─|

40 ─|- - - - - - - -

20 ─|

 0 ─|_____
        F₀ 50              100
        Area per molecule  [Å²]
```

Figure 1.21. Graph for the compression of a film of stearic acid spread over water, showing the dependence of the area on the shearing. At the steepest gradient of the curve the area per molecule, F_0, can be read off the *x* axis.

Example showing the magnitude of the film pressure: For a film of thickness 10^{-7} cm, a film pressure of 1 mN/m corresponds to an internal pressure of 10^{7} mN/m, or 10 atm.

Example for the calculation of the area of a tenside molecule (source: B. Franklin): 1 teaspoonful (5 cm³) of fatty acid of molecular weight 300 and relative density 0.9 spread over half an acre of water (~2 × 10^{7} cm²) creates a monomolecular layer of:

$$\begin{array}{ll} 4.5 \text{ g fatty acid} & = 0.015 \text{ mol } (M_r = 300) \\ 0.015 \cdot 6.02 \cdot 10^{23} & = 9 \cdot 10^{21} \text{ molecules} \end{array}$$

$$\frac{2 \cdot 10^{7} \text{ cm}^{2}}{9 \cdot 10^{21}} = 22 \cdot 10^{-16} \text{ cm}^{2}/\text{molecule} = 22 \text{ Å}^{2}/\text{molecule}$$

1.5.3 Tensides: Structure, Typical Examples, Essential Physical Properties, Degradable Tensides

Tensides may be classified as *anionic, cationic, nonionic,* or *zwitterionic.*

Their principal uses are as wetting agents, detergents, foam formers, dispersants, and emulsifiers.

Dispersants and emulsifiers can be classified only with difficulty; they vary greatly in chemical composition, and are treated in detail in the chapters on dispersions and emulsions.

Wetting agent: a) branched chain with central
                         hydrophilic group

                      b) short hydrophobic chain with
                         hydrophilic end group

Foaming agent: medium-length hydrophobic chain with
                        hydrophilic end group

Detergent: long hydrophobic chain with hydrophilic end group

Figure 1.22. Relationship between structure and potential applications of tensides (analogous to reference [11]).

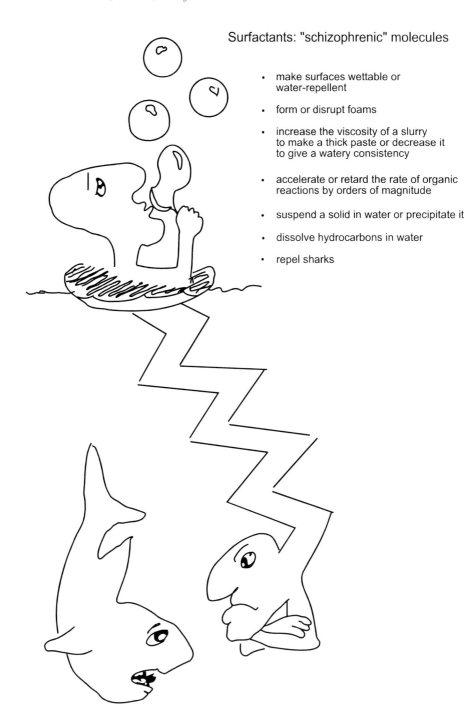

Figure 1.23. Surfactants: "schizophrenic" molecules [12].

A good survey of commercially available emulsifiers, dispersants, and detergents, with their names and suppliers, is given in McCutcheon's book, published annually [13].

Tensides belong in the class of association colloids, which are introduced in Section 1.8.

Figure 1.22 depicts the general relationship between the structures and properties of tensides. This relationship is based on theoretical principles which cannot be dealt with here. However, the structures in the diagram can be used in the assessment of tensides on principle for their suitability as wetting agents or detergents.

Tensides have many uses. They have been described as Janus compounds (for their double faces) on account of their possession of both lipophilic and hydrophilic character, and even as "schizophrenic" molecules (Figure 1.23).

The chemical structures of the various classes of tensides are listed in Table 1.9.

Table 1.9. Structures of surface-active substances.

Type	Hydrophilic substituent	Example	Name
anionic	$-COO^-$	$CH_3-(CH_2)_n-COO^-\ Na^+$	soaps, salts of fatty acids
	$-O-SO_2-O^-$	$CH_3-(CH_2)_n-O-SO_2-O^-\ Na^+$	sulfates of fatty alcohols
	$-SO_2-O^-$	$CH_3-(CH_2)_n-SO_2-O^-\ Na^+$	alkyl sulfonates
		$CH_3-(CH_2)_n-\langle\bigcirc\rangle-SO_2-O^-\ Na^+$	alkyl aryl sulfonates
	$-CO-\underset{R}{N}-CH_2-CH_2-SO_2-O^-$	$CH_3-(CH_2)_m-CO-\underset{R}{N}-CH_2-CH_2-SO_2-O^-\ Na^+$	fatty acylated aminoethyl sulfonates
cationic	$-\underset{CH_3}{\overset{CH_3}{N^+}}-CH_3$	$CH_3-(CH_2)_n-\underset{CH_3}{\overset{CH_3}{N^+}}-CH_3\ Cl^-$	alkyltrimethyl-ammonium chloride
zwitterionic	$-\underset{CH_3}{\overset{CH_3}{N^+}}-CH_2-COO^-$	$CH_3-(CH_2)_n-\underset{CH_3}{\overset{CH_3}{N^+}}-CH_2-COO^-$	N-alkylbetaine
nonionic	$-O-(C_2H_4-O)_nH$	$CH_3-(CH_2)_m-O-(C_2H_4-O)_nH$	polyethylene oxide alkyl ether
		$CH_3-(CH_2)_m-\langle\bigcirc\rangle-O-(C_2H_4-O)_nH$	polyethylene oxide alkyl aryl ether
H-bond active	$-CO-NH-\underset{CH_2OH}{\overset{CH_2OH}{C}}-CH_2OH$	$CH_3-(CH_2)_n-CO-NH-\underset{CH_2OH}{\overset{CH_2OH}{C}}-CH_2OH$	fatty acyl derivatives of trimethylolamino-methane

Some typical wetting agents are listed in Table 1.10. The following properties must be considered in the assessment of wetting agents:

- the minimum surface tension attainable, regardless of the amount of tenside required;
- the depression of surface tension achieved with a specified concentration of tenside;
- the time required for a tenside to achieve equilibrium.

The last property is of particular importance for wetting agents. Selected tenside types that excel in this respect are shown in Figure 1.24. These tensides permit the depression of γ in water by up to about *25 mN/m* in a very short period (15 s). Formulation chemists are very interested in such tensides, since the concentration necessary to reach the minimum γ, the critical micelle concentration (*CMC*; see Section 1.8.1), is small, in the case of Aerosol OT 0.7 g/L.

Table 1.10. Some typical wetting agents.

Structure	Name
i-C$_4$H$_9$, i-C$_4$H$_9$ N-C-(CH$_2$)$_7$-CH$_2$ CH$_3$-(CH$_2$)$_7$-CH-SO$_3$Na O	sodium *N,N*-diisobutyloleamide sulfonate
C$_4$H$_9$, C$_4$H$_9$ SO$_3$Na	sodium dibutylnaphthalenesulfonate
R$_1$-O-C-CH-SO$_3$Na R$_2$-O-C-CH$_2$ O O	sodium sulfosuccinic acid ester
C$_2$H$_5$ C$_4$H$_9$CHCH$_2$-O-C-CH-SO$_3$Na C$_4$H$_9$CHCH$_2$-O-C-CH$_2$ C$_2$H$_5$ O O	sodium di(2-ethylhexyl)sulfosuccinate Aerosol OT
CH$_3$ CH$_3$ CH$_3$ CH$_3$ CH$_3$-CH-CH$_2$-C-C = C-C-CH$_2$-CH-CH$_3$ OH OH	*tert*-acetylene glycol 2,4,7,9-tetramethyl-5-decyne-4,7-diol Surfynol 104

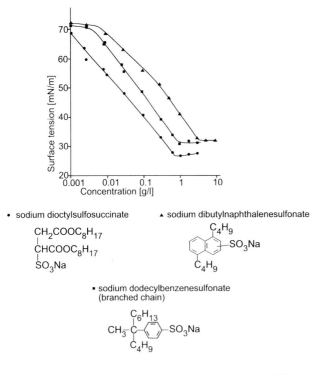

- • sodium dioctylsulfosuccinate
- ▲ sodium dibutylnaphthalenesulfonate

- ▪ sodium dodecylbenzenesulfonate
  (branched chain)

Figure 1.24. Some tensides which reach equilibrium rapidly in aqueous solution [14].

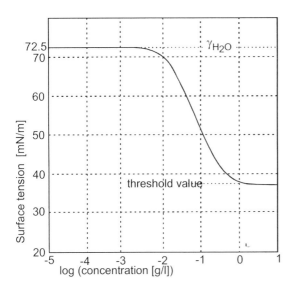

Figure 1.25. Typical concentration dependence of the surface tension of aqueous tenside solutions.

The concentration dependence of the surface tension of a tenside is shown in Figure 1.25; the curve is typical for tensides. The set of curves in Figure 1.26 illustrates the change of γ over time for a number of wetting agents at the same concentration.

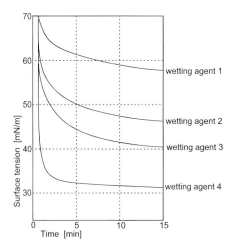

Figure 1.26. Change of surface tension over time for solutions of various wetting agents of the same concentration (0.1 g/L).

Since reference books usually deal only with water-soluble tenside types, we have compiled an equivalent list of tensides for use in organic solvents, Table 1.11.

Table 1.11. Tensides for organic solvents.

Anionic tensides
–   linear alkylbenzenesulfonates with 12 carbons
–   sodium salts of petroleum sulfonates with $M_r < 450$
–   sulfosuccinate esters ROOCCH$_2$CH-(SO$_3^-$M$^+$)COOR, for example sodium dioctyl-sulfosuccinate
–   dinonylnaphthalenesulfonate

Cationic tensides
–   long-chain amines with 12–18 carbons
–   tetraalkylammonium salts, for example di(2-ethylhexyl)ammonium chloride

Nonionic tensides
–   polyoxyethylated alkylphenols, for example *p*-nonylphenols with a maximum of five ethylene oxide (EO) groups
–   sorbitan fatty acid esters, for example sorbitan monooleate
–   polyoxyethylated sorbitan fatty acid esters
–   alkanolamine–fatty acid condensation products, for example alkyldiethanolamide

---

In general, tensides are only poorly degradable, and insufficiently so to meet the strict requirements of current legislation. In addition, they are usually toxic to fish as a result of their surface activity, and they reduce oxygen uptake in water treatment plants. As a result, the development of biodegradable tensides has become a matter of some importance. One possible alternative takes the form of biotensides, compounds produced by microorganisms. Structurally, these are extremely varied. Some biotensides which have attracted particular attention in the recent literature are illustrated in Figure 1.27.

Using sophorolipid as a starting material, Kaosoap of Japan has covered almost the entire HLB area by transesterification, so that in principle emulsifiers, wetting agents, and washing and cleaning preparations may all be obtained from biotensides. The compound *surfactin* is of great interest for its low surface tension. This compound was reported as early as 1984 by D. G. Cooper [15].

As an emulsifier, *emulsan* deserves attention: it is completely adsorbed at the water/oil interface, it is nontoxic and biodegradable. Details are given in reference [16].

Finally, *rhamnolipid* deserves a mention, with its values of $\gamma = 25$ mN/m, minimum interfacial tension of 0.06 mN/m, and 100% biodegradability. It has good wetting properties but is not suitable as a washing agent.

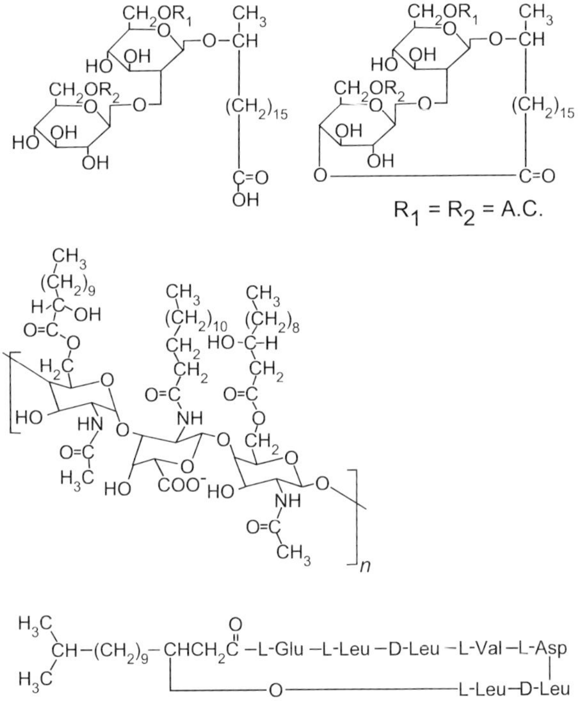

sophorolipid

emulsan (*n*: 840)

surfactin (γ < 25 mN/m)

Figure 1.27. Biotensides.

Biotensides are four to five times more expensive than ordinary tensides. Usually, however, they have a greater specific efficacy, so the cost of their use is reduced. An alternative to biotensides, in terms of their biodegradability, is offered by tensides derived from saccharides: *sorbitan fatty acid esters, saccharose fatty acid esters,* and *fatty alcohol polyglycosides.* Henkel has recently started producing *alkyl polyglucosides* commercially (Figure 1.28).

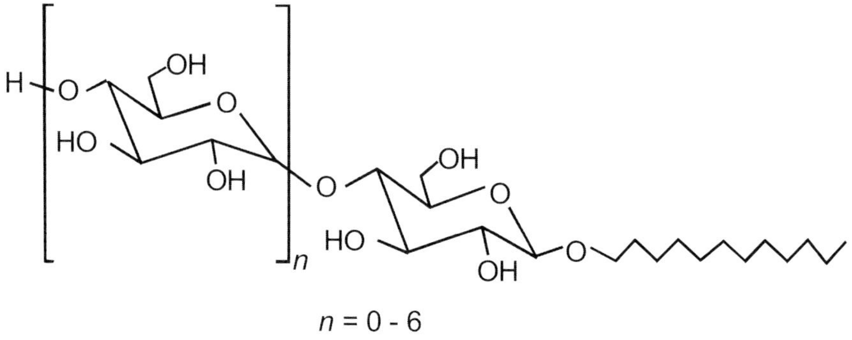

$$n = 0 - 6$$

Figure 1.28. A new class of tensides: alkyl polyglucosides (APG), synthesized from fatty alcohols (8–14 C) and glucose. Advantages: excellent cleaning action and skin compatibility (no R38 notice), biodegradable.

Another class of tensides is the *glucamines*. Paraffinsulfonic acid glucamine salt is skin-compatible and is not an eye irritant. According to Hüls, the topic of mono-saccharide-based tensides is far from exhausted.

Reduction of tenside concentration by synergy: Certain tenside combinations are eff-ective at lower concentration than either tenside alone, for example a fatty alcohol ether sulfate and a linear alkyl sulfate.

1.6 Cohesion, Adhesion, and Spreading

These phenomena of interface chemistry play a decisive role in wetting, rewetting, washing, and cleaning. Formulation chemists therefore need to have good knowledge of them.

1.6.1 Cohesion in a Liquid

Cohesion in a liquid is a measure of the attraction between similar molecules (*A* and *A*); adhesion is a measure of that between dissimilar molecules (*A* and *B*).

To divide a cylinder of liquid A with cross-section $S = 1$ cm^2 into two halves so that two new surfaces are created (the two hatched surfaces in Figure 1.29), each with surface tension γ_A, work must be done, specifically, the work of cohesion W_c (Equation 1.18).

$$W_c = 2\gamma_A \tag{1.18}$$

The work of adhesion, which is necessary to overcome the attraction between two dissimilar molecules A and B in a liquid, is equal to the newly generated surface tensions $\gamma_A + \gamma_B$ minus the interfacial tension lost in the separation, γ_{AB}. Thus, the adhesion work W_a is as given in Equation 1.19.

$$W_a = \gamma_A + \gamma_B - \gamma_{AB} \qquad \text{Dupré Equation} \tag{1.19}$$

The adhesion work for a liquid and a solid is:

$$W_a = \gamma_S + \gamma_L - \gamma_{SL} \tag{1.20}$$

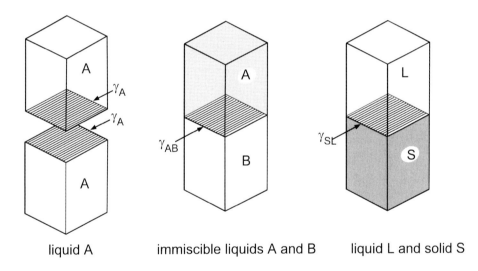

liquid A	immiscible liquids A and B	liquid L and solid S

Figure 1.29. Cohesion and adhesion.

1.6.2 Spreading of One Liquid over Another

If a drop of a water-insoluble substance such as oleic acid is layered onto the surface of water, it may behave in one of three ways:

1. It may remain in the form of a lens (Figure 1.30a).
2. It may spread into a thin film until the entire surface is covered with a "duplex film". A duplex film is thick enough for the two interfaces of film/air and liquid/film to be independent of one another and for each to have its characteristic interfacial tension.
3. It may spread into a monolayer in equilibrium with lenses of excess oil (Figure 1.30b).

a)   a drop of nonspreading oil on water

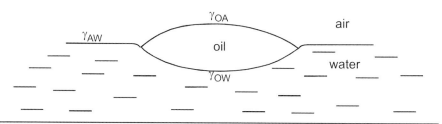

b)   *n*-hexanol spreading on water

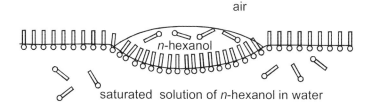

Figure 1.30. Spreading of a liquid (e.g. oil) on a second liquid (e.g. water) [1].

If the affinity of the oil molecules for one another is greater than that of the oil for water, then no spreading occurs; otherwise, the oil spreads out over the surface of the water. In other words, spreading occurs when the adhesion work is greater than the cohesion work: $W_a - W_c > 0$.

This difference is known as the spreading coefficient S.

$$S = W_a - W_c = (\gamma_A + \gamma_B - \gamma_{AB}) - 2\gamma_A \qquad (1.21)$$

or

$$S = \gamma_B - \gamma_A - \gamma_{AB} = \gamma_B - (\gamma_A + \gamma_{AB}) \qquad (1.22)$$

Spreading occurs when $S > 0$; that is, a substance spreads out if the sum of the free surface energies of the new surface and of the new interface is smaller than the free energy of the old surface.

When a liquid L spreads over a solid S, Equation 1.23 applies:

$$S = \gamma_S - (\gamma_L + \gamma_{SL})$$ (1.23)

This spreading coefficient is of particular interest to us, since it is of decisive importance in the problem of wetting for dispersions, which will be described in Chapter 5, and also plays a role in cleaning with detergents, treated in Chapter 10.

The above discussion of the spread of a liquid over another liquid is limited to the *initial* spreading; Table 1.12.

Table 1.12. Initial spreading coefficients on water at 20°C [17].

Substance	Spreading coefficient S [mN/m]
ethanol	50.4
propionic acid	45.8
diethyl ether	45.5
acetic acid	45.2
acetone	42.4
oleic acid	24.6
undecylenic acid	32
chloroform	13
benzene	8.9
hexane	3.4
octane	0.22
dibromoethene	–3.19
liquid paraffin	–13.4

Before equilibrium is reached, the surface of the water becomes saturated with the spreading liquid, which itself becomes saturated with water. When mutual saturation has occurred, γ_{BA} and γ_B alter. S may become smaller, or even negative. This means that, even after initial spreading, a liquid may gather together into a lens. γ_{AB} does not change when this occurs, since interfacial tension is always measured under conditions of mutual saturation.

Example 1: Figure 1.30b: *n*-hexanol on water
S_i (initial) = 72.8 – (24.8 + 6.8) = 41.2 mN/m;
S_f (final) = 28.5 – (24.7 + 6.8) = –3.0 mN/m.

Example 2: benzene γ_o = 28.9 mN/m; γ_{AB} = 35.0 mN/m;
S_i (initial) = 72.8 – (28.9 + 35.0) = +8.9 mN/m → spreads on water.
However, after equilibration, the spreading coefficient becomes negative:
γ_A = 28.8 mN/m; γ_B 62.2 mN/m,
S_f(final) = 62.2 – (28.8 + 35.0) = –1.6 mN/m.

The fact that benzene spreads over water is nothing to do with polarity but rather a result of the fact that the cohesive forces between benzene molecules are much weaker than those of their adhesion to water.

1.6.3 Adsorption of Tensides at Liquid Interfaces

Amphiphilic tensides are adsorbed at interfaces. Through their accumulation at the o/w interface they act as a bridge between the polar and nonpolar phases.

The adsorption of tensides in binary systems has been quantified in the fundamental Gibbs Equation (1878), Equation 1.24:

$$\Gamma = -\frac{c}{RT} \cdot \frac{d\gamma}{dc} \tag{1.24}$$

R: gas constant $= 8.314$ J·K^{-1}·mol^{-1}
c: concentration [mol/L]
Γ: excess concentration of the tenside at the interface [mol/cm^2].

The differential quotient $\left(\dfrac{d\gamma}{dc}\right)$ indicates the change in γ with the concentration of the tenside in the bulk of the liquid. A decrease in the surface tension per unit tenside concentration leads to a positive value of Γ; in other words, the tenside accumulates at the interface. If we know this excess concentration Γ for a given tenside concentration in the bulk of the liquid, then we can use it to calculate the area occupied by each molecule at the interface.

If tensides that form soluble monolayers are added to the liquid in such quantities that they saturate not only the interfacial layer but also the bulk of the liquid, the surplus particles aggregate in the form of *micelles*. Micelles are of colloid size and are not surface-active. These association colloids are described in Section 1.8.

1.7 The Solid–Liquid Interface

1.7.1 Wetting and the Contact Angle

When a drop of liquid is placed on a level solid surface, it can either spread completely over the surface or remain as a droplet on the surface with a distinct contact angle θ. Figure 1.31 represents five different contact angles θ schematically.

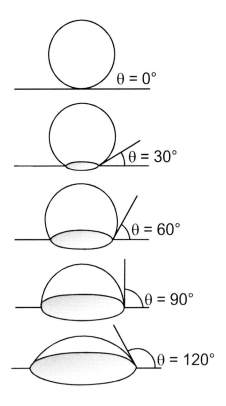

Figure 1.31. Contact angle between a drop of liquid and a solid.

If we assume that the various surface forces can be represented by surface tensions acting in the direction of the surfaces, we obtain Figure 1.32.

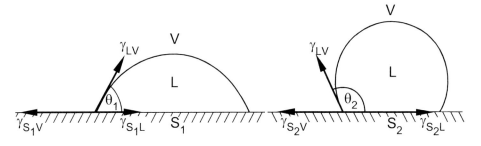

Figure 1.32. The Young–Dupré Equation: summation of forces at solid/liquid/gas phase boundaries, depicted with the same liquid but different solids.

For the three phases S, A (V), and L meeting at a point in the projection (Figure 1.32), vector addition yields the Young Equation (1.25).

$$\gamma_{SV} = \gamma_{SL} + \gamma_{LV} \cos \theta \tag{1.25}$$

or

$$\cos \theta = \frac{\gamma_{SV} - \gamma_{SL}}{\gamma_{LV}} \tag{1.26}$$

S = solid
L = liquid
A or V = air, gas, vapor

The largest values for θ so far measured between water and a smooth surface lie in the range 105°–110° (for example for paraffin wax and water). When $\theta > 100°$, the water gathers in droplets that run over the surface and are easily shaken off.

Some authors set a threshold of $\theta < 90°$ for wetting and $\theta > 90°$ for nonwetting.

Roughness in the surface increases the deviation of the natural value of θ from 90°. For a roughness factor r (r is the ratio of the actual to the apparent surface), the mean contact angle θ' for a rough surface is defined in Equation 1.27:

$$\cos \theta' = r \cdot \cos \theta \tag{1.27}$$

When this is combined with the Dupré Equation (first quoted as Equation 1.19)

$$W_{SL} = \gamma_{SV} + \gamma_{LV} - \gamma_{LS} \tag{1.28}$$

we obtain the Young–Dupré Equation, 1.29:

$$W_{SL} = \gamma_{LV} \cdot (1 + \cos \theta) \tag{1.29}$$

When $\theta \leq 90°$, Equation 1.26 predicts that a reduction in γ_{LV} will result in a reduction in θ, that is, an improvement in wetting. Addition of a wetting agent decreases γ_{LV}; thus, better wetting occurs.

Examples of the influence of θ on wetting:
– Soot floats on water ($\theta \sim 80°$); TiO_2 sinks ($\theta \sim 30°$). When stearic acid is adsorbed onto the TiO_2, θ is changed to $\sim 110°$; the hydrophobized TiO_2 floats on water.
– Quartz sand (SiO_2) is completely wetted by water; however, sand hydrophobized by adsorption of an organic silicon compound is completely water-repellent. It clumps together in water and is plastic (can be formed into any shape). When the water is drained off, the sand is once more freeflowing.
– Calcium carbonate is employed as a filler for paper. For use as a filler in polyolefins, it must first be hydrophobized by adsorption of fatty acids such as stearic acid; a monomolecular layer on the surface of the mineral suffices.

Figure 1.33 shows the position of the contact angle θ for various degrees of immersion of a sphere in a fluid.

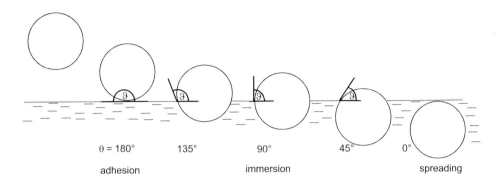

Figure 1.33. Contact angles for immersion of a sphere in a fluid.

1.7.2 Wetting of Solids as a Function of Their Surface Energy

There are many old rules of thumb concerning the wetting of solids, such as: "Polar solids are wetted by polar liquids", or "Liquids with a low surface tension wet solids with a higher surface energy". However, despite their general application, these rules have many exceptions: for example, water does not spread over iron or many other metals. Table 1.13 gives further details of surface energies of solids.

Table 1.13. Surface energy of some solids.

Solid	Surface energy [mJ/m^2]
copper	1100
silver	900
aluminum	500
iron	1700
glass	73
nylon	46
PVC	40
polystyrene	33
polyethylene	31
Teflon	18.5

Rules of thumb for wetting of solids:

- Liquids of lower surface tension wet solids of higher energy, but liquids of higher surface tension do not wet solids of lower energy. There are exceptions to this rule.

- A rule without exceptions, in contrast, is: a liquid of low surface tension can only wet and adhere to a solid of high surface energy if the intermolecular forces between the solid and the liquid resemble those within the liquid.

- The specific free surface energy can be split into terms for the nonpolar (γ_d, dispersion forces) and polar (γ_p, polar forces) interactions. On contact of a liquid and a solid, the two terms can be treated additively. For example, water ($\gamma_L = 72.8$ mN/m, $\gamma_d = 21.8$ mN/m, $\gamma_p = 51.0$ mN/m) does not spread on graphite, iron, copper, or silver, although their surface energies are all considerably greater than that of water. For the liquid to spread, the free surface energy of the solid must have a dispersion term γ_d of at least 243 mN/m. This is the case for aluminum, but not for the solids mentioned above. The spreading of water on aluminum may be predicted from the Zisman–Fowkes–Girifalco theory.

- A liquid with the same specific free surface energy as the solid on which it rests will only wet that solid completely ($\theta = 0$) if both the liquid and the solid have the same polar and nonpolar interaction terms.

Determination of the surface energies of solids can be problematic. The contact angles θ_{SVL} with the solid of a series of liquids of decreasing surface tensions are measured. The surface tension of those liquids which wet the solid completely ($\theta_{SVL} = 0$) is called the critical surface tension of the solid γ_C and is assumed to be equal to the surface tension of the solid (Figure 1.34).

However, if different series of liquids (polar and nonpolar liquids) are used to measure the surface energy of solid paraffin, the results for the two series differ: 15 mN/m for polar and 22 mN/m for nonpolar liquids. This is because *the surface energy of a condensed phase comprises different energy terms* (Equation 1.30):

$$\gamma = \gamma_d + \gamma_p \tag{1.30}$$

γ_d = term for London forces (dispersion forces)
γ_d = term for polar forces (dipole–dipole, hydrogen bonds, etc.)

Example: Water has $\gamma_d = 21.8$ mN/m, $\gamma_p = 51.0$ mN/m, $\gamma = 72.8$ mN/m $\qquad$ (1.31)

In contrast, hydrocarbons experience dispersion forces only.

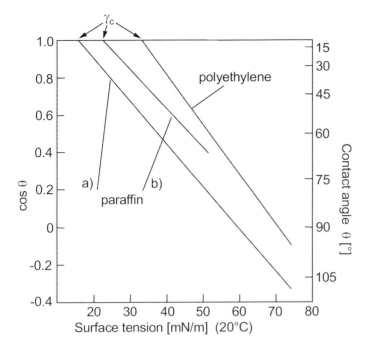

Figure 1.34. Critical surface tension γ_c of solids: a) measured with polar liquids, b) measured with apolar liquids [18].

These facts are of great technical significance for the problems of wetting and adhesion of plastics.

The determination of the surface energies of solids bristles with errors; it is difficult to find reliable values in the literature.

1.7.3 Penetration of a Liquid into a Capillary

This process plays a role in the wetting of rough surfaces that contain capillaries, such as that of paper. The gaps in powders and agglomerates may also be modeled as capillaries. The penetration of liquid into capillaries (Figure 1.35) is treated in Equation 1.32.

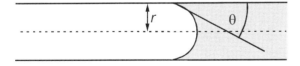

Figure 1.35. Penetration of a liquid into a capillary.

$$P = -\frac{2\gamma_{LV} \cdot \cos\theta}{r} \tag{1.32}$$

Penetration occurs only when $\theta < 90°$.

For $\theta = 0°$:

$$P = -\frac{2\gamma_{LV}}{r} \tag{1.33}$$

The time course of wetting is described by the *Washburn Equation* (l = depth of penetration within the time t, r = pore radius, η = viscosity):

$$\frac{dl}{dt} = \frac{r \cdot \gamma_{LV} \cdot \cos\theta}{4\eta \cdot l} \tag{1.34}$$

or, after integration:

$$\frac{l^2}{t} = \frac{r \cdot \gamma_{LV} \cdot \cos\theta}{2\eta} \tag{1.35}$$

The capillary pressure P is negative under a concave surface. The liquid is drawn spontaneously into the capillary if the latter is wetted, i.e. if $\theta < 90°$. The rate of penetration of the liquid into the capillary $\left(\dfrac{dl}{dt}\right)$ is important in the wetting process (see the Washburn Equation, 1.34). Equation 1.34 has a consequence of significance for formulation chemists that is not empirically obvious: rapid penetration occurs when $\gamma_{LV}\cos\theta$ is large, θ and η (viscosity) are small, and r is large (loose packing in powders and agglomerates). A high value of γ_{LV} and a small contact angle are mutually exclusive. For good spreading, however, a low value for θ is more important. Formulation chemists must therefore try to minimize the contact angle to near $0°$ without reducing γ_{LV} more than necessary to do so. Thus the frequent practice of improving wetting that already suffices by the addition of further wetting agent is, in fact, not only unnecessary but actually counterproductive. Equation 1.38 gives the penetration time for agglomerates or powders, as used in research.

With the definition of porosity ε from Equation 1.36

$$\varepsilon = \frac{void\ space}{bulk\ volume} = \frac{V_b - V_p}{V_b} = 1 - \frac{V_p}{V_b}, \tag{1.36}$$

and V_b = bulk (total) volume, V_p = true volume of the particles, and O_v = specific surface area with respect to volume, the mean hydraulic radius r of a porous system can be calculated from Equation 1.37 as follows:

$$r = \frac{\varepsilon}{2 \cdot (1 - \varepsilon) \cdot O_v}$$

(1.37)

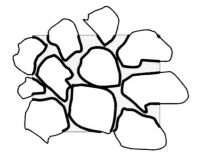

☐ : 2 **◡** = mean hydraulic radius of the porous system

(powder volume : 2x particle surface area)

Figure 1.36. Powder porosity and mean hydraulic pore radius represented as a two-dimensional projection.

Since the specific surface area is inversely proportional to the mean particle size, the capillary radius in the Washburn Equation may be replaced by the mean particle size x and the porosity ε. This yields the following equation, Equation 1.38 (C is a constant).

$$t = \frac{C \cdot \eta \cdot l^2}{x \cdot \gamma_{LV} \cdot \cos \theta \cdot \varepsilon^{2.5}}$$

(1.38)

1.7.4 Measurement of Rate and Extent of Wetting

Contact angles of liquids may be measured by light microscopy. Various techniques are necessary to reduce the error to under 2°. Advancing and receding contact angles can be measured (*contact angle hysteresis*). The measurement of the contact angle of a drop of liquid on a fiber requires a special technique that makes use of photomicrography.

The wetting of materials such as powders, porous aggregates, paper, and textiles can be determined by measurement of the speed at which the liquid front soaks a bed or a column of the material and the application of the Washburn Equation (1.34). The *Enslin cell* (Figure 1.37) can be used as the apparatus [14].

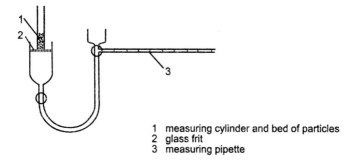

1   measuring cylinder and bed of particles
2   glass frit
3   measuring pipette

Figure 1.37. Measurement of contact angles on powders with the Enslin apparatus [14].

Wetting may be assessed qualitatively by the use of a number of simple tests:

The wetting time of woven textiles is measured with the *dip-wetting test* (DIN standard 53 901). The best known test for textile wetting is the *Draves test* (1939), another method in which the time taken for a scrap of the textile to sink into the tenside solution is measured.

The wetting properties of pigment powders and many other fine powders or porous materials such as cellulose powder can be studied by the preparation in a suitable suspension medium of a paste of the material, which is then left to dry out in a Petri dish. Drops of liquids with various surface tensions are placed on this homogeneous layer of material and the diameter of the resultant spots measured after a specified time. The diameter is plotted as a function of the surface tension of the liquid. Such plots yield a critical surface tension that indicates the *integral wettability* of the powder.

This is only a selection of the numerous wetting tests that exist and for which specific apparatus is available.

1.8 Association Colloids, Basic and Secondary Structures

As was explained in Sections 1.1 and 1.2.1, in the strict sense colloidal systems are systems in which gas bubbles, droplets of liquid, solid particles, macromolecules, or associations of amphiphilic molecules in the size range of about 1–1000 nm are distributed throughout a continuous phase. The continuous phase may be a gas, a liquid, or a solid. Thus, in general, colloidal particles are ensembles of molecules, or, in the case of ionic crystals, of ions. In contrast, macromolecules can make up colloidal particles on their own, or in groups (think of the quaternary structure of polypeptides; see Chapter 13, "Food Formulation"), if the structures formed are within the defined size range. However, when such macromolecular systems are disturbed to the extent that the macromolecules associate to a greater degree, such as in gelation or flocculation, they are no longer considered to be colloids. Biocolloids too are therefore only to be treated as colloids under certain conditions.

   This narrow definition of colloids is nevertheless a little too restrictive for modern colloid chemistry. Certain structures ranging from, for example, sponge phases and bi-continuous microemulsions to gels, and also planar, cylindrical, and spherical lamellar phases and emulsions, are of pre-eminent importance in practical colloid chemistry, but lie outside the area covered by the strict definition of colloids.

   The *association colloids*, however, do fall within this area. Certain substances can form colloidal associations spontaneously, owing to their particular molecular structure. These association colloids (e.g. spherical micelles) are distinct from other substances in that they are in a state of unrestricted thermodynamic equilibrium. Unlike the thermo-dynamically stable association colloids (and solvated macromolecules), other colloids are thermodynamically unstable. So, for example, colloidal dispersions of gases, liquids, or solids undergo changes in particle size through Ostwald ripening (see Chapter 9).

1.8.1 Spherical Micelles

Solutions of surface-active substances have unusual characteristics. Dilute solutions behave like those of normal molecules. At a certain concentration, some of their physical properties change abruptly, for example osmotic pressure, electrical conductivity, and surface tension (Figure 1.38).

   If all the physical properties of a particular system are compared with one another, it will be observed that these changes in properties occur in a very narrow concentration range. The abnormal behavior of the solution can be explained in terms of the formation of *organized aggregates, or micelles*. The concentration at which micelle formation becomes observable is known as the *critical micelle concentration, CMC*. Above the *CMC*, the micelle is the thermodynamically stable form, although it is in equilibrium with single molecules and oligomers. As a result of micelle formation, the thermal behavior of tensides differs from that of other low molecular weight compounds. At low temperatures, the solubility changes little initially, but as the temperature rises to a point characteristic for that substance, the *Krafft point*, the solubility increases sharply as a result of micelle formation. The *CMC*s of some typical tensides are given in Table 1.14.

Table 1.14. Critical micelle concentrations for various tensides in water [20].

Tenside	*CMC* [g/L] in water
alkyl benzenesulfonate (C_{10-13} alkyl group)	0.65
C_{12-17} alkanesulfate	0.35
C_{15-18} α-olefinsulfonate	0.30
C_{12-14} fatty alcohol-2 EO-sulfate	0.30
nonylphenol + 9 EO	0.049
oleyl/cetyl alcohol + 10 EO	0.035

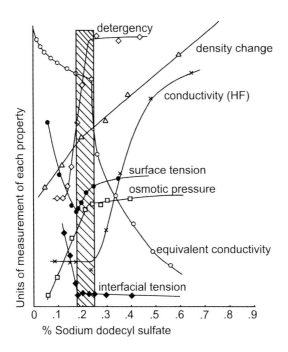

Figure 1.38. Changes in the physical properties of tenside solutions in the range of the *CMC* (after reference [19]).

Figure 1.39 is a sketch of spherical micelles in water and a nonpolar solvent. It should be noted, though, that in reality the flexible alkyl chains are curled up on themselves.

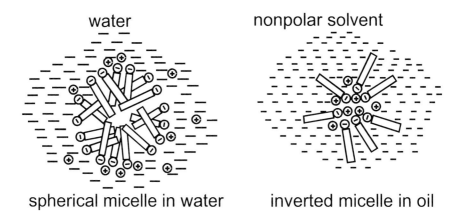

Figure 1.39. Sketch of spherical micelles.

In aqueous solution the lipophilic hydrocarbon chains point into the center of the micelle and the hydrophilic groups are in contact with the water (Figure 1.39). Part of the molecule has a tendency to come out of solution, the other part tries to stay in solution. The factors affecting this balance are:

1. The interaction of the HC chains with water.
2. The interaction of the HC chains with each other.
3. The solvation of the head group.
4. The interaction between solvated head groups and the ionic environment.

The balance is also significantly influenced by the relative sizes of the hydrophilic and hydrophobic portions of the molecules, the HLB value (see Chapter 2). Micelles form structures that minimize the contact between hydrocarbon and water while maximizing contact between hydrophilic head groups and water.

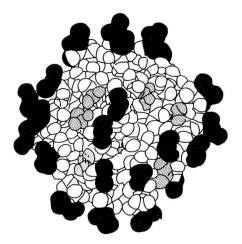

Figure 1.40. Computer-generated representation of a spherical micelle of dodecanoate; black: carboxy groups, shaded: terminal methyl groups; from reference [21].

As can be seen from Figure 1.40, considerable exposure of the hydrocarbon parts to water nevertheless still occurs. Because of the density of distribution of the hydrocarbon chains in the core of the micelle, the head groups on the surface cannot be packed together tightly. In a spherical micelle, a maximum of one third of the surface is occupied by head groups.

Micelles should not be regarded as solid particles. The individual molecules exist in a dynamic equilibrium with the other tenside molecules in the disperse and continuous phases and the boundary layer. *Association colloids form spontaneously when the tenside concentration of the solution exceeds the CMC.*

In a micellar dispersion the aggregate sizes are distributed as shown in Figure 1.41, which plots the frequency of occurrence against the aggregation number S (the aggregate size).

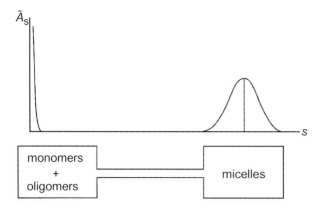

Figure 1.41. Distribution curve for aggregates of aggregation number *S* (source: H. F. Eicke, University of Basel).

Rather than steadily increasing aggregation, a relatively narrow band is observed for the size distribution of the aggregates. In small aggregates the alkyl groups cannot be sufficiently screened from the water for micelles to form. Large aggregation numbers, on the other hand, are also not favored because of the repulsion between the polar head groups. As a result, there is a narrow range of preferred sizes for aggregates. Aggregate numbers for common tensides are about 50–100.

The great technical importance of micelles stems from the fact that they act as *reservoirs for tensides*. If association and micelle formation did not occur, the solubility of tensides would be smaller than the *CMC*. Micelles serve as reservoirs for the provision of further tensides to newly formed accessible surfaces. Thus they are useful as *cleaning agents and stabilizers*. Their solutions form stable foams when shaken. They also play an important role in *emulsion polymerization*, for example in the manufacture of latices.

Their solubilizing effect on insoluble substances means the number of their potential applications is enormous. *Nonionic surfactants are about ten times more effective as solubilizers than anionic surfactants.* An inverse micelle in oil dissolves water in the same way in which normal micelles in aqueous solution dissolve insoluble organic substances. In dry-cleaning, for example, the solvent removes oily dirt but not water-soluble soiling. Therefore, tenside and water are added to the solvent, so the water-soluble dirt is lifted by the water held in the inverse micelles. Another example of the use of micelles is their great technical potential in tertiary oil recovery from beds of porous rock.

The interesting feature is that micelles offer the possibility of "dissolving" water-insoluble substances in water, or water itself in organic solvents. The solubilization of a nonpolar compound in the core of a micelle is depicted in Figure 1.42. *The properties of the micelle interior are the same as those of a hydrocarbon.* Thus, micelles can dissolve considerable quantities of nonpolar organic substances; in a homologous series of tensides, the free energy of solubilization is proportional to the hydrocarbon chain length [22, 23].

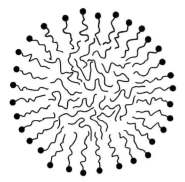

Figure 1.42. Solubilization of a nonpolar compound in the core of a micelle.

1.8.2 Basic Structures

It is not unusual for unexpected results to be obtained in practical formulation work in colloid chemistry. Unless unsuitable tensides have been used, for example those which fail to stabilize the emulsion sufficiently, this is often related to the formation of certain special structures. Such special forms are treated in sections 1.8.2–1.8.6, but these sections should serve only as a warning that in formulation, especially when high concentrations of tensides are used, more complex structures than emulsions or dispersions of solids may need to be taken into account.

The geometric form of the structure adopted by the associated tenside is heavily influenced by steric factors. The *critical packing parameter* (*R* value) introduced by Israelachvili will serve to give us our bearings; it relates the volume of the hydrocarbon chain v_H to the area of the polar group a_0 multiplied by the chain length l' [24] (Equation 1.39).

$$R = \frac{v_H}{a_0 \cdot l'} \tag{1.39}$$

Table 1.15 shows the relationship between the *R* value and the structure of the associated tenside.

Table 1.15. Structures of associated tensides.

R Value	Associate structure
0 – 1/3	spheres
1/3 – 1/2	cylinders
1/2 – 1	vesicles, lamellae

In practical usage, the first and third of these structures are of particular interest. The monomolecular shells of the droplets in o/w and w/o emulsions (first structure) stabilize the droplets against coalescence by their physical properties. When coalescence occurs, shell material turns back into micelles.

In vesicles, the stabilizing effect of the shell relies on two mechanisms, a) an increase in the free energy when the surface grows by deformation, and b) resistance to conversion into lamellae.

1.8.3 Phase Equilibria

The formation of the supramolecular structures is determined not only by steric factors, but also by the composition of the tenside solution. Depending on the additives present, tensides can associate in different ways (for example, to form cubic or hexagonal structures). The states of the corresponding solutions are referred to as phases. Multicomponent mixtures of amphiphilic molecules in solution can demonstrate extremely complicated phase equilibria. Different phases have very varied properties and potential applications, and the changes that occur on heating, cooling, or dilution and consequent phase change are by no means trivial [25, 26]. The assorted possible structures adopted by the associated tenside molecules, such as micelles, bilayers, or cylinders, may form numerous phases. There are *isotropic micellar* phases, *lamellar smectic*, *hexagonal*, and *cubic* phases. Addition of an immiscible liquid alters the phase behavior. Thermodynamically stable microemulsions, which are of considerable practical importance, may develop, as may macroemulsions, which are only kinetically stable, or more complex systems such as vesicles or other membrane systems.

These mesophases are structures which do not have the sort of three-dimensional lattice structure found in crystalline solids but which possess a lesser order in the form of one, two, or three degrees of translational freedom. They retain, however, a certain long-range orientation from the solid state. Accordingly, their physical properties are sometimes typical of crystalline solids and at other times of liquids. Thus, liquid crystals generally display viscous flow properties, yet behave anisotropically when it comes to electrical and electromagnetic properties like conductivity, dielectric permittivity, and magnetic susceptibility.

Liquid crystalline structures are classified into two principal groups, the *thermotropic* and *lyotropic* phases.

Thermotropic liquid crystals are single-component systems that pass through one or more liquid-crystalline phases on melting of the solid phase before they become normal liquids with isotropic distribution of the molecules above the clearing temperature T_c. In *nematic* phases, the molecules rotate freely around their long axis and adopt a favored alignment on average. *Smectic* phases display a higher degree of order: in addition to their ordered orientation, with free rotation about the long axis, the centers of gravity of the molecules are, on average, ordered in layers.

Figure 1.43 a. Lamellar phase.   **Figure 1.43 b.** Hexagonal phase.

In the tenside context, lyotropic liquid-crystalline systems are of particular interest. These consist of at least two components, an amphiphile and a solvent. In contrast to thermotropic systems, for certain compositions of lyotropic systems, two phases can co-exist, for example an isotropic phase with a low concentration of tenside alongside a liquid-crystalline, tenside-rich phase (in a test tube, these would appear as two separate liquid layers). For systems with more than two components, more than two phases are also possible. The lamellar and the hexagonal phase are considered to be the most important liquid-crystalline phases. In the lamellar phase, tenside molecules are arranged in bilayers (Figure 1.43a); the hydrocarbon chains and the water between the polar groups exist in a disordered, liquid-like state. In the hexagonal phase, cylindrical micelles adopt a parallel, hexagonal arrangement (Figure 1.43b). The reverse arrangement is also possible, that is, with the polar groups pointing into the center of the cylinder and the hydrocarbon chains directed outwards. Such a phase is known as an inverted phase; the inverted micelles can take up as much as 40% water. In addition, there are also cubic phases, which can form, for example, from monoglyceride/water systems in which the hydrocarbon chain contains more than fourteen carbon atoms, and which are very viscous. Although lateral intermolecular forces have an effect, the molecules in the mesophases still retain a great deal of mobility about their long axis.

In addition, there are isotropic liquid phases containing disordered, connected bi-layers, known as sponge phases (Figure 1.44) [27, 28] (reviews: references [29–32]; clear description of phase behavior: reference [33]). A more recent view considers these systems as phases of fluctuating surfaces rather than as phases containing particles. From this point of view, *the existence of concentrated microemulsions can be attributed to rapid thermal fluctuations of tenside films* (reviews on microemulsions: [33–40]).

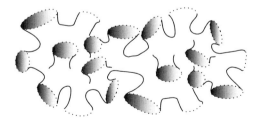

Figure 1.44. Sketch of a sponge phase.

The following example shows how completely different structures can form when only the concentration of components in a mixture is varied, even for simple tensides. The phase diagram for the ternary water/pentanol/sodium dodecyl sulfate (SDS) system is reproduced in Figure 1.45. As well as the biphasic area *LL'* there are four single phases, the isotropic phase *L* and the three mesophases L_α (lamellar), H_α (hexagonal), and *R* (cubic). If oil is added to this system, further phases are created: alongside an assortment of multiphasic regions, there are microemulsions and an oil-rich sponge phase (Figure 1.46).

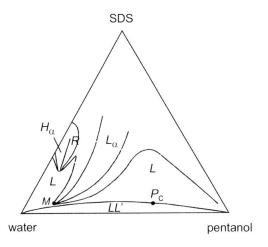

Figure 1.45. Phase diagram for the ternary water/pentanol/SDS system at 25°C. *L*: isotropic phase, H_a: hexagonal phase, *R*: cubic phase, L_a: lamellar phase; *M* is an azeotropic-type point; P_c is a critical point. From reference [26], p. 184.

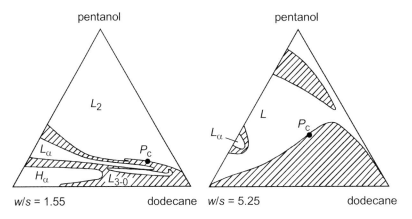

Figure 1.46. Addition of dodecane: phase diagram for the water/dodecane/pentanol/SDS system at 21°C. Water/tenside ratio 1.55. and 5.25 (left and right). The hatched areas are multiphasic. *L* and L_2 are microemulsions, H_a: hexagonal phase, L_a: lamellar phase, L_{3-O}: oil-rich sponge phase; from reference [26], p. 194.

1.8.4 Lamellar L_α Phases

The distance between layers in lamellar phases is not constant, but can be varied continuously by addition of a solvent from molecular dimensions up to 100 nm. Water, oil, or salt can be utilized. As is also the case for macroemulsions, there is strong electrical repulsion between the lamellar structures stretched by water uptake. The stability of lamellar phases, which do not coalesce even when large quantities of oil are added, is attributed to the flexibility of the bilayers. In very dilute solutions these undulate, resulting in a repulsive entropic effect between neighboring lamellae that approach one another, the *undulation interaction*. The stability of the oscillating lamellae can be considerably improved by the addition of a suitable cosurfactant, much as is the case for tenside shells in macroemulsions.

1.8.5 Sponge Phases and Bicontinuous Microemulsions

Sponge phases, like bicontinuous, concentrated microemulsions, are examples of disordered, connected, fluctuating membranes. Thermal fluctuation is the principal influence.

Sponge phases can be obtained from lamellar systems by the addition of alcohols. The lamellar bilayers are converted into a continuous surface dividing the volume into two regions of equal size. There are thermodynamic reasons for this, since both regions contain liquid of the same composition (for example, an oil/alcohol mixture in the case of an L_{3-O} sponge). The bilayer (inverted in this case) is swollen with water. In an L_{3-W} sponge, water is the liquid and the oil is present in the tenside bilayer.

In *bicontinuous microemulsions*, the two regions differ in size; one contains water, the other oil.

1.8.6 Spherical Lamellar Systems

If the elasticity constants κ and $\bar{\kappa}$ for mean curvature and Gaussian curvature of the bilayer are such that the energy required for the formation of a round shell is small, unilamellar and multilamellar spherical or ellipsoidal shapes – vesicles – are formed instead of bicontinuous phases [41].

Vesicles and liposomes are now of particular importance in the pharmaceutical and cosmetic industries. Active organic substances can be enclosed in these tiny compartments (the diameter of unilamellar vesicles is around 0.01–0.1 µm; for multilamellar liposomes, it is a few µm) and thus introduced into biological systems.

When C_5–C_{10} alcohols (pentanol–decanol) are added to sodium dodecyl sulfate/aqueous salt solution, the micelle phases convert into vesicle phases and then, on addition of further alcohol, into lamellar and sponge phases. The sequence when alcohol is added is thus $L_4 \rightarrow L_\alpha \rightarrow L_{3-W}$.

When little tenside is present, dilute, polydisperse vesicles dispersions form. The vesicles are unilamellar and vary in size between 150 Å (hexanol) and 1250 Å (decanol). At higher concentrations of tenside, smaller vesicles containing *multiple lamellae* (up to four layers) are created. These phases are highly viscous and extremely viscoelastic.

In all the systems so far studied, the tenside/alcohol ratio is critical. This means that $\bar{\kappa}$ (the Gaussian elasticity constant; see also Gaussian curvature, Equation 3.6) must be such that stable vesicles can form. This stabilizing elasticity contribution is counteracted by entropic contributions to the free energy.

Multilamellar vesicles can arise in lamellar L_α phases. Smaller multilamellar vesicles (diameter 0.2–0.5 μm) may also occur in the L_4 phase of a biphasic L_α/L_4 system. The high viscosity of the L_4 phase ensures that no phase demixing occurs. Figure 1.47 depicts the relationship of the tenside concentration ϕ and the Gaussian elasticity constant $\bar{\kappa}$ to the formation of these "membrane phases".

Detailed information about liquid-crystalline phases, with particular reference to biological membranes, may be found in references [25, 42, 43].

In this context, it should be mentioned that vesicle formulations in common use are not thermodynamically stable, but rather kinetically metastable, and can be manufactured, for example, from lamellar phases by energy input.

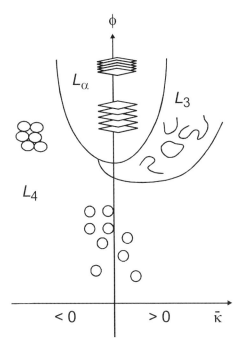

Figure 1.47. Sketch phase diagram for the sodium dodecyl sulfate/aqueous salt solution/alcohol system as a function of tenside concentration ϕ and Gaussian elasticity constant $\bar{\kappa}$. L_α: lamellar phase; L_3: sponge phase; L_4: vesicle phase (from reference [44]).

References for Chapter 1:

[1]   D. J. Shaw, Introduction to Colloid and Surface Chemistry, Butterworths, London, 1980.
[2]   C. M. Hansen, A. Beerbower, Solubility Parameters, in Kirk–Othmer, Encyclopedia of Chemical Technology, Supp. Vol., 2nd ed., 1971, p. 889.
[3]   A. F. M. Barton, CRC Handbook of Solubility Parameters and Other Cohesion Parameters, CRC Press, Boca Raton, FL, 1983.
[4]   D. H. Everett, Basic Principles of Colloid Science, Royal Soc. of Chemistry, London, 1988.
[5]   E. Matijevic, Chem. Technol. 1973, p. 656.
[6]   R. Menold, Chem.-Ing.-Tech. *58*, 533 (1986).
[7]   R. Polke, Farbe + Lack *90* (6), 457 (1984).
[8]   M. Glor, Dechema-Kurs Formulierungstechnik 1987.
[9]   H. C. Hamaker, Physica *4*, 1058 (1937).
[10] A. Martin, J. Swarbrik, A. Camerata, Physikalische Pharmazie, Wiss. Verl. Ges., Stuttgart, 1980.
[11] M. J. Rosen, J. Am. Oil Chem. Assoc. *49*, 295 (1972).
[12] M. J. Rosen, Chemtech *15*, 292 (1985).
[13] McCutcheon's Detergents and Emulsifiers Annual, Allured, Ridgewood, N.Y.
[14] L. Carino, H. Mollet, Ber. VI. Int. Kongr. grenzflächenakt. Stoffe, Zürich, 1972, p. 563.
[15] D. G. Cooper, American Oil Chemists Society Monogr. *11*, 281 (1984).
[16] U.S. Patent No. 4 395 353.
[17] W. D. Harkins, The Physical Chemistry of Surface Films, Reinhold Publ., New York, 1952.
[18] H. W. Fox, W. A. Zisman, J. Colloid Sci. *7*, 428 (1952).
[19] W. C. Preston, J. Phys. Colloid Chem. *52*, 84 (1948).
[20] H. Andree, P. Krings, Chem. Ztg. *99*, 168 (1975).
[21] K. A. Dill et al., Nature, *309*, 42 (1984).
[22] K. S. Birdi, T. Magonisson, Colloid Polym. Sci. *254*, 1059 (1976).
[23] K. Shinoda, Principles of Solution and Solubility, Marcel Dekker, Inc., Basel, New York, 1978.
[24] J. N. Israelachvili, J. Michell, B. W. Ninham, J. Chem. Soc. Faraday Trans. II *76*, 1525 (1976).
[25] P. Ekwall, in Advances in Liquid Crystals, Vol. 1 (D. H. Brown, Ed.), Academic Press, New York, 1975.
[26] A.-M. Bellocq, in Surfactant Sci. Ser., Vol. 61: Emulsions and Emulsion Stability (J. Sjöblom, Ed.), Marcel Dekker, Inc., New York, 1996.
[27] G. Porte, J. Marignan, P. Bassereau, R. May, J. Phys. (Paris) *49*, 511 (1988).
[28] D. Gazeau, A. M. Bellocq, D. Roux, T. Zemb, Europhys. Lett. *9*, 447 (1989).
[29] A. M. Belocq, D. Roux, in Microemulsions: Structure and Dynamics, (S. E. Friberg, P. Bothorel, Eds.), CRC Press, Boca Raton, FL, 1987.
[30] M. Kahlweit, R. Strey, G. Busse, J. Phys. Chem. *94*, 3881 (1990).

[31]   A. M. Bellocq, D. Roux, in Progress in Microemulsions (S. Martellucci, A. N. Chester, Eds.), Plenum Press, New York, 1985.

[32]   H. Kunieda, K. Nakamura, A. Uemoto, J. Colloid Interface Sci. *150*, 235 (1992).

[33]   W. M. Gelbart, A. Ben-Shaul, D. Roux (Eds.), Micelles, Membranes, Microemulsions and Monolayers, Springer Verlag, New York, 1994.

[34]   A. M. Bellocq, J. Biais, P. Bothorel, B. Clin, G. Fourche, P. Lalanne, B. Lemaire, B. Lemanceau, D. Roux, Adv. Colloid Interface Sci. *20*, 167 (1984).

[35]   L. M. Prince (Ed.), Microemulsions. Theory and Practice, Academic Press, New York, 1976.

[36]   I. D. Robb (Ed.), Microemulsions, Plenum Press, New York, 1982.

[37]   K. L. Mittal (Ed.), Micellization, Solubilization and Microemulsions, Plenum Press, New York, 1977.

[38]   S. Friberg, P. Bothorel (Eds.), Microemulsions: Structure and Dynamics, CRC Press, Boca Raton, FL, 1987.

[39]   H. L. Rosano, M. Clausse (Eds.), Microemulsion Systems, Marcel Dekker, Inc., New York, 1987.

[40]   M. Corti, V. Degiorgio (Eds.), Physics of Amphiphiles: Micelles, Vesicles and Microemulsions, North Holland, Amsterdam, 1987.

[41]   B. D. Simons, M. E. Cates, J. Phys II (Paris) *2*, 1439 (1992).

[42]   V. Luzzati, in Biological Membranes (D. Chapman, Ed.), Academic Press, New York, 1968.

[43]   K. Fontell, Prog. Chem. Fats Lipids *16*, 145 (1978).

[44]   D. Roux, S. Candaux, Images de la recherche: les systèmes moléculaires organisés, ed. CNRS, 1994, p. 29.

2.1.1 Phase Volumes and Emulsion Types

A collection of spheres of equal radius packed as closely as possible fills 74% of the volume it takes up; the remaining 26% is empty or, in the case of an emulsion, is the outer phase (Figure 2.6). According to W. Ostwald's theory, at a ratio between the phases $\phi > 0.74$, an emulsion is too densely packed and either phase inversion or disruption of the emulsion will occur. For a given system, between the phase volume ratios 0.26 and 0.74 both o/w and w/o emulsions are possible; below and above these ratios, only one type can exist if the spheres are homogeneous.

If the spheres are not homogeneous, a greater packing density than 74% is possible, since the smaller spheres can fit between the larger ones (Figure 2.7). If the droplets can be deformed, still greater packing densities can be achieved. Figure 2.8 represents a densely packed emulsion of polyhedral droplets analogous to a polyhedral foam.

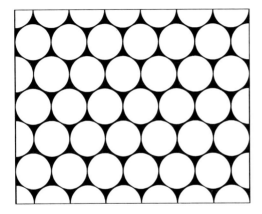

Figure 2.6. Close packing of spheres filling 74% of the available space; cf. [3].

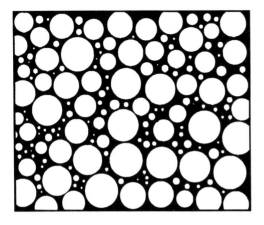

Figure 2.7. Packing densities > 74% are possible for particles of different sizes; cf. [3].

Figure 2.8. Polyhedral foam with lamellae; cf. [3].

2.1.2 Emulsion Viscosity

If the inner phase of an emulsion occupies less than 30% of its volume, the individual droplets seldom interfere with each other. The change in viscosity as a function of concentration can be described approximately with the *Einstein Formula*, Equation 2.4:

$$\eta = \eta_0 \left(1 + 2.5 \cdot \phi\right) \tag{2.4}$$

(ϕ = ratio of volumes of the inner to the outer phase; η_0 = viscosity of the outer phase)

For the manufacture of a highly viscous emulsion under these conditions, Equation 2.4 shows that the outer phase (the aqueous phase in the case of an o/w emulsion, a solvent immiscible with water in the case of a w/o emulsion) must already have a high viscosity. Thickeners soluble in the outer phase can be used to alter the viscosity as required. This aspect is of particular importance in the case of emulsions for use in cosmetics.

When the inner phase occupies more than ϕ_i 30% of the volume, the droplets begin to influence one another, and the viscosity increases up to a ϕ_i value of 50–52%. At higher values it escalates sharply, accompanied by non-Newtonian behavior. At $\phi_i \cong 68\%$, the emulsion is usually unstable unless a special emulsifier is added; this is the *inversion point*. At still higher inner- to outer-phase volume ratios, either the spheres adopt an even denser packing arrangement (honeycomb) or flatten out. The maximum possible occupation of volume is achieved at $\phi_i = 74\%$. At $\phi_i > 74\%$, the emulsion becomes poly-disperse. If water is added, the viscosity–concentration plot does not follow the previous course but makes a hysteresis curve (Figure 2.9).

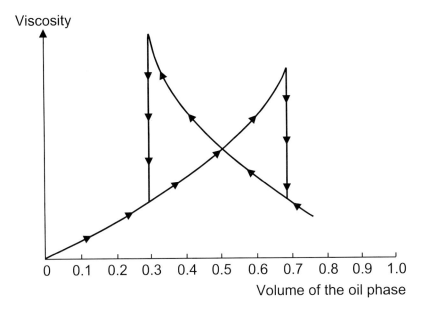

Figure 2.9. Viscosity hysteresis on dilution of a concentrated emulsion [3].

2.1.3 Emulsifiers

The correlations between the chemical structure of surface-active substances and their emulsifying activity are complex, since the composition of oil and water phases varies. As both the composition of the two phases and the concentration of the dispersant must be considered, *it is not possible to classify specific tensides as general emulsifiers.* Nevertheless, there are some general guidelines that are helpful in the selection of surfactants to act as emulsifiers.

For a tenside to act as an emulsifier, it must have the following properties:

1. It must show good surface activity and create a low surface tension. If this is not the case, the emulsifier can be combined with a suitable tenside. It must tend to migrate to the surface rather than remaining in solution in the bulk phase. Therefore, it must have both hydrophilic and hydrophobic groups. Too great a solubility in either one of the two phases impairs the efficacy.

2. It must form a film at the interface, either on its own or in conjunction with other molecules also adsorbed there. The film must be a *condensed film*; that is, *in the case of o/w emulsions, the hydrophobic groups in the interfacial film should interact strongly in a lateral direction, i.e. with their neighbors in the film; in the case of w/o emulsions, the hydrophilic groups should interact thus.*

3. It must migrate to the interface fast enough to ensure that the interfacial tension is lowered sufficiently while the emulsion is being manufactured.

4. Emulsifiers that are preferentially soluble in oil give w/o emulsions; low molecular weight, hydrophilic emulsifiers such as soaps induce o/w emulsions, as do water-soluble, macromolecular emulsifiers.

5. A mixture of a preferentially oil-soluble tenside with a water-soluble tenside creates more stable emulsions than a single tenside.

6. The more polar the oil phase, the more hydrophilic the emulsifier should be; the less polar the oil to be emulsified, the more lipophilic the emulsifier.

The adsorbed emulsifier molecules are oriented at the phase boundary such that their hydrophobic parts are in the oil and their hydrophilic parts are in the water. *Although the reduction in interfacial tension is indeed significant for the dispersion process and for spontaneous emulsification at very low interfacial tension, it is not of decisive importance for the stability of the emulsion.* It has been found that the most effective emulsifiers occupy little space in the interfacial layer.

The adsorbed quantity of the emulsifier can be calculated from the Gibbs Equation (Equation 2.5):

$$\Gamma = -\frac{c}{RT} \cdot \frac{\mathrm{d}\gamma}{\mathrm{d}c} \tag{2.5}$$

(c: concentration of the tenside in the bulk of the liquid [mol/L]; Γ: excess tenside per unit area of the interface compared with the quantity in the bulk of the liquid [mol/cm²]).

Once the excess tenside concentration at the interface is known, the area occupied by each individual molecule in the interfacial layer can be calculated.

Example: $\Gamma = 4.0 \cdot 10^{-10}$ mol/cm²
    Number of molecules/cm² $= \Gamma \cdot N_L = 4.0 \cdot 10^{-10} \cdot 6.023 \cdot 10^{23} = 24.09 \cdot 10^{13}$ cm⁻²
    Area per molecule $= 1/(\Gamma \cdot N_L) = 41 \cdot 10^{-16}$ cm² $= 41$ Å
    (N_L: Loschmidt number).

2.1.3.1 Mode of Action of an Emulsifier

Emulsion formation by chemical means (emulsifiers) occurs as follows:

To start with, the emulsifier must be present at the interface between the phases to be emulsified in a sufficient quantity to ensure that rapid adsorption lowers the interfacial tension so much that the phases separate into droplets. The resulting flows and turbulence cause further division, which is reinforced by the spreading of molecules of the emulsifier in the interface. This spreading itself contributes to the rapid coverage of the interface. The supply of more emulsifier to the interface must also be assured; that is, the rate of diffusion from the bulk solution to the interface must be adequate. If the energy supplied to the system is no longer sufficient to increase the surface area of the droplets further, then the division of the inner phase stops.

The adsorbed layers around the oil or water droplets should prevent their coalescence. A thin boundary layer made up of the continuous phase, that is, the dispersion medium, forms between approaching droplets (Figure 2.2). The physical properties of this layer are determined by the type of the adsorbed emulsifier.

The molecules of emulsifier diffuse continuously out of the outer phase into this intervening layer and decrease the interfacial tension. New emulsifier molecules are supplied from the continuous phase outside the boundary layer by diffusion; that is, the diffusion path is relatively long. As a result, the surface tension decreases more slowly in the intervening layer than it does elsewhere at the interface between the droplet and the outer phase. The adsorbed layer moves physically along the oil/water interface into the area between the two droplets; as it does so, it carries a thin layer of liquid with it, thus preventing the intervening layer from thinning and, therefore, the droplets from coalescing. This process is called the *Marangoni effect*. It also plays an important role in other areas, for example the stability of foams (see Chapter 4).

For stable emulsions, the thickness of the intervening layer has been observed to decrease fairly quickly to a limiting value of about 50–200 Å. The fact that such a thin liquid layer resists further thinning shows that the layer exerts an overpressure, known as the gap pressure, which protects the particles from coalescence.

2.1.3.2 Survey of Common Emulsifiers

It can be seen from the list below that common emulsifiers are of the most varied chemical types. A good overview of the products available commercially, especially in the American and British markets, can be obtained from the international edition of "McCutcheon's Detergents and Emulsifiers" published annually [4].

Emulsifiers

A. *Low molecular weight emulsifiers of principally hydrophilic nature. Preferred type as o/w emulsifiers*

1. *Anionic*
   soaps (Na, K, NH$_4$, and morpholinium salts of fatty acids), Na lauryl sulfate, Na cetyl sulfate, Na mersolate, Na 2-ethylhexyl sulfate, Na xylenesulfonate, Na naphthalenesulfonate, Na sulfosuccinate, R–COOC$_2$H$_4$SO$_3$Na, R–CONHC$_2$H$_4$SO$_3$Na (R = C$_{17}$H$_{33}$), Na oleyl lysalbinate, Na oleyl protalbinate, Turkey-red oil, natural sulfonated oils, Na salts of dialkyl sulfosuccinate esters, bile salts, resin soaps.

2. *Cationic*
   laurylpyridinium chloride, lauryltrimethylammonium chloride, laurylcolamine formylmethylpyridinium chloride.

3. *Nonionic*
   polyoxyethylene fatty alcohol ethers, polyoxyethylene fatty acid esters.

B. *Low molecular weight emulsifiers of principally lipophilic nature. Preferred type as w/o emulsifiers*
   Mg stearate, Mg oleate, Al stearate, Ca oleate, Ca stearate, Li stearate, di-, tri- etc. esters of fatty acids with polyols, cholesterol, lanolin, oxidized fats and oils.

C.   Low molecular weight emulsifiers with less pronounced properties
fatty acid esters of polyols and polyoxyethylene, polyoxypropylene fatty alcohol ethers,
polyoxypropylene fatty acid esters, lecithin, monoesters of fatty acids and polyols,
triethylcetylammonium cetyl sulfate, laurylpyridinium laurate, chloronitroparaffins.

D.   High molecular weight emulsifiers
albumin, casein, gelatin, products of protein degradation (glue), gum arabic, tragacanth,
carrageenan, saponin, cellulose ethers and esters, polyvinyl alcohol, polyvinyl acetate,
polyvinylpyrrolidone.

In Table 2.1 below, further examples of common emulsifiers are listed (see also
references [4–6]).

Table 2.1. Examples of emulsifiers.

C_9H_{19}—⬡—$(O\text{-}C_2H_4)_n OH$ lipophilic hydrophilic	nonionic emulsifier
$C_{12}H_{25}$—⬡—$SO_3^- 1/2 Ca^{++}$ lipophilic hydrophilic	anionic emulsifier
Oleyl—N$\big\langle$ $(C_2H_4O)_n H$ / $(C_2H_4O)_m H$ lipophilic hydrophilic	cationic emulsifier $n, m = 1 \text{ - } 40$
R—C(=O)—NH-CH$_2$-CH$_2$-N$^{(+)}$(CH$_3$)$_2$—CH$_2$-COO$^{(-)}$ lipophilic hydrophilic	amphoteric (zwitterionic) emulsifier $R = Alkyl\ C_7 \text{ - } C_{17}$

We should also mention AMP (= 2-amino-2-methyl-1-propanol) here; many authors
refer to it as the most effective cationic (in protonated form) emulsifier.

2.1.3.3 Systems for the Selection of Emulsifiers

The most carefully constructed system is that of Griffin, based on the concept of hydrophile–lipophile balance (HLB). Each emulsifier is assigned a dimensionless number between 0 and 20:

Numbers between 0 and 9:   oil-soluble hydrophobic emulsifiers
Numbers between 11 and 20: water-soluble hydrophilic emulsifiers

Examples are given in Table 2.2.

Table 2.2. HLB values for emulsifiers.

Name	Chemical name	Type	HLB
Span 85	sorbitan trioleate	N	1.8
Tegin 0	glycerin mono/dioleate	N	3.3
Span 80	sorbitan monooleate	N	4.3
Brij 72	ethoxylated stearyl alcohol (2 mol [†])	N	4.9
Catinex KB-10	ethoxylated nonylphenol	N	6.6
Triton X-35	ethoxylated octylphenol	N	7.8
* Atlox 4861 B	alkyl aryl sulfonate	A	8.6
* Eumulgin RT20	ethoxylated technical castor oil	N	9.6
* Tween 85	ethoxylated sorbitan trioleate	N	11
* Igepal CA-630	ethoxylated nonylphenol (9 mol [†])	N	12.8
* Atlox 4851B	mixture of nonionic and anionic	N/A	13.2
* Synperonic OP11	ethoxylated octylphenol (11 mol [†])	N	14
* Synperonic NP15	ethoxylated nonylphenol (15 mol [†])	N	15
* Renex 720	ethoxylated C_{13}–C_{15} alcohols	N	16.2
*	Na oleate	A	18
Myrj 59	polyethoxyethanol(100) stearate	N	18.8
Ethomeen T/25	ethoxylated stearamine	C	19.3
	Na lauryl sulfate (pure)	A	40

*   emulsifier for o/w emulsions
N nonionic
A anionic
C cationic
[†] number of moles of ethylene oxide per mole of tenside

Emulsifiers with a HLB value of 10 are said to be balanced hydrophilically–lipophilically. The emulsifier is oriented such that its hydrophobic hydrocarbon residue dips into the oil phase while the hydrophilic groups are in water. If the HLB concept is applied to ionic tensides, additional dissociation effects can result in HLB values over 20; for example, sodium dodecyl sulfate has a value of 40.

The HLB System

HLB = Hydrophile-Lipophile Balance

In Griffin's system [7], each tenside is assigned an HLB value (from 1 to 20) on the basis of the hydrophilic portion of its molecule. Griffin at first developed the HLB system only for nonionic substances, such as fatty acid esters, polyglycol ethers of polyols, fatty acids, fatty alcohols, etc., and then extended it later to include other tensides (Table 2.3).

Table 2.3. Tensides and HLB values.

	HLB value	Application
lipophilic	0 – 3	defoamers
	3 – 8	w/o emulsions
	7 – 9	wetting agents
	8 – 18	o/w emulsions
	11 – 15	detergents
hydrophilic	15 – 18	solubilizers

Determination of HLB value:

– Calculation from the theoretical composition (for nonionic ethylene oxide adducts):

$$HLB \ \ value = \frac{\text{molar mass of the hydrophilic part}}{\text{molar mass of the emulsifier}} \cdot 20 \ \ = \frac{M_H}{M} \cdot 20 \qquad (2.6)$$

– Calculation from the saponification number S of the ester and the acid number A of the recovered acid (for fatty acid esters):

$$HLB = 20 \cdot \left(1 - \frac{S}{A}\right) \qquad (2.7)$$

– by NMR (for nonionic ethoxylates)
– by calorimetry (enthalpies of mixing)
– with reference substances (oils) of known HLB value

As is clear from Equation 2.6, the HLB value reflects the proportion of the molecule in the aqueous phase with respect to the relative molecular mass.

The HLB values of a mixture of emulsifiers, HLB_{Mi}, are obtained by adding the components together (Equation 2.8):

$$HLB_{Mi} = HLB_1 \cdot g_1 + HLB_2 \cdot g_2 + ...$$ (2.8)

$g_1, g_2, ...$ are the mass fractions of the components.

For fatty acid esters, the HLB value can be calculated from the saponification number of the ester and the acid number of the fatty acid, Equation 2.7.

There are other formulae for ethoxylated fatty acid esters and other classes of compounds. The HLB values obtained for ionic emulsifiers from Equation 2.6 are too low; in other words, the emulsifier is more hydrophilic than expected from the ideal distribution. The equation can be corrected with an additional term, C (Table 2.4), as in Equation 2.9. When C is positive, the effective HLB value is greater than expected in the ideal case. This means that the tenside molecules are dipping more deeply into the aqueous phase than was assumed. The hydrophilic groups must therefore be pulling the attached hydrophobic residues into the aqueous phase somewhat. If C is negative, the HLB value becomes smaller; the tenside molecules are more hydrophobic than predicted. They are not so deeply engaged with the aqueous phase as expected in the ideal case. The hydrophobic groups pull their hydrophilic neighbors into the oil phase a little.

Correction of HLB values for ionic emulsifiers and polyglycol ethers:

$$HLB = 20 \cdot \frac{M_H}{M} + C$$ (2.9)

Table 2.4. Correction factors C according to Griffin [7].

Emulsifier	C
aliphatic polyglycol ethers	−1.2
aromatic polyglycol ethers with one alkyl group	−1.9
aromatic polyglycol ethers with two alkyl groups	−4.4
ethanolamine salts of *n*-dodecylbenzenesulfonic acid	+2.1
sodium salts of *n*-alkyl sulfonates	+5.5
sodium salts of *n*-alkyl sulfates	+6.0

Since in general HLB values have been recorded for nonionic emulsifiers above all, and few for anionic emulsifiers, this aspect has been refined further for use in the application of anionic tensides/emulsifiers. In this case, the HLB value can be estimated by means of an incremental method (Equation 2.10, Table 2.5).

The HLB values thus calculated for nonionic emulsifiers generally agree with values obtained by Griffin's method. Ionic tensides, on the other hand, yield values well over 20, so greater than Griffin's.

Incremental method for calculation of HLB values:

$$HLB = 7 + \sum H + \sum L \qquad\qquad (2.10)$$

Table 2.5. HLB increments for hydrophilic and hydrophobic groups.

Hydrophilic groups		Lipophilic groups	
group	H value	group	L value
NaSO$_4$-	38.7	-CH<	0.47
KOOC-	21.1	-CH$_2$-	0.47
NaOOC-	19.1	-CH$_3$-	0.47
HOOC-	2.1	-CF$_2$-	0.87
HO- (free)	1.9	-CF$_3$	0.87
-O-	1.3	benzene ring	1.66
-OH··· (sorbitan ring)	0.5	-(CH$_2$CHCH$_3$O)-	0.11
N (tertiary amine)	9.4		
ester (free)	2.4		

Examples:

oleic acid C$_{17}$H$_{33}$COOH	$7 + 2.1 - (17 \cdot 0.47) = 1.1$
Na oleate	$7 + 19.1 - (17 \cdot 0.47) = 18.1$
sorbitol monooleate	$7 + (5 \cdot 1.9) - (17 \cdot 0.47) = 8.5$
Na lauryl sulfate	$7 + 39 - (12 \cdot 0.47) = 40$
propylene glycol monolaurate	$7 + 1.9 - (12 \cdot 0.47) = 3.3$

Numerous other methods exist for the determination of HLB values which cannot be described here. We will merely mention the measurement of HLB values from the *phase inversion temperature PIT* (for which see later). Certain relationships between the PIT value and the HLB value can be defined.

The advantage of using the PIT rather than the HLB is that the PIT is directly measurable and all conditions such as type of oil, composition of aqueous phase, etc., are automatically taken into consideration. *This method is only suitable for nonionic emulsifiers.*

The HLB concept is an empirical principle. HLB values have no physical basis:

– they do not permit prediction of the actual stabilizing action of interface-active compounds;
– the structure and composition of the oil phase and the composition of the aqueous medium are not taken into consideration (e.g. addition of electrolytes);
– if there are several polar groups in the molecule, their relative positions are not considered;
– mixtures of interface-active substances cannot be treated additively.

In contrast, a physically based model for the characterization of emulsifiers with respect to type of emulsion and stability can be derived from the modes of action of adsorbed layers [8].

C. M. Donnald [9] was able to find a physical relationship between HLB values and solubility parameters for alkyl polyglycol ethers. R. G. Laughlin [10] treated HLB values from a thermodynamic point of view.

2.2 Formulation of Emulsions

The practical significance of the HLB concept lies in the fact that every substance that is to be emulsified has its own "*required HLB value*". To emulsify a particular substance, an emulsifier or a mixture of emulsifiers with the same HLB value must therefore be used. This can be determined as follows.

2.2.1 Determination of the Requisite HLB Value of an Oil

Mixtures are prepared of a pair of emulsifiers of the same chemical type, one lipophilic and one hydrophilic, for example Span 60 (sorbitan monostearate) and Tween 60 (polyoxyethylene(20) sorbitan monostearate), with HLB values covering a range from around 5 to 14. The HLB value of a mixture is calculated from the sum of the individual percentages of the components (Table 2.6).

From these ten mixtures of emulsifiers, test emulsions are produced, each with the following composition:

    20 % oil
     4 % mixed emulsifiers (20 % with respect to the oil)
    76 % water

The oil is placed in a glass measuring cylinder and the emulsifier dissolved therein. The water is added to this homogeneous mixture, which is then emulsified by stirring, homogenization, etc. It is important that each of the 10 test emulsions be made in exactly the same way. After the emulsions have been left to stand for a time depending on the effectiveness of the emulsification, they are compared for transparency, creaming, sedimentation, and turbidity. The entire test procedure must be carried out at a constant room temperature.

If there is no obvious difference between the different emulsions, the series of tests is repeated with less emulsifier. If, on the other hand, all the emulsions are poor and there is little difference between them, the test is carried out once more with a larger quantity of emulsifier.

Once an optimum mixture has been identified in the test series, for example at an HLB value of 10.3, the value can be determined more exactly by means of further experiments with mixtures of emulsifier in a narrow HLB range, say between 9.5 and 11.

Table 2.6. HLB values of test mixtures of emulsifiers.

No.	Emulsifier mixture		Calculated HLB value
	Span 60 [%]	Tween 60 [%]	
1	100	–	4.7
2	87	13	6.0
3	68	32	8.0
4	50	50	9.8
5	45	55	10.3
6	40	60	10.8
7	35	65	11.3
8	30	70	11.8
9	25	75	12.2
10	20	80	12.9

A selection of the requisite HLB values for commonly emulsified substances is given in Tables 2.7 and 2.8 (from the Atlas specialty chemicals brochure, which contains seven chapters on the HLB system; undated).

Table 2.7. Requisite HLB values (w/o) for commonly emulsified substances (±1).

gasoline	7
mineral oil	6
petroleum	6

As has already been mentioned, the HLB value of a mixture of emulsifiers can be easily calculated from those of the component fractions.

Example: HLB value of a mixture of 70% Tween 80 (HLB = 15) and
30% Span 80 (HLB = 4.3) →
HLB = 0.7 · 15 + 0.3 · 4.3 = 11.8.

Here we will highlight a range of nonionic emulsifiers with an HLB scale from 1.8 to 16.7. This is the HLB system available from Atlas Chemie; it comprises lipophilic sorbitan esters (Spans) with HLB values from 1.8 to 8.6, and their polyoxyethylene derivatives, hydrophilic Tweens, which have high HLB values between 9.6 and 16.7. Span 20 and Tween 20 are two emulsifiers of the same chemical class, one hydrophobic and one hydrophilic laurate ester. The emulsifiers with the number 40 are palmitate esters, those with 60 are stearates, and those with 80 are oleates.

The HLB values of a mixture of two emulsifiers and the requisite HLB value of an oil are related as described in Equation 2.11 (W_A: % emulsifier A, W_B: % emulsifier B).

Table 2.8. Requisite HLB values (o/w) for frequently emulsified substances (±1).

acetophenone	14	ethyl benzoate	13
acid, dimer	14	ethylaniline	13
acid, isostearic	15–16	fenchone	12
acid, lauric	16	glycerin monostearate	13
acid, linoleic	16	heavy gasoline	14
acid, oleic	17	isopropyl lanolate	14
acid, ricinoleic	16	isopropyl myristate	11–12
alcohol, cetyl	15–16	isopropyl palmitate	11–12
alcohol, decyl	15	jojoba oil	6–7
alcohol, hexadecyl	11–12	lanolin, anhydrous	9
alcohol, isodecyl	14	lanolin, liquid	9
alcohol, isohexadecyl	11–12	laurylamine	12
alcohol, lauryl	14	menhaden oil	12
alcohol, oleyl	13–14	methylphenylsilicone	11
alcohol, stearyl	15–16	methylsilicone	11
alcohol, tridecyl	14	mineral oil, paraffin base	10
arachidyl propionate	7	mineral oil (light), "	10–11
Arlamol E	7	mineral oil (medium), "	9
beef suet	5	mineral oil, naphthenic (light)	11–12
beeswax	9	mink oil	5
benzene	15	*N,N*-diethyltoluamide	7–8
benzonitrile	14	nitrobenzene	13
bromobenzene	13	nonylphenol	14
butyl stearate	11	*ortho*-dichlorobenzene	13
cake paraffin	10	palm oil	10
carbon tetrachloride	16	peanut oil (hydrogenated)	6–7
carnauba wax	15	petroleum	14
castor oil	14	pine oil	16
ceresin wax	8	polyethylene wax	15
chlorobenzene	13	polyoxypropylene(30) cetyl ether	10–11
chloroparaffin	12–14	propylene, tetramer	14
cocoa butter	6	rapeseed (canola) oil	6
corn oil	10	silicone oil (volatile)	7–8
cottonseed oil	5–6	soya oil	6
cyclohexane	15	styrene	15
decahydronaphthalene	15	toluene	15
decyl acetate	11	trichorotrifluoroethane	14
diethylaniline	14	tricresyl phosphate	17
diisooctyl adipate	9	Vaseline	7–8
diisooctyl phthalate	13	white spirit	14
diisopropylbenzene	15	xylene	14
dimethylsilicone	9		

$$HLB_{Oil} = \frac{W_A \cdot HLB_A + W_B \cdot HLB_B}{W_A + W_B} \tag{2.11}$$

In order to determine the amount of an emulsifier (A) that must be mixed with some other emulsifier (B) to achieve an HLB value of X, the equation below can be used (2.12):

$$\% \ (A) \ = \ \frac{100 \cdot \left(X - HLB_B \right)}{HLB_A - HLB_B} \quad \textit{where} \quad \% \ (B) = 100 - \% \ (A) \tag{2.12}$$

Example: How much Span 80 (HLB = 4.3) and how much Tween 80 (HLB = 15) are required to give an HLB value of 12?

$$\% \ \text{Tween 80} = \frac{100 \cdot \left(12 - 4.3 \right)}{15 - 4.3} = 72 \ \% \tag{2.13}$$

$$\% \ \text{Span 80} = 100 - 72 \ \% = 28 \ \% \tag{2.14}$$

It may also be that the preliminary experiments yield a good emulsion at an HLB value of, say, 4.7, and another at a value of 12. The emulsion with the low HLB value is a w/o emulsion (cannot be diluted with water), that with the high value an o/w emulsion (easily diluted with water).

2.2.2 Effect on the Emulsion of the Class of the Emulsifier

An HLB value of a given magnitude can be obtained by mixing emulsifiers of various chemical types. The "correct" chemical type is just as important as the "correct" HLB value. If we know that a mixture of Span 60 and Tween 60 (stearates) with an HLB of 12 yields a better emulsion than any other mixture of these emulsifiers, this HLB value will also be fairly close to the best for other chemical classes of emulsifiers, and we must now check whether Span–Tween mixtures of other types (such as laurate, palmitate, or oleate) with an HLB value of 12 are more effective than the stearates, or whether emulsifiers from other classes give even better results.

Figure 2.10 a shows an emulsifier of the polyoxyethylene (POE)–sorbitan oleate ester type with its unsaturated oleate residue in oil. An unsaturated chain of this type seems to prefer unsaturated oils.

A similar emulsifier is depicted in Figure 2.10 b, but this time with a stearate residue instead of oleate. Saturated chains such as this, or laurate or palmitate, seem to prefer saturated oils. Although both types of oil may require an emulsifier with an HLB value of 12, the emulsifier that has a chemical structure closer to that of the oil is considerably more effective.

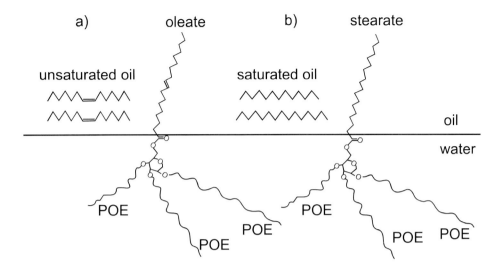

Figure 2.10. Selection of Tween type to correspond to the type of the oil to be emulsified.

The influence of the chemical class of the emulsifier on the emulsion stability is plotted in Figure 2.11. Three mixtures, A, B, and C, of emulsifiers of different chemical structure were used, all suitable for the oil to be emulsified and all used in the same concentration. The graph shows that all three mixtures yield good emulsion stability in the area around an HLB value of 12.

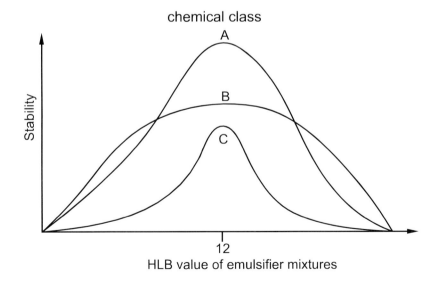

Figure 2.11. Stabilization of emulsion by different classes of emulsifiers as a function of the HLB value of the emulsifier mixtures.

However, mixture A is by far the best. Nevertheless, it is possible to achieve similarly good results by increasing the concentration of B or C. The choice of emulsifier is often dictated by economic and practical considerations. Mixture B may be preferred over mixture A if it is cheaper or if fair emulsion stability over a wide HLB range is more advantageous than excellent stability over a very narrow range.

It may well be found that oleates in particular give good smoothness, while stearates are particularly suited to adjust the consistency. In the same system, the laurates might give excellent emulsion stability at very low emulsifier concentrations, and thus permit economical formulation.

2.3 Stabilization by Solid Particles

Very finely divided particles that are smaller than the droplets in an emulsion and are well wetted by the water or oil phase can stabilize an emulsion. They accumulate at the water/oil phase boundary in the form of a solid layer which prevents the inner phase from coalescing (Figure 2.12).

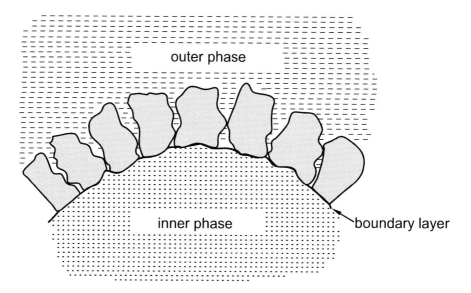

Figure 2.12. Accumulation of solid particles at the o/w interface.

If the solid is preferentially wetted by one of the two phases, it can gather at the interface if this curves away from the wetting phase. Bentonites, which are preferentially wetted by water, therefore form o/w emulsions, in contrast to gas soot, which is preferentially wetted by oil and yields w/o emulsions (Figure 2.13).

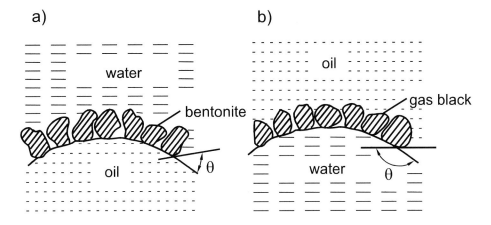

Figure 2.13. Stabilization of an emulsion by adsorption of solid a) hydrophilic and b) hydrophobic particles.

The fine particles at the interface prevent both coalescence of the droplets and also, if the solid particles repel one another, aggregation. Solid particles with a contact angle of 90° form the most stable emulsions (Figure 2.14).

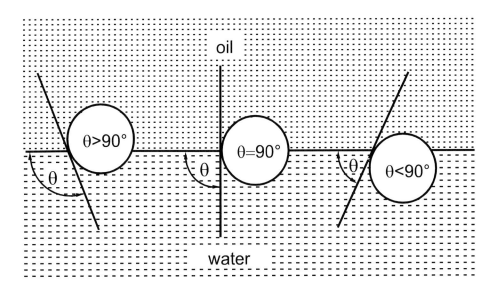

Figure 2.14. Adsorbed particles of solid with a contact angle of 90° give the most stable emulsions.

In fact, the particles are wetted principally by the outer phase. This results in the minimization of the interface between the two phases and thus in further stabilization of the droplets in the emulsion.

If only inorganic solids are used to stabilize an emulsifier, that is, if no actual emulsifier is employed, the energy necessary for the formation of the droplets must come from vigorous mixing, for example with a high-speed impeller, and the particles must be added gradually. Alternatively they can be added together with the inner and outer phases.

Benzene/water emulsions are stabilized by $CaCO_3$, toluene/water emulsions by pyrite, water/benzene with powdered charcoal.

The accumulation of the solid at the oil/water phase boundary creates an interfacial "film" of high stability and firmness, which increases the stability of the emulsion. Indeed, an emulsion stabilized with tenside does become still more stable if the interfacial properties of the disperse phase can be made to resemble those of a solid. It is also possible that high zeta potentials build up due to the presence of the powder in the interface, and consequently the emulsion is stabilized.

2.4 Phenomenology of Emulsions

The emulsion type, o/w or w/o, can be identified by various observations and experimental methods:

1. o/w emulsions have a creamy consistency, whereas w/o emulsions are oily or greasy (the viscosity of an o/w emulsion is often little different from that of a true aqueous solution; in contrast, w/o emulsions often have an unctuous or buttery consistency, which is generally a result of a liquid crystalline gel structure);

2. an emulsion mixes immediately with any liquid that is miscible with its dispersion medium;

3. an emulsion can be colored with dyes that are soluble in the dispersion medium (for example, methylene blue for o/w emulsions, Sudan blue for w/o emulsions);

4. o/w emulsions are usually fair electrical conductors.

2.5 Stability of Emulsions

Since emulsions are thermodynamically unstable, the word "stable" is used with reference to the emulsion lifetime. In this context, three important concepts should be mentioned (Figure 2.3):

1. Creaming and sedimentation. These phenomena occur as a result of disparities in density. The rising or settling of the dispersed droplets is not necessarily associated with aggregation and is generally not considered as instability. The droplets can be redispersed.

2. Flocculation. Flocculation or coagulation of the dispersed liquid particles is a type of emulsion instability. However, as long as the individual droplets exist, the emulsion has not been destroyed, as the droplets can be redispersed.

3. Breaking of the emulsion; coalescence. The emulsion is only disrupted when the droplets coalesce, and thus the phases separate and the emulsified system is destroyed. Therefore, the rate of coalescence of the droplets was chosen as the only quantitative measure for the stability of an emulsion.

2.6 Rate-Determining Factors in Coalescence

The kinetic stability or instability of an emulsion depends on various factors.

Rate-determining factors in coalescence

1. Nature of the interfacial film
2. Electrical and steric barriers
3. Viscosity of the dispersion medium
4. Volume ratio of the disperse phase and the dispersion medium
5. Droplet size distribution
6. Temperature

2.6.1 Nature of the Interfacial Film

The stability of the emulsion is heavily dependent on the *mechanical strength* of the interfacial film. The adsorbed tensides should be "condensed" by *strong lateral intermolecular forces*, but the *film should also be very elastic*. A combination of water- and oil-soluble tensides is best suited to achieving this aim. An oil-soluble tenside with long, straight hydrocarbon chains and weakly polar head groups incorporated into the film reduces electrical repulsion between the water-soluble tenside molecules and consolidates the film by dispersion forces. A *complex film* has the densest packing.

Mixed interfacial films and structural effects.

1. Highly pure emulsifiers yield loosely packed and thus mechanically poorly stable interfacial films. Therefore, good emulsifiers usually consist of a mixture of two or more tensides. The usual combination is one water- and one oil-soluble emulsifier. As described above, the oil-soluble type, with a long, straight HC chain and an only weakly polar head group, amplifies the lateral interactions between the surface-active molecules in the interfacial film and consolidates the latter, resulting in greater mechanical strength. For example: the combination of lauryl alcohol and sodium lauryl sulfate yields a densely packed monomolecular film and improved stability compared with an emulsion with only one emulsifier. Electrostatic repulsion between the ionic head groups is reduced and the hydrophobic HC chains can approach one another more closely. The most stable o/w emulsions are obtained when both components of the mixture of emulsifiers have the same chain length and are present in the same molar concentration. Further examples of emulsifier combinations are given in Table 2.9.

2. *Emulsions protected by macromolecules are not sensitive to electrolytes.* Various polymers spread over the water/air and water/oil interfaces to form coherent and mostly viscous films, and result in good emulsion stability. The molecular configuration of such polymers is flat. Examples are polyesters, poly(vinyl acetate), poly(vinyl benzoate), and poly(methyl methacrylate).

Table 2.9. More examples of combinations of emulsifiers.

Na cetyl sulfate / cetyl alcohol	good
Na cetyl sulfate / oleyl alcohol	poor
Na cetyl sulfate / cholesterol	good
cholesterol alone	poor
sorbitan ester (Span) / POE sorbitan ester (Tween)	spontaneous formation of o/w emulsions with mechanically stable interfacial films
tall oil fatty acids / alkyl aryl sulfonates (petroleum sulfonate)	ditto

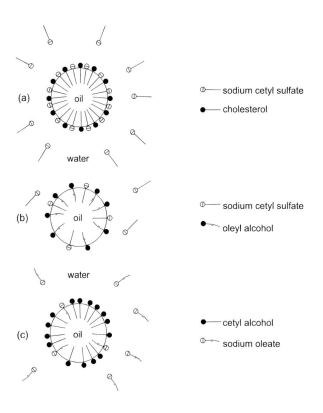

Figure 2.15. Sketch of the adsorption of emulsifier combinations at the o/w interface [11].

Figure 2.15 gives examples of a dense film (a), a film with loose packing (b), and a film with insufficient packing (c), and the corresponding emulsion stabilities.

A frequently used mixture of emulsifiers consists of oil-soluble sorbitan esters (Span) and water-soluble POE–sorbitan esters (Tween). In Figure 2.16, a possible structure is depicted for the stabilizing film built up at the phase boundary out of these molecules. The stronger interaction of the POE-sorbitan with the aqueous phase causes the hydrophilic group of this molecule to extend further into the water than does the ester without oxyethylene groups, and this permits the hydrophobic groups of the two different types of molecules to get closer to each other in the interfacial layer and to interact better than would be the case if only one of the tensides were present.

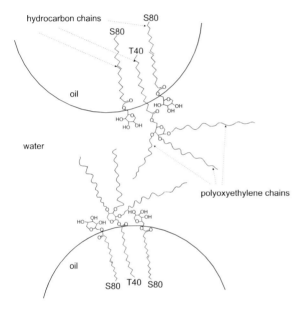

Figure 2.16. Interaction of Span 80 and Tween 40 at the oil/water interface [12].

A thick, tough film is particularly important in the case of w/o emulsions. Since the water droplets are usually electrically neutral (or weakly negatively charged), there is no electrical barrier to coalescence. The stability of the emulsion relies entirely on the mechanical effect of the protective layer. The thickness of the surface film is about 0.01 µm for most emulsions.

2.6.2 Electrostatic and Steric Barriers

In o/w emulsions, the charged, hydrophilic part of the tenside faces the water, and the electrical charge on the droplet acts as a barrier to prevent coalescence. For ionic tensides, the sign of the charge on the droplet is the same as that on the tenside.

In emulsions stabilized with nonionic tensides, the disperse phase is charged either by adsorption of ions from the aqueous phase, or by motion and friction of the droplets in the dispersion medium separating electrical double layers. When the latter occurs, the phase with the higher dielectric constant is the positively charged one.

High molecular weight emulsifiers stabilize emulsions principally by steric repulsion.

2.6.3 Viscosity of the Dispersion Medium

This aspect has already been partially treated in Section 2.1.2. The stability of the emulsion depends on the viscosity of the continuous phase because of the influence of the viscosity on the diffusion of the droplets. A low diffusion constant reduces the number of collisions, so the rate of coalescence becomes smaller. For this reason, *concentrated emulsions are often more stable than dilute ones*, since the viscosity of the continuous phase rises with the number of droplets and therefore diffusion declines. In practice, emulsion viscosity and thus stability are increased by addition of *thickeners*. Common thickeners include cellulose derivatives, gelatin, casein, starch, dextrins, carob (locust bean) flour, PVA (polyvinyl alcohol), PVP (polyvinylpyrrolidone), xanthan gum (Kelzan), carbopols (acrylic acid polymers), tragacanth, microcrystalline cellulose, and alginates. In certain proportions, mixtures of oil, water, and tenside can form *liquid crystalline structures that stabilize the system*.

2.6.4 Volume Ratio of the Disperse Phase and the Dispersion Medium

An increase in the volume of the disperse phase in relation to the volume of the continuous phase leads to the enlargement of the interfacial film area and thus to a decline in stability. If the volume of the disperse phase exceeds that of the continuous phase, the emulsion becomes unstable with respect to the inverted emulsion. The tenside layer around the disperse phase is now larger than the one that would be necessary to surround the continuous phase; therefore, it is unstable with respect to the smaller emulsifier film (which has a lower free surface energy). If both types of emulsion are possible with the emulsifier used, then a phase inversion can occur.

2.6.5 Size Distribution of the Droplets

Larger droplets are thermodynamically more stable than smaller ones, since the ratio of interface (surface area) to volume is lower. As a result the larger droplets grow at the expense of the smaller ones until the emulsion breaks. *The narrower the size distribution of the droplets, therefore, the more stable the emulsion.* The change in size distribution of droplets in an emulsion over time is represented in Figure 2.17.

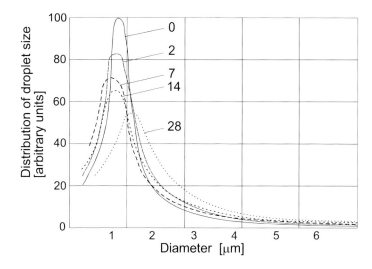

Figure 2.17. Size distribution of droplets in a stabilized emulsion at different times (in days); from reference [13].

2.6.6 Temperature

The rate of coalescence of an emulsion depends heavily on the temperature. A change in temperature alters the interfacial tension between the phases. For most liquids, γ decreases linearly with increasing temperature, as in the empirical formula of Ramsay and Shields or Eötvös. In addition, the viscosity of the interfacial film and the homogeneous phases, the solubility of the emulsifier in both phases, and the thermal motion of the particles all change.

2.7 Inversion of Emulsions

Phase inversion makes it possible for us to change an o/w emulsion into a w/o emulsion or vice versa.

The type of emulsion depends, for example, on the order in which the phases were added, the sort of tenside used, the ratio of phases, the temperature, and the presence of electrolyte or other additives.

If water is added to an apolar tenside solution, a w/o emulsion usually results, whilst addition of oil to an aqueous solution of tenside yields an o/w emulsion.

Temperature-related change in the type of emulsion is caused by the change in tenside hydrophobicity with temperature. For nonionic tensides, elevated temperatures encourage conversion from o/w to w/o emulsion type; for ionic tensides a change from w/o into o/w tends rather to occur on cooling. The behavior of the interfacial tension during the phase change is interesting. *In a narrow temperature range $\gamma_{o/w}$ tends to 0, and the emulsion droplets cease to be stable.*

Addition of electrolyte can also change the hydrophobicity of the interfacial film and thus cause phase inversion. Strong electrolytes lower the electrochemical potential of the particles, and the interactions between tenside ions and counterions are amplified. This can reduce the stability of o/w emulsions (salting out).

The inversion of an o/w emulsion stabilized by an interfacial film of sodium cetyl sulfate and cholesterol is depicted in Figure 2.18. Addition of strong electrolytes, that is, polyvalent cations (in the case represented in Figure 2.18, Ba^{2+} or Ca^{2+}), neutralizes the charge on the droplets. Small quantities of water are trapped inside the aggregating oil droplets.

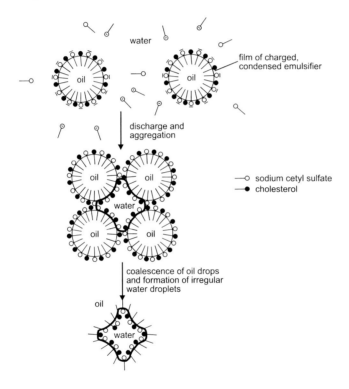

Figure 2.18. Inversion of an o/w emulsion stabilized by Na cetyl sulfate and cholesterol through addition of polyvalent cations. The negative surface charges are neutralized when the cations are adsorbed, so the oil drops can coalesce [14].

The molecules in the interfacial film align themselves such that irregularly shaped water droplets are formed, which are stabilized by a rigid, uncharged film, and dispersed in the oil. The coalescence of the oil droplets into a continuous phase completes the process of inversion. The phenomenon of coalescence is important as the start of both the creaming process and the inversion process.

2.7.1 Phase Inversion Temperature

The rate of coalescence increases as the temperature increases and the stability decreases. The rate is influenced by a variety of factors, such as the size and type of the hydrophilic and hydrophobic groups of the emulsifier, emulsifier concentration, length of hydrophobic chains, and so on.

When the temperature is raised, the degree of hydration of the hydrophilic groups of the tenside declines and the tenside becomes less hydrophilic. Its HLB value decreases. So, if an emulsion is o/w type at lower temperatures, it can invert to a w/o emulsion when the temperature is increased. Similarly, a w/o emulsion created at a high temperature can invert to an o/w type when the temperature drops. *The temperature at which this inversion occurs is called the phase inversion temperature, PIT.* At this temperature the hydrophilic and lipophilic properties of the emulsifiers are balanced. The PIT can be determined by visual observation.

Since the interfacial tension of the emulsifier reaches a minimum in the region of the PIT, it is a good idea to emulsify a mixture near the PIT, since there very small droplets can be formed with little energy expenditure.

As has already been mentioned in Section 2.1.3.3, there is an almost linear relationship between the PIT and the HLB value.

2.7.2 Emulsion Inversion Point (EIP)

The EIP is the point at which a w/o emulsion turns into an o/w emulsion when water is added. It is given in units of cm^3 water added to $1 cm^3$ oil. This assumes that the emulsion consists of water, oil, and emulsifier. The EIP value declines with increasing HLB value of the emulsifier; it reaches a minimum that corresponds to the optimum stability of an o/w emulsion. The EIP method is a fast way of finding out the physical stability of emulsions. *The most stable o/w emulsions are often manufactured starting from w/o emulsions that are then inverted.*

2.8 Emulsification Techniques

2.8.1 Procedure and Apparatus

In principle, there are two different ways to disperse a liquid finely in another immiscible liquid:
a) by condensation starting from seeds;
b) by division of large drops into smaller ones.

The condensation method is limited to a few specific applications (for example, the injection of vapor into liquid with condensation).

The usual technical procedures rely on the dispersion principle and the introduction of mechanical energy to the system. The disperse phase is subjected to a steep velocity gradient, so that the drops break up into many smaller ones. The inertial forces breaking up the droplets ($\sim v^2 \cdot d_P^2 \cdot \rho$) are counteracted by the surface forces ($\sim \gamma \cdot d_P$; d_P: particle diameter; ρ: density).

The shearing effect can be achieved either by an agitator (the stator–rotor principle) or by pressure-induced flow through a constriction. Emulsifying equipment is accordingly classified into two classes:

a) "colloid mills" (stator–rotor) for wide viscosity ranges;
b) homogenizers (pressure-induced flow) for lower viscosities.

The following equipment is used to make emulsions:

- simple mixers
- turbine mixers
- colloid mills
- homogenizers
- ball mills
- agitator ball mills
- perforated disk mills
- high-pressure homogenizers
- ultrasound apparatus

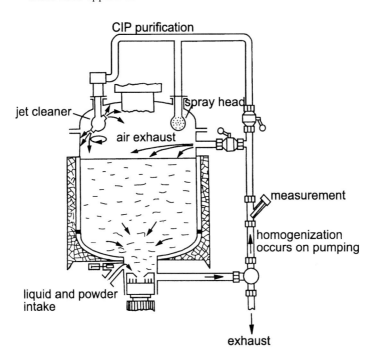

Figure 2.19. Emulsification apparatus for use in the laboratory.

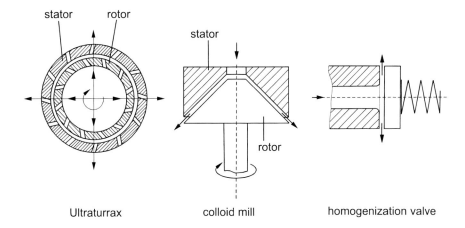

stator   rotor

stator

rotor

Ultraturrax   colloid mill   homogenization valve

Figure 2.20. Examples of emulsification apparatus.

In the choice of emulsification apparatus, after product-specific properties and viscosity the most important consideration is the mean particle size produced. Figures 2.19 and 2.20 depict some emulsification apparatus.

The characteristics of three types of emulsification equipment, the colloid mill, the ball mill, and the homogenizer, are collected in Table 2.10.

Table 2.10. Characteristics of different emulsification apparatus (from reference [15]).

	Colloid mill		Ball mill	Homogenizer
	25 m/s	50 m/s		
viscosity range [mPa·s]	10^3–10^5	1–5000	300–8000	1–20 000
optimal viscosity [mPa·s]	15 000	2000	600–2400	1–200
particle size [µm]	2–100	1–100	0.5–100	0.5–20
optimal particle size [µm]	2–4	1–3	1	0.1–2
power requirement [kW]	2–150	2–150		2–220
premixing necessary	yes	yes	no	yes
may be used with volatile solvents	yes	yes	yes	yes

Ultrasound apparatus and the high-pressure homogenizer are described in Chapter 5.

Since the break-up of large droplets into smaller ones can not only be caused by the addition of mechanical energy to the system but can also be facilitated by reduction of the interfacial tension between the two phases by means of tensides, and the fine droplets must be prevented from coalescing, *optimal formulation of the mixture with tensides, emulsifiers, and so on is at least as important as mechanical comminution*. At very low interfacial tensions, systems may even self-emulsify spontaneously (if $\gamma_{o/w}$ < 1 mN/m).

2.8.2 Process Technology: Mixing of the Components

The order and method of addition of the components have a decisive influence on the course of emulsification and its results [16, 17]. Of the four standard techniques listed below, only the first requires a large input of energy.

1. Agent in water
2. Agent in oil
3. Nascent soap method
4. Alternating addition

Method 1 is a *forced emulsification*. Note on Method 4: In the manufacture of cosmetics, two general emulsification methods are used, the "English" and "continental" methods. The former is alternating addition, the latter the agent-in-oil method, which is thought to give better emulsions but needs a higher emulsifier concentration.

Detailed descriptions of the standard techniques:

1. AGENT IN WATER
The emulsifier is dissolved directly in water and the oil then added while the mixture is stirred vigorously ("forced emulsification"). This creates o/w emulsions directly. If w/o emulsions are required, more oil is added until inversion occurs. These coarse inverted emulsions must be refined in a colloid mill.

2. AGENT IN OIL
The emulsifier is dissolved in the oil phase. The emulsion can be formed in two ways:
a) By direct addition of the mixture to water, with spontaneous formation of an o/w emulsion. This process usually creates droplets of a consistent size, about 0.5 µm.
b) By direct addition of water to the mixture, slowly and in small quantities (in order to encourage the emulsion to "start up"). Addition of further amounts of water while the mixture is slowly stirred results in inversion to give an o/w emulsion.

3. NASCENT SOAP METHOD
In situ soap formation. Only suitable for emulsions that are stabilized with soaps. Permits formation of both o/w and w/o emulsions. A fatty acid is dissolved in the oil, the alkali in the water. The soap forms at the interface when the two phases are mixed. Spontaneous formation of stable emulsions.

4. ALTERNATING ADDITION
Water and oil are added alternately in small quantities to the emulsifier. Particularly suitable for food emulsions, for example mayonnaise and other emulsions of vegetable oil.

The particle size distributions for four emulsions, each consisting of 10% olive oil, 0.5% soap, and 89.5% water, and made by four different methods, are listed in Table 2.11. In comparison, the agent-in-oil method yields more uniform emulsions (uniform particle size) with good stability.

Table 2.11. Droplet size distribution for various methods of manufacture (from reference [3]).

Particle size range	Percentage of particles in size range			
[μm]	Emulsion 1	Emulsion 2	Emulsion 3	Emulsion 4
0 – 1	47.5	71.8	68.5	80.7
1 – 2	41.1	26.4	28.4	17.1
2 – 3	7.4	1.4	2.0	2.0
3 – 4	2.1	0.3	0.5	0.2
4 – 5	0.7	–	0.1	–
5 – 6	0.1	0.1	0.3	–
9 – 10	0.2	–	–	–

Emulsion 1:   Agent in water
Emulsion 2:   Agent in water with additional homogenization after stirring
Emulsion 3:   In situ soap formation with simple stirring
Emulsion 4:   Ditto, but followed by homogenization
Emulsion composition: 10% olive oil, 0.5% soap, 89.5% water

2.8.3 Self-Emulsifying Systems and Spontaneous Emulsification

Self-emulsifying systems are concentrated formulations that form emulsions when poured into water and, if necessary, lightly agitated. The nascent soap method prefigures this phenomenon. Self-emulsifying oils are known too; in these, water-soluble soaps are solubilized with a phenol, amyl alcohol, or benzyl alcohol. Instead of the soap, sulfonated fatty acids or sulfonated aromatic and hydroaromatic hydrocarbons may be used.

Many sulfonated oils are adequately soluble in oils that are to be emulsified, even in the form of their sodium salts; the best in this respect are the triethanolamine salts of fatty acids. PEO tensides are particularly suitable for use in self-emulsifying oils, since various types are soluble in both oil and water. They can also be mixed with oil-soluble sulfonates; for example sodium dodecylbenzenemonosulfonate and octylphenol, condensed with 10–12 moles of ethylene oxide, are mixed into the oil to be emulsified.

More recently formulated self-emulsifying systems contain mixtures of emulsifiers that form complexes at the interface, for instance mixtures of sorbitan fatty acid esters and their condensation products with ethylene oxide. Self-emulsifying glycerin monostearates are mixtures of, for example, lipophilic glycerin monostearate with a hydrophilic emulsifier such as sodium stearate.

Certain individual tensides make it possible to reach extremely low $\gamma_{o/w}$ values (<1 mN/m). With these, the spontaneous formation of emulsions with very fine droplets (<0.1 μm) can be achieved. However, the stability of emulsions with this composition is unsatisfactory.

Examples of such tensides:

–   Petroleum sulfonates have a very low o/w interfacial tension: $\gamma_{o/w} \sim 10^{-3}$ dyn/cm.
–   Aerosol OT, Na di(2-ethylhexyl)sulfosuccinate, gives a very low $\gamma_{o/w}$ value.

These tensides take us into the realm of micellar emulsions.

Example of spontaneous emulsification (from Bowcott and Schumann): A coarse o/w emulsion is prepared, such as benzene in water stabilized with a soap such as potassium oleate. A long-chain alcohol like hexanol is added to this emulsion, upon which the particle size decreases until the emulsion becomes transparent, that is, until the particle size is in the range 100–500 Å. The mixture is now made up of swollen micelles; for various reasons that cannot be discussed here, it is an emulsion and not a microemulsion and is therefore not thermodynamically stable (whereas microemulsions are stable).

Three mechanisms have been postulated for spontaneous emulsification:

a) *Interfacial turbulence*
If a solution of lauric acid in oil is carefully layered over a solution of sodium hydroxide in water, an emulsion forms in the aqueous phase. Quincke supposes that the spontaneous emulsification is caused by local depression of the interfacial tension resulting from uneven soap formation. Oil droplets are torn from the interface by the subsequent vigorous spreading and then stabilized by the soap.

b) *Diffusion and stranding*
Spontaneous emulsification can be triggered by diffusion alone, for example when a solution of ethanol and toluene is gently brought into contact with water. When the alcohol diffuses out of the oil into the water, it carries some oil with it. The oil is ejected from the solution and "stranded" in the water in the form of small emulsified droplets.

c) *Negative interfacial tension*
If the interfacial tension in a small area becomes negative, the interface tends to expand spontaneously, making the interfacial tension less negative.

As a general explanation of spontaneous emulsification, it may be said that it results from hydrodynamic instability in the phase boundary when the interfacial tension is low. In such a case droplets of very small diameter (10–50 nm) are created. These have kinetic energy of about the same order of magnitude as the energy of interaction between particles. Such emulsions are therefore of high stability. Spontaneous emulsification is of considerable practical importance in the manufacture of, for example, fine latices.

2.8.4 Emulsifiable Concentrates

For some purposes it is useful to manufacture o/w emulsions just before use from a concentrate which should, if possible, be emulsifiable in water. These preparations are of particular importance in crop protection, in which active substances are often prone to hydrolysis and are therefore sold in anhydrous formulations. The user simply mixes them with the required quantity of water on the spot before use. Another area in which emulsifiable concentrates are used is that of cutting oils or cooling lubricants for metal-working; these concentrates, however, usually need intensive agitation to emulsify them in water, rather than emulsifying spontaneously like the emulsifiable concentrates used in agrochemistry.

Since the use of emulsifiable concentrates for crop protection demands spontaneous emulsification without the input of a great deal of mechanical energy, the emulsifier composition is decisive. The correct HLB value to suit the active substance must be selected. As well as this, the other requirements for the product must be fulfilled, such as spontaneity of emulsification, compatibility with the other constituents of the spray formulation, and long storage life. These requirements will certainly not be fulfilled by every emulsifier with the right HLB value. For good spontaneous emulsification, the following points are necessary:

- creation of turbulence at the interface, which is encouraged by low viscosity in the emulsifiable concentrate;
- solubilizers that encourage the mutual solubility of the phases (e.g. alcohols);
- emulsifier constituents that travel rapidly through the interface by convection.

In practice, mixtures of at least two and usually three to five emulsifiers are used. The ionic components are most frequently calcium or amine salts of alkyl aryl sulfonates, and the nonionic emulsifiers are ethylene oxide adducts of fatty alcohols, alkyl phenols, castor oil, or sorbitan esters with different HLB values. For a carefully selected mixture of ionic and nonionic emulsifiers, 5% or so in the formulation may suffice; in other cases, 10–20% may be required.

The stability of an emulsion depends on the temperature and the salt content of the water. Warm, hard water needs emulsifier mixtures with a more hydrophilic HLB value, achieved by an increased proportion of ionic emulsifiers, and vice versa for colder, softer water.

Emulsifiable concentrates are manufactured by dissolution of the liquid or solid active ingredient, along with the necessary emulsifiers and other additives, in a water-immiscible solvent or mixture of solvents. The resulting solution must be clear and must emulsify spontaneously in water. The most commonly used stabilizers are epoxides such as epoxidized soya oil. If the active substance is of low viscosity, there is no need for the solvent.

Example of an emulsifiable insecticide concentrate (from reference [18]):

    264 g/L active substance (95 %)

      38 g/L nonylphenol polyethoxyethanol (10 mol ethylene oxide, HLB 13.5)

      15 g/L castor oil polyethoxylate (32 mol ethylene oxide, HLB 16)

      25 g/L calcium dodecylbenzenesulfonate (HLB 8.6)

    250 g/L xylene

    388 g/L aromatic petroleum fraction (boiling range 170–210°C)

Apparatus: heated/cooled vessel with agitator, supply vessels with dosing meters and outlets.

2.8.5 Emulsifying Aids

Emulsifying aids help in achieving and retaining the requisite emulsion stability. They are added before or during emulsification. The following additives are among the most important auxiliaries:

Emulsion thickeners

Thickeners are added to emulsions in order to increase the viscosity of the outer phase. Since they also act as emulsifiers to some extent, like lecithin from egg yolk in the manufacture of mayonnaise, they are also known as "quasiemulsifiers". Amongst the large number of such substances are methylcellulose, sodium alginate, waxes, proteins such as gelatin and casein and gums like locust bean gum, polysaccharides such as starch, dextrins, pectin.

High viscosity alone is not sufficient to stabilize an emulsion, but viscous emulsions are more stable than runny ones. High viscosity hinders coalescence and flocculation. However, it should be pointed out that viscosity has only a small effect in o/w emulsions [19].

In the production of inks for textile printing, large quantities of alginates or locust bean gum are used to manufacture thick pastes. Synthetic polymers from styrene and acrylic acid derivatives, or butadiene and acrylonitrile, are also used for this purpose. The carbopol resins (made by Goodrich) and Lyoprint (Scott Bader) should also be mentioned.

Sometimes it is necessary instead to decrease the viscosity of an emulsifier system. Plasticizers serve this purpose, for example added to margarine so that it can be spread satisfactorily even after storage in the refrigerator. Plasticizers include glycerin esters of fatty acids (HLB 2–4), cholesterols, and modified fats and oils.

Solubilizers

These are added to emulsions to increase the solubility of an emulsifier in a phase. For example, in the emulsifiable concentrates used in crop protection (Section 2.8.4), small amounts of aliphatic alcohol are added to encourage spontaneous emulsification. Other solubilizers are urea and its derivatives, added to textile or dye auxiliaries, or the small quantities of water added to chlorinated hydrocarbons in dry-cleaning. Solubilizers are also used to adapt the density and viscosity of both phases. In principle, any solvent miscible with the outer phase may be used. Price is often the determining factor.

Protective colloids

These are added to the outer phase in order to hinder coalescence of the droplets; they achieve this by holding the droplets formed in suspension, surrounding them as do emulsifiers but without penetrating the inner phase. Household laundry detergents usually contain carboxymethylcellulose for this reason. This holds the dirt released from the fibers in solution and prevents it from being reabsorbed by the substrate.

Other protective colloids are lecithins, cholesterol, gelatin, proteins, casein, gums, saponins, lignin sulfonates.

Preservatives

Preservatives prevent fungal and bacterial growth. They are unnecessary in the presence of cationic emulsifiers, as these usually have bactericidal or bacteriostatic properties.

Preservatives are normally more soluble in oil than in water (so, for example, *p*-hydroxybenzoate esters). This should be borne in mind when deciding on the dosage of preservative.

Suitable preservatives for emulsions are aldehydes such as formaldehyde and chloral hydrate, and phenol derivatives like *o*-phenylphenol, *p*-chloro-*m*-cresol, and pentachlorophenol (only when unlikely to cause physiological problems). Heterocycles such as *o*-chloroaniline dichlorotriazine and derivatives of the 1,2-benzisothiazole ion are also used.

Defoaming agents
Most common trade defoamers are suitable for use in emulsions. Since emulsifiers are now designed to produce as little foam as possible, the addition of antifoaming agents is only occasionally necessary.

A good type of defoamer is a nonionic polyalkylene glycol ether with mixed ethylene and propylene chains.
Additives to prevent incrustation on the lid of the container
The most suitable additives are glycerin and sorbitol.

2.9 Some Important Consequences of the Theory of Emulsion Stability

2.9.1 The Electrical Double Layer in Emulsions

The most important difference between electrical double layers in emulsions and in dispersions is that in emulsion droplets there is a double layer on each side of the interface. Figure 2.21 shows the structure of the electrical double layer in the aqueous phase at the w/o phase boundary with adsorbed emulsifier. The picture is not entirely true to the oil/water system; the electrical double layer in the oil phase has been omitted.

In Figure 2.22, the course of the potential on both sides of the phase boundary is sketched (source: van den Tempel). If a salt is present, cations and anions are distributed unevenly between the two phases, since anions are more oil-soluble than cations. The oil phase therefore becomes negatively charged. In the absence of surface-active ions, the potential changes as in Figure 2.22a. The electrical double layer in the oil phase is several μm broader than the one in the aqueous phase, which is 10^{-3} to 10^{-2} μm thick. This means that the interactions between droplets in a w/o emulsion extend much further and the repulsion energy is small, so the droplets can easily run together. The electrical potentials of adjacent water droplets overlap, decreasing the stability and giving the system a strong tendency to coalesce (in the absence of an emulsifier). The rapid decline in potential in an o/w emulsion, on the other hand, stabilizes it.

The addition of tensides, which collect at the interface, has a drastic effect on the course of the potential (Figure 2.22b). The charge is now concentrated in the aqueous phase, so that the zeta potential is sufficiently large to stabilize the emulsion.

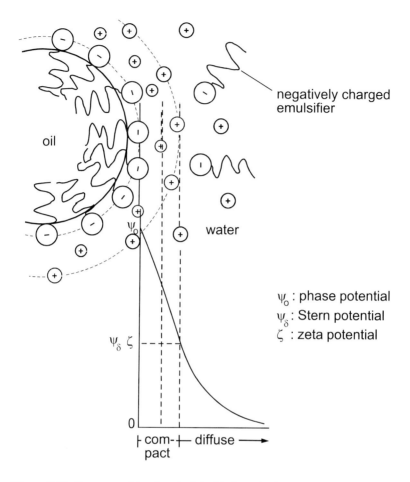

oil

negatively charged
emulsifier

water

ψ_0 : phase potential
ψ_δ : Stern potential
ζ : zeta potential

$\psi_\delta\ \zeta$

0

⊢ com-⊢ diffuse ⟶
pact

Figure 2.21. Structure of the electrical double layer in the aqueous phase at the o/w interface, with adsorbed emulsifier [20].

According to Verwey, *a zeta potential on the order of ±100 millivolts is required for stabilization*. When there is a large amount of electrolyte in the aqueous phase in addition to the tenside, the potential changes as shown in Figure 2.22c. The added electrolyte reduces the thickness of the diffuse double layer. The mutual repulsion of the charged polar groups causes the film to expand and to have a low π value (film pressure). When electrolyte is added, the repulsion is decreased and the film becomes denser; π increases. The role of electrolyte addition in lowering the interfacial tension and thus increasing emulsion stability, initially rather difficult to understand, can be qualitatively explained in these terms. *The stabilizing effect of electrolytes on emulsions seen in certain cases is due to consolidation at the interface*. The repulsion potential V_R and the van der Waals attraction potential V_A for the electrical double layer are given in Equations 2.15–2.17 without further explanation.

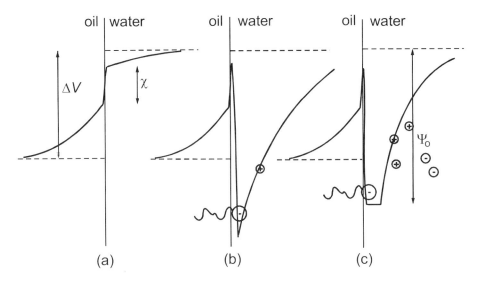

Figure 2.22. Course of the potential at the oil/water interface in emulsions: a) without emulsifier, b) with emulsifier, c) with emulsifier at high electrolyte concentration [20].

$$V_R = 4.62 \cdot 10^{-6} \cdot \frac{r}{v^2} \cdot \gamma^2 \cdot e^{-\kappa \cdot H_0} \tag{2.15}$$

$$V_A = -\frac{A \cdot r}{12 H_0} \tag{2.16}$$

$$V = V_R + V_A \tag{2.17}$$

(r: particle radius; v: valency of counterion; γ: function of $v \varepsilon \psi_0 / kT$; ε: dielectric constant; ψ_0: double layer potential; k: Boltzmann constant; A: Hamaker constant; H_0: distance between particles; κ: Debye–Hückel function; $1/\kappa$: Debye length)

The flocculation of droplets in an emulsion depends on the valency of the counterions, as described in the Schulze–Hardy rule (see also Section 5.6):

Mono-, di-, trivalent counterions flocculate as 100 : 1.6 : 0.13.

The mechanism differs from that in the case of sols. A doubly charged ion causes the same kind of flocculation as a singly charged ion would at 1/10 of the concentration. In general, the higher the concentration of the potential-determining ions (defined below), the lower the concentration of indifferent ions (ionic strength), and the lower the charge on the counterions, then the greater the stability of an emulsion.

Potential-determining ions play a particular role in the flocculation of emulsions and dispersions. For example: for an aluminum sol, H^+ and OH^- ions are potential-determining. At low pH, the sol is positively charged because of adsorption of Al^{3+} ions; at high pH it is negative on account of adsorption of OH^- or AlO_2^- ions.

In the case of oxides (e.g. SiO_2, Al_2O_3, TiO_2), protons act as potential-determining ions; that is, the surface charge and potential depend on the pH. In the case of aluminum oxide the isoelectric point (electrical charge = 0) is at pH = 9. Below this value, aluminum oxide systems make positively charged colloid particles; above it, negatively charged ones. H^+ is generally a potential-determining ion for proteins and biological substances owing to the dependence of dissociation of acidic or basic groups on the pH of the solution.

2.9.2 Electrical Double Layer in w/o Emulsions

In w/o emulsions the continuous phase (oil) has a low dielectric constant. These organic phases can only carry small charges; in benzene, for example, ion concentrations of 10^{-10} mol/L are possible. The electrical double layer is thus stretched out, with Debye lengths $1/\kappa$ of a few µm. The capacitance of the double layer is very low, so small charges result in high zeta potentials (50 mV and more). The forces of interaction between the charged droplets in a w/o emulsion extend much further than those in an o/w emulsion. Because the repulsion energy between water droplets is so low, they can coalesce with little energy input. The high zeta potentials have no stabilizing effect (Overbeek). As a consequence, w/o emulsions have to be stabilized sterically (with macromolecules or polymers).

2.9.3 Emulsion Zeta Potentials

The equations for the zeta potentials of dispersions do not apply to emulsions. Instead, the complicated equations derived by Boots must be employed. Hunter developed corrections for the calculation of ζ. Electrophoretic mobility depends on particle shape and size. The effect of the latter is given by a function κr, calculated by Henry.

For making comparisons or drawing qualitative conclusions the electrophoretic mobility is sufficient information. In general, the zeta potential is an acceptable measurable parameter for the assessment of emulsion stability, *as long as the repulsive forces in the emulsion are exclusively or predominantly electrostatic in nature.* For the reasons already given, rather than use the zeta potential, which is very difficult to calculate, we will adopt the underlying measurable variable V (electrophoretic velocity) or the electrophoretic mobility V/E (E: electric field strength). Measurements of V/E under various experimental conditions can be used to give information about the emulsion stability.

2.9.4 Steric Stabilization

If the adsorbed layers of two droplets are about to touch, steric repulsion may be present alongside the electrostatic or dispersion force interactions. When the adsorbed layer consists of short-chain molecules like most tensides, there are four possible causes for the steric hindrance that occurs on the approach of two droplets (Figure 2.23a–d). In the case of macromolecules, the free movement of the chains is hindered by interpenetration of the adsorbed macromolecular layers (Figure 2.24). The result is entropic repulsion.

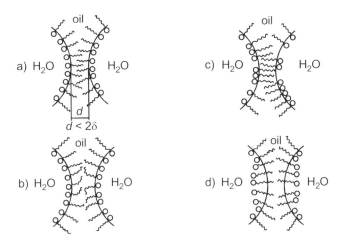

Figure 2.23. Steric hindrance in adsorbed layers of short-chain molecules: a) interpenetration of interacting adsorbed layers; b) crowding of the molecules; c) compression of the molecules within the layers and formation of gaps; d) partial desorption due to pressure (cf. [8]).

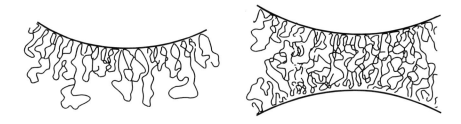

Figure 2.24. a) Schematic view of the adsorption of macromolecules; b) interpenetration of adsorbed macromolecular layers (cf. [8]).

The intentional steric stabilization of emulsions is achieved, however, with polymeric emulsifiers, in particular hydrocolloids, that is, water-soluble macromolecules and polyelectrolytes containing groups that dissociate in water to give macroions and counterions, for example PVA, polysaccharides ("gums"), alginates, and cellulose derivatives. These are used with o/w emulsions. The stabilization at the o/w interface due to hydrocolloids results from:

a) the creation of an elastic interfacial film;
b) the modification of the rheological properties of the dispersion medium;
c) the repulsive barrier of the electrical double layer.

A combination of hydrocolloid and tenside often gives a more stable emulsion than the individual components. A mixed film can only exist if the hydrocolloid and the emulsifier have the same charge.

Besides hydrocolloids, a wide variety of polymer types can effect steric stabilization. The mobility of hydrocarbon or polyglycol ether groups tends to prevent the approach of emulsion droplets to one another. All the factors that favor the mobility of the emulsifier molecules within the interfacial layer contribute to stability. Thus, emulsifiers with long hydrocarbon chains are more effective emulsifiers for w/o systems than those with shorter chains.

Hydrogen bonding and the formation of complexes are also steric factors that can increase the emulsion stability. Hydration of the hydrophilic groups keeps the droplets apart. Liquid-crystalline structures may form, which likewise raise the stability of the emulsion.

It has already been mentioned that emulsions protected with macromolecules are not sensitive to addition of electrolytes. The reason is that it is not electrostatic repulsion that stabilizes the emulsion, but the low level of attraction due to dispersion forces. At high salt concentrations, however, dehydration of the macromolecules may result in flocculation.

Macromolecules not only prevent flocculation but also cause it. Flocculation occurs when only minor amounts of the macromolecule are added to the dispersion: in this case, one macromolecule can be adsorbed on two droplets simultaneously and thus act as a link holding them together and causing flocculation. This occurs when the droplet surfaces are not completely covered with emulsifier molecules so that the tail of a macromolecule that is adsorbed on one droplet can be adsorbed by the bare surface of a neighboring droplet. This process is known as *sensitization*, and plays an important role in the flocculation of colloid particles in waste-water treatment.

We can only make rough generalizations about the concentration of macromolecules at which this effect occurs. In polystyrene suspensions, all the primary particles were agglomerated at a 2% level of addition, whereas at 3% the suspension was once more without any agglomeration. For w/o emulsions containing monoglycerides and soap as emulsifiers, creaming occurred when the tenside concentration was below about 3%. The critical region is therefore ~<3%. If, on the other hand, the emulsifier is present in relatively large quantities, mixed phases of water, tenside, and oil may form that are largely responsible for the behavior of the emulsion.

The optimum emulsifier quantity can only be determined by series of trial emulsifications similar to those described for the determination of the requisite HLB value for an oil in Section 2.2.1.

2.9.5 Kinetics of Coalescence; Determination of Emulsion Stability

The stability of an emulsion is usually measured from the rate of decline in the number of droplets during the initial phase of aggregation. Von Smoluchowsky calculated the rate of diffusion-controlled coalescence of spherical droplets as a result of collisions in the absence of energy barriers against coalescence [21]. If every collision between particles results in a decrease in the number of particles, then the change in number over time is given by a second-order kinetic equation (2.18):

$$-\frac{dn}{dt} = 4\pi \cdot D \cdot r \cdot n^2 \tag{2.18}$$

D: diffusion coefficient
r: collision radius = maximum distance between the centers of the particles at which
collision still occurs
n: number of particles/cm^3

If we assume that in fact there is an energy barrier to be overcome before coalescence can occur, we obtain the following equation (2.19) for the decline in particle numbers:

$$-\frac{dn}{dt} = 4\pi \cdot D \cdot r \cdot n^2 \cdot e^{-E_A/kT} \tag{2.19}$$

E_A: activation energy
T: temperature
k: Boltzmann constant
t: time

Integration yields

$$\frac{1}{n} = 4\pi \cdot D \cdot r \cdot e^{-E_A/kT} \cdot t + const \tag{2.20}$$

The Einstein Equation (2.21) holds for the diffusion coefficient D of spherical particles (a: particle radius; η: viscosity):

$$D = \frac{kT}{6\pi \eta a} \tag{2.21}$$

If the particles reacting are in contact with one another, that is, if $r = 2a$, then the reciprocal of the number of particles can be written as a function of t:

$$\frac{1}{n} = \left(\frac{4kT}{3\eta}\right) \cdot e^{-E_A/kT} \cdot t + const \tag{2.22}$$

All the variables in this equation bar the activation energy E_A can be determined experimentally. The number of particles can be counted with a microscope. E_A can be read from a plot of $1/n$ against t. The dependence of coalescence on the volume ratio of the two phases can be taken into account as follows:

Let the mean volume of a droplet be defined as a new variable:

$$\overline{V} = \frac{V}{n} \tag{2.23}$$

where V = the volume of the disperse phase per cm^3 emulsion. Thus:

$$\overline{V} = \frac{4}{3} \cdot \left(\frac{V \cdot k \cdot T}{\eta} \right) \cdot e^{-E_A/kT} \cdot t + const \tag{2.24}$$

Differentiation gives the rate of coalescence and thus a measure for emulsion stability:

$$\frac{d\overline{V}}{dt} = \frac{4}{3} \cdot \left(\frac{V \cdot k \cdot T}{\eta} \right) \cdot e^{-E_A/kT} \tag{2.25}$$

The effectiveness of the emulsifier system is discernible from the activation energy barrier E_A. This barrier includes both electrical and mechanical screening of the droplet from coalescence. It is usually dependent on temperature.

Davies [22] reported that the type of emulsion formed depends on the kinetics of coalescence of the two liquid phases mixed in the presence of an emulsifier. The rate of coalescence G_1 of oil droplets dispersed in water can be calculated from the following equation (Equation 2.26):

$$G_1 = C_1 \cdot e^{-E_{A1}/kT} \tag{2.26}$$

The collision factor C_1 depends on the volume ratio of the oil phase to the aqueous phase and on the reciprocal of the viscosity of the aqueous phase. E_A is the energy barrier to be overcome before coalescence can occur. In an o/w emulsion, E_{A1} is a function of the electrical potential and the hydration energy of the emulsifier. For the coalescence of water droplets in a w/o emulsion, Equation 2.27 applies:

$$G_2 = C_2 \cdot e^{-E_{A2}/kT} \tag{2.27}$$

E_{A2} is the energy barrier to coalescence. Most importantly, it is a function of the number of -CH$_2$- groups in an emulsifier molecule between two water droplets approaching one another. Before these droplets can coalesce, the energy barrier due to the lipophilic part of the emulsifier molecule must be overcome.

If a mixture of oil and water is combined with an emulsifier, initially both o/w and w/o dispersions form. The type of emulsion finally created depends on which rate of coalescence is greater, G_1 or G_2.

According to Davies, an o/w emulsion forms if G_2 is very much larger than G_1. If not, a w/o emulsion forms [23].

o/w emulsions are the more stable if $G_2 = 1$ or $G_2/G_1 \gg 1$.
w/o emulsions are more stable, if $G_1/G_2 \gg 1$.

In practice, emulsions tend to be stable only when or G_1 or G_2 is smaller than about $10^{-6}\,C_1$ or $10^{-6}\,C_2$.

References for Chapter 2:

[1]   International Union of Pure and Applied Chemistry, Manual on Colloid and Surface Science, Butterworths, London, 1972.
[2]   S. E. Friberg, J. Yang, in Surfactant Science Series Vol. 61: Emulsions and Emulsion Stability (J. Sjöblom, Ed.), Marcel Dekker, Inc., New York, 1996.
[3]   P. Becher, Emulsions: Theory and Practice, Reinhold Publishing Corporation, New York, 1966.
[4]   McCutcheon's Detergents and Emulsifiers, Int. Ed., Ridgewood, New York.
[5]   H. Stache, Tensid-Taschenbuch, Carl Hanser Verlag, München, 1979.
[6]   N. Schönfeldt, Grenzflächenaktive Aethylenoxid-Addukte, Wissenschaftliche Verlagsgesellschaft, Stuttgart, 1976.
[7]   W. C. Griffin, J. Soc. Cosmet. Chem. *1*, 311 (1950); *5*, 249 (1954).
[8]   H. Sonntag, Lehrbuch der Kolloidwissenschaft, VEB Deutscher Verlag der Wissenschaften, Berlin, 1977.
[9]   C. M. Donnald, Can. J. Pharm. Sci., Vol. 5, No. 3, 81 (1970).
[10] R. G. Laughlin, J. Soc. Cosmet. Chem. *32* (1981), p. 371–392.
[11] J. H. Schulman, E. G. Cockbain, Trans. Faraday Soc. *36*, 651 (1940).
[12] J. Boyd, C. Parkinson, P. Sherman, J. Colloid Interface Sci. *41*, 359 (1972).
[13] J. Stauff, Kolloidchemie, Springer Verlag, Berlin, 1960.
[14] J. H. Schulman, E. G. Cockbain, Trans. Faraday Soc. *36*, 661 (1940).
[15] L. H. Rees, Chem. Eng. *81*, 86 (1974).
[16] A. L. Smith (Ed.), Theory and Practice of Emulsion Technology, Academic Press, New York, 1976.
[17] K. J. Lissant (Ed.), Emulsions and Emulsions Technology, Marcel Dekker, Inc., New York, 1974.
[18] H. Helfenberger, in DECHEMA-Kurs Formulierungstechnik 1987, Thema 7.
[19] E. L. Knoechel, D. E. Wurster, J. Am. Pharm. Assoc. Sci. Red. *48*, 1 (1959).
[20] B. Dobias, Tenside Detergents *15*, 228 (1978).
[21] M. Von Smoluchowsky, Physik. Z. *17*, 557, 585 (1916); Z. Phys. Chem. *92*, 129 (1917).

[22]   J. D. Davies, in 2nd Int. Congress of Surface Activity, Vol. 1, Butterworths, London, 1957, p. 426.

[23]   J. D. Davies, E. K. Rideal, Interfacial Phenomena, Academic Press, New York, 2nd ed. 1963, p. 366 ff.

3 Microemulsions, Vesicles, and Liposomes

Microemulsions and liposome preparations are used in cosmetics and other applications. What is the difference, from a practical point of view, between a microemulsion and a vesicle or liposome system? Microemulsions, being thermodynamically stable systems, are easy to manufacture reproducibly, whereas metastable vesicles can vary in type and size depending on the exact manner of their production. On use (application to the skin, in the case of cosmetics), microemulsions convert to another phase (because of solvent evaporation and interactions with the epidermis), while vesicles remain in their kinetically metastable state and change much more slowly.

3.1 Microemulsions

3.1.1 Phenomenology

O/w and w/o systems with interfacial tension ~0, which can be created by means of specialized tensides or mixtures of tensides and cotensides, can create structures with periodic order out of lamellae, cylinders, or more complex objects, as described in Section 1.8. These preparations possess some of the characteristics of solids as well as high viscosity, which can cause problems in some industrial applications.

In contrast, in microemulsions, another system with interfacial tension ~0, the interfaces are disordered. This type of system behaves as a transparent liquid with low viscosity. At low oil or water concentration, swollen micelles are present, known as microemulsion droplets. At other concentrations, the structures created are a great deal more complex; bicontinuous structures such as sponge phases form (see Section 1.8.5). All these different and thermodynamically stable structures are known by the single (and rather unfortunately chosen) name of microemulsions, coined by Schulman [1].

Hoar and Schulman characterized the first microemulsions in 1943 [1]. They described transparent 1:1 (v/v) mixtures of water and oil containing large quantities of an ionic tenside and a nonionic cotenside. Alcohols of medium chain length are suitable for use as cotensides. For example, equal volumes of toluene and water can be mixed with an anionic tenside such as potassium oleate until an emulsion forms. This emulsion is titrated with hexanol until it becomes transparent.

A microemulsion described by Shinoda has the following composition (Table 3.1):

Table 3.1. Microemulsion [2], discussed in [3].

Oil	Water	Tenside	Cotenside
60% benzene	27%	6.5% K oleate	6.5% butanol

The starting material is neat benzene, to which alcohol is added, followed by water and the soap. These transparent solutions have a mixture of the properties of hydro-carbons and water. This immediately suggests the great range of possibilities for their application.

Thus, for example, an oil-soluble dye gives the same color in benzene as in a microemulsion of benzene, water, tenside, and cotenside. Similarly, a water-soluble dye behaves identically whether in water or in a microemulsion of benzene, water, tenside, and cotenside.

Light and X-ray scattering studies show that these mixtures comprise a dispersion of aqueous droplets with a diameter of about 8 nm, and that most of the potassium oleate and the hexanol is at the oil–water interface. The smallness of the droplets explains the low turbidity of the microemulsion.

The principal distinction between a microemulsion and an ordinary emulsion is neither the size of the droplets nor the degree of cloudiness, but the facts that microemulsions form spontaneously, that their properties are independent of the manner in which they were produced, and that they are thermodynamically stable. Ordinary emulsions, in contrast, require mechanical or chemical work for their production, so that at best they are *kinetically stable*, that is, the droplets are protected from coalescence. At their simplest, microemulsions are a special case of emulsions: colloidal solutions of normal or inverted, swollen micelles [4].

As a consequence of their large water–oil interface, microemulsions are only thermodynamically stable if their interfacial tension is so low that the positive free interfacial energy γA (A = interfacial area w/o) is negated by the term for the free energy of mixing (= $-T \cdot S$). The entropy of mixing for a single droplet is of the order of k_B according to the Boltzmann Equation (k_B = Boltzmann constant, $1.3806 \cdot 10^{-23}$ J/K; S: entropy, W: thermodynamic probability):

$$S = k_B \cdot \ln W \qquad \left[\frac{J}{mol \cdot K} \right] \qquad (3.1)$$

It follows that under these conditions the interfacial energy is on the same order as the thermal energy (γ = interfacial tension):

$$k_B \cdot T = 4\pi R^2 \cdot \gamma \qquad (3.2)$$

If the radius of a droplet R = 10 nm, γ is ~0.03 $\frac{mN}{m}$!

The interfacial tension between oil and water is ~50 mN/m. The task of the tenside is the reduction of this value right down to $\gamma \rightarrow 0$. Tensides adsorb spontaneously at the w/o interface because of their amphiphilicity. This facilitates the spreading of the interface and thus the reduction of the interfacial tension. According to the Gibbs Equation (Section 1.6.3) the gradient $-\dfrac{d\gamma}{dc}$ is proportional to the extent of adsorption (c: tenside concentration).

Figure 3.1 depicts the interfacial tension as a function of the logarithm of the tenside concentration. At low concentrations there is little adsorption and the interfacial tension declines only slowly. When further tenside is added, γ falls rapidly until, within a small concentration range, it reaches a steady value. This behavior can be explained by micelle formation ($\rightarrow$ critical micelle concentration, *CMC*). Formation of aggregates often prevents γ from sinking very low. However, if a second tenside is added, a cotenside, of a different chemical type from the first tenside, the effects of both tensides are magnified. The interfacial tension can then reach very low levels. A water-soluble tenside and a more oil-soluble cotenside, such as sodium dodecyl benzenesulfonate in water and pentanol in cyclohexane, can bring γ so low that any further addition of tenside would make it negative (which is impossible). This would result in spontaneous expansion of the interface from uptake of the excess tenside and cotenside and a simultaneous increase in γ to positive values once more. *This process is the spontaneous formation of a microemulsion.*

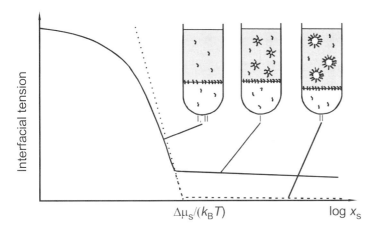

$\Delta\mu_S/(k_B T)$ $\log x_S$

Figure 3.1. Interfacial tension as a function of the tenside concentration for a biphasic system with two different tensides (I, II). When the concentration is lower than the *CMC* (discontinuity in curve I), the tensides are present as monomers and γ drops rapidly owing to adsorption at the o/w interface. At concentrations higher than the *CMC*, tenside I forms micelles. Since the interface is almost entirely covered with tenside molecules, γ decreases only gently after this. Tensides like I are used for macroemulsions. Tensides like II can be used to make microemulsions. At the *CMC*, nanodroplets (microemulsion droplets) form, resulting in an extremely low interfacial tension (from [5]).

In the region in which microemulsion formation occurs, a molecule of sodium dodecyl sulfate covers an area of 80–90 Å^2 in the presence of 0.3 M NaCl. This area depends only slightly on the concentration, the nature of the cotenside, and the electrolyte concentration. One gram of sodium dodecyl sulfate covers about 1780 m^2. Such an enormous interfacial area can be accommodated in microemulsions in a number of ways (see Figure 3.2), specifically, as water droplets in oil, oil droplets in water, or irregular bicontinuous structures.

Measurements of light scattering, and other experiments, can be used to show that many microemulsions contain spherical droplets. The size of the droplets can be estimated from the composition of the microemulsion and the area covered by one molecule (measured from the interfacial tension). *Almost all the tenside molecules are located at the interface.* Therefore the total interfacial area *A* is roughly equal to the number of tenside molecules multiplied by their molecular area.

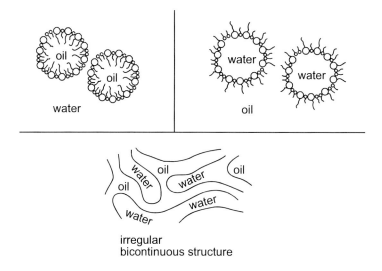

Figure 3.2. Structures in microemulsions; cf. [6].

To manufacture microemulsions with ionic emulsifiers or tensides, it is necessary to add a more hydrophobic cosurfactant, as has already been stated. One explanation for this fact is that the ionic tenside is not sufficiently hydrophobic to permit the solubilization of the oil phase. This suggestion is in agreement with experimental findings which show that longer-chain ionic tensides require less cosurfactant (higher alcohols) than those with shorter chains. The particle size in a microemulsion varies with temperature, with the balance between hydrophilic and hydrophobic groups or components of the emulsifier, and with the oil content.

Droplets in a microemulsion undergo large and reversible changes in size when small alterations are made to temperature or pressure; this is not true of micelles.

Whether a microemulsion is of the w/o or o/w type depends on the tendency of the interfacial region to bend in one direction or the other (Bancroft's rule) [8], which is determined by factors such as the difference in interfacial tension or interfacial pressure on the two sides of the interface, and the difference in the volumes and compressibilities of the hydrophilic and hydrophobic groups. If the hydrophilic side of the interfacial film has a greater tendency to expand than does the hydrophobic side, an o/w microemulsion forms; in the converse case, if the hydrophobic side of the interfacial film has a greater tendency to expand than does the hydrophilic side, a w/o microemulsion forms.

Microemulsions are often studied with the aid of phase diagrams, since these are a great help in understanding the manner in which the phases coexist at a given temperature (Figure 3.3).

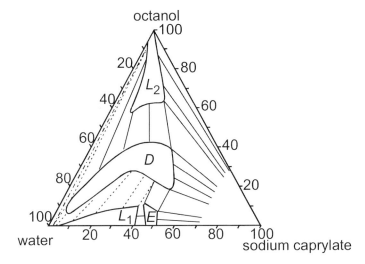

Figure 3.3. Phase diagram for the system water/sodium caprylate/octanol. Isotropic phases: $L1$ (normal micelles), $L2$ (inverse micelles); liquid-crystalline phases: E (hexagonal packing of cylinders), D (lamellar) [7].

Microemulsions made with ionic tensides need large quantities of tenside. For a microemulsion with a water:hydrocarbon ratio of 1:1 about 15% tenside must be added. For many uses, this is prohibitive. For certain nonionic tensides, smaller amounts are sufficient; for a polyethylene glycol alkyl ether, 5% is enough. However, the temperature range for both o/w and w/o microemulsions is very narrow.

Friberg [7] believes that, if 75% of a solvent can be replaced by an aqueous tenside solution containing 10% ionic tenside (an optimistic figure), the economic break-even point is reached, if the ratio of the cost of the tenside to that of the solvent is 10. Nowadays, cost ratios of 4 are considered acceptable, which shows that the concept of microemulsion use is an economically healthy one! It should also be mentioned that in some cases, a single tenside can be used to achieve very low interfacial tensions and thus to form microemulsions. This is possible for ionic tensides that have two HC chains, like ethyl hexyl sulfosuccinic acid (Aerosol OT), and for some nonionic tensides within a narrow temperature range in which their solubility in oil is similar to their solubility in water (PEG alkyl ethers).

It is generally simpler to formulate w/o microemulsions than o/w systems; it is harder to find the correct proportions of oil and emulsifiers for the o/w type than for the w/o type. For this reason, o/w microemulsions are often manufactured by phase inversion of w/o microemulsions [9].

3.1.2 Theory of Microemulsions

The *Schulman condition* for microemulsions, $\gamma \approx 0$, means that the Langmuir surface pressure of the tenside molecules in the molecular layer is equal to the interfacial tension between water and oil (see Section 1.5.2). When this condition is fulfilled, the tenside molecules take up the equilibrium area Σ_0 at the oil/water boundary and have energy μ_s. The Schulman condition now enables us to relate the geometry of the tenside molecule to the ratio of the oil/water surface area S containing the tenside molecules and the volume V (v_s: volume, l_s: length, N_s, number, Φ_s: volume fraction of tenside molecules):

$$\frac{S}{V} = \frac{\Sigma_0 \cdot N_s \cdot v_s}{V \cdot v_s} = \frac{\Phi_s}{l_s} \tag{3.3}$$

In addition, on the basis of theory we define a variable called the persistence length, ξ, the length over which a property does not change noticeably (for example, the lipophilic interior of o/w microemulsion droplets). Consider a microemulsion as divided into cubes of side ξ. Each cube contains either oil or water; in other words, the volume of the cube is roughly equivalent to that of a microemulsion droplet. Let the volume fractions of oil and water be Φ_O and Φ_W, respectively. Then the ratio of the total oil/water interfacial area S to the volume V is, by statistical reasoning:

$$\frac{S}{V} = \frac{6 \cdot \xi^2 \cdot \Phi_O \cdot \Phi_W}{\xi^3} \tag{3.4}$$

Combination of Equations 3.3 and 3.4 yields the interesting relationship 3.5.

$$\xi = 6 \cdot l_s \cdot \frac{\Phi_O \cdot \Phi_W}{\Phi_s} \tag{3.5}$$

Equation 3.5 is regarded as the definition of the so-called Schulman line, which connects compositions of the same persistence length in the Gibbs phase triangle (Figure 3.4).

As has already been mentioned, microemulsions have very low interfacial energies (Equation 3.2). The type of structures present in a microemulsion therefore depends upon the energy of curvature of the interfaces, amongst other factors [10]. The energy of curvature E_c per unit area of the unimolecular tenside layer can be represented as shown in Equation 3.6:

$$E_c = \frac{\kappa}{2} \cdot \left(\frac{1}{R_1} + \frac{1}{R_2} - \frac{2}{R_s} \right)^2 + \overline{\kappa} \cdot \frac{1}{R_1 \cdot R_2} \tag{3.6}$$

R_1 and R_2 are the radii of curvature of the interface (they also appear in the Young–Laplace Equation for soap bubbles, Figure 4.2). R_s is the radius of spontaneous curvature, which occurs because of the asymmetry of the lipophilic and hydrophilic regions of the tenside molecule (see also the oriented wedge theory, Figure 2.5). κ and $\bar{\kappa}$ are the elasticity (curvature) constants for the stiffness and the Gaussian curvature.

The significance of the stiffness can be illustrated by comparison with the thermal energy. We expect disordered microemulsion phases for $\kappa < k_B \cdot T$ and ordered lamellar phases for $\kappa > k_B \cdot T$. If $R_s < \xi$, a bicontinuous, spongey structure is indicated, and if $R_s > \xi$, a phase of dispersed droplets is seen (see above for details of the persistence length ξ). The sign of κ/R_s, according to Bancroft's rule (see Chapter 2 on emulsions), is also related to ordinary emulsions. In the case of ionic tensides, the solvating ability of water increases with increasing temperature. As this happens, the radius of spontaneous curvature may change sign and the phases may invert.

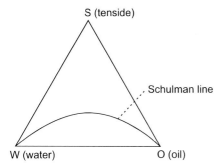

Figure 3.4. Phase triangle for a three-component system. The curve is a Schulman line, which connects all points of the same persistence length in a microemulsion.

In microemulsions with a low oil content in an aqueous continuous phase, or a low water content in a continuous phase of oil, isolated nanodroplets are present. If the concentration of the disperse phase is increased, then phase inversion must occur, for example from an o/w to a w/o microemulsion. The disperse phase becomes the continuous phase and vice versa. In an intermediate concentration range, though, aqueous and oily continuous phases coexist. This is a bicontinuous microemulsion. According to percolation theory, these conversions occur around concentrations of 20% and 80% [11].

3.1.3 Uses of Microemulsions

The general application of aqueous tenside solutions as replacements for nonpolar solvents is particularly interesting from the points of view of toxicity and environmental acceptability. Nevertheless, the most important application for microemulsions is found in the tertiary recovery of crude oil. At present, conventional methods can economically exploit no more than 50% of oil deposits (primary and secondary oil recovery). By means of microemulsions and other methods, yields could be increased by 10–20%. For the time being, oil that cannot be extracted by these methods may be considered unobtainable.

Cleaning: Industrial cleaning processes usually involve the simultaneous removal of both hydrophobic (fats, oils, lipids) and hydrophilic soiling (inorganic salts, pigments, sugars, proteins). The use of emulsion systems has advantages compared with classical methods for textiles and hard surfaces; microemulsions are best, as they have no stability problems. The more commonly used type is the o/w microemulsion. However, these are harder to formulate with standard tensides and tend to foam. While the high tenside concentration results in more effective cleaning, it is also more expensive. Recent developments and patents in this field deal mostly with the dry-cleaning of textiles. Detergent solutions for hard surfaces, which contain high concentrations of pyro-phosphate and tripolyphosphates, can be reformulated as microemulsions by the addition of aromatic or other organic solvents, resulting in better stability and cleaning ability.

Cosmetics: Microemulsions are very widely used in this field. The requirement for stable and transparent emulsions is perfectly met by microemulsions. It is more common to employ o/w microemulsions, though the w/o type is also sometimes utilized. Fragrances and vitamins are solubilized with polyoxyethylene derivatives of sorbitan fatty acid esters and ethers; pentanol and hexanol or the less aggressive glycols serve as cosurfactants, and cholesterol and nonionic tensides as additives to the alcohol.

Further areas in which microemulsions are used include:
– lubricants
– emulsion polymerization
– coatings
– fuels.

G. Gillberg [12] cites 245 references on the topic.

Finally it should be remarked that the first microemulsions were manufactured by chance in 1928; they consisted of carnauba wax in water. These emulsions were developed from research into household cleaning products. One such formulation was "Aerowax", a self-polishing floor wax comprising carnauba wax, oleic acid, potassium hydroxide, borax, and water. When warmed, the borax decomposes to give sodium hydroxide and boric acid, which releases fatty acid from the soap to act as a cotenside in the interfacial layer (L. Prince). Other "forerunners" of microemulsions include cutting oils, fragrance oils, polymers, and pesticides in the form of microemulsions.

3.2 Vesicles and Liposomes

3.2.1 Nomenclature

Phospholipids play an important role in both plant and animal cells, and are used in formulation of very many foodstuffs (think, for example, of soya lecithin). As will be seen in Chapter 13, phospholipids are esters of glycerin with two different fatty acids and a phosphoric acid derivative. The diversity of phospholipids means that they and their mixtures have a very wide range of colloidal properties. With water, these amphiphiles are able to form different lyotropic mesophases such as lamellar, cubic, and hexagonal phases, as well as vesicles and liposomes (cf. Section 1.8).

Vesicles are a special case of the lamellar phase in which the bilayer membranes make closed uni- or multilamellar spherical or ellipsoidal structures. Vesicles based on amphiphilic lipids made from biological substances are known as *liposomes*. Less strictly, however, the term "liposome" is often used synonymously with "vesicle".

Bangham et al. [13] showed by electron microscopy that when lamellar phases of membrane phospholipids are treated with ultrasound, structures with closed membranes resembling cells form – "microvesicles" (small bubbles or vessels), dubbed "bangosomes" at the time. Just as these names are derived from the name of the discoverer or the form of the discovery, the word "liposome" points to the basic building blocks of the structure, the lipids.

3.2.2 Phenomenology

An electron micrograph of the onion-layer structure of a multilamellar liposome is reproduced in Figure 3.5.

Such vesicles are classified according to structure and size by the scheme shown in Figure 3.6.

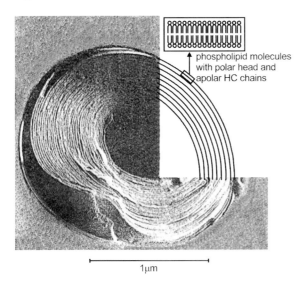

Figure 3.5. Electron micrograph of a multilamellar liposome (prepared by freeze-fracture). Top right: sketch of a phospholipid bilayer. The liposome shown here contains a series of concentrically arranged bilayers.

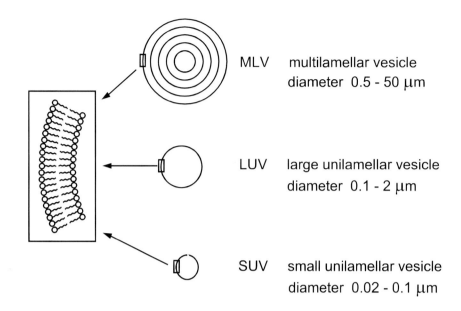

MLV   multilamellar vesicle
       diameter 0.5 - 50 μm

LUV   large unilamellar vesicle
       diameter 0.1 - 2 μm

SUV   small unilamellar vesicle
       diameter 0.02 - 0.1 μm

Figure 3.6. Vesicle types.

What is the relationship between vesicles and other structures such as micelles and lamellae? Let us consider the structures that can be adopted by a surface-active compound when its concentration is raised; when the critical micelle concentration (*CMC*) is passed, micelles form. Hydrophobic interactions pack the hydrocarbon chains closely together, whereas the strong repulsion between the ionic head groups pushes them as far apart as they can go. These effects result in spherical micelles. When the concentration of surfactant is increased further, ionization and thus the mutual repulsion of the hydrophilic head groups is repressed. The reduction in the area required by these groups permits reorganization into cylindrical structures and thence into lamellae consisting of bilayers of amphiphilic molecules in which the hydrophobic chains are in contact and the hydrophilic parts are oriented towards the water layer sandwiched between the lamellae.

If the amphiphile has a hydrocarbon part that takes up a large surface area, for example if the molecules has two HC chains, then some phases in the series may be "leapfrogged", not passed through, as the concentration is increased. In general, the lamellar phase is observed for simple straight-chain tensides, but not if the HC chain is too short (caprylates) or if the hydrophilic part is very large.

Since the spatial arrangement of molecules is primarily a question of space requirements (though we must not forget the effects of the intermolecular interactions), it is obvious that the addition of water- or oil-soluble substances can affect this arrangement.

So far, we have discussed thermodynamically stable systems. However, depending on the route by which these states are reached, intermediate metastable states are possible; they can also be created from thermodynamically stable systems by addition of energy.

Most vesicle systems are metastable forms, which can be obtained from lamellar phases, for example, by mechanical shearing.

3.2.3 Preparation

There are various techniques for the production of vesicles. Some are only suited to the laboratory scale, others can also be used for large-scale manufacture.

MLV – agitation of lipid films or lyophilized foams
– size fractionation by filter extrusion
LUV – reverse-phase evaporation
SUV – ultrasound
– high-pressure homogenizer
– chelate dialysis ("Lipoprep", laboratory apparatus)
– injection of alcohol
– injection of ether
– spontaneous formation on the addition of certain lipids.

Although amphiphiles with two hydrocarbon chains like phosphatidylcholine are best, vesicles can also be made from unsaturated and saturated fatty acids in pH ranges specific to their chain length, as long as the temperature is above the Krafft point (for details of the Krafft temperature, see Figure 13.7). Above the Krafft temperature, which in this case is 7–10° lower than the melting point of the fatty acids, the HC chains in the solid are able to rotate freely about their long axis. In addition, about half of the molecules of fatty acid are undissociated. Alternatively, the corresponding alcohols may be added, since this has a similar effect to that seen in the stabilization of emulsions. Comparable results are obtained with a mixture of sodium dodecyl sulfate and dodecanol.

Not only biological but also synthetic amphiphiles with two HC chains have been used in studies of vesicles. Mortara et al. [14] produced vesicles from dihexadecyl phosphate (DHP) by ultrasonication at 55°C; they were stable below the Krafft temperature (35°C), but more sensitive to salts than are vesicles made from naturally occurring phospholipids.

Analogously, cationic vesicles were made by K. Deguchi and J. Mino [15] from dioctadecyldimethylammonium chloride (DODAC). Like the DHP vesicles, these are sensitive to electrolyte; at electrolyte concentrations greater than 0.1 M, the surfactant vesicles are destroyed.

T. Kunitake [16] studied a large number of cationic, anionic, amphoteric, and nonionic amphiphiles with two HC chains. It was found that all amphoteric molecules with two HC chains 10–20 carbon atoms in length form lamellae in aqueous media. The two chains need not be the same length. The amphiphiles with shorter chains, with Krafft points below room temperature, form multilamellar vesicles spontaneously.

Further examples are collected in Table 3.2. They will also serve to provide an overview of the type and size of the vesicles thus formed.

On the laboratory scale, the most common method of preparing vesicles is the ultrasonication of the amphiphile solution. If the time of the ultrasound treatment is kept short, multilamellar vesicles are created, while longer treatment converts these into small unilamellar vesicles. It is nevertheless also possible to produce vesicles by simple shearing in mixers such as homogenizers or colloid mills, which are also used in emulsification (Chapter 2). These techniques are of particular interest on the manufacturing scale.

Table 3.2. Examples of vesicle preparation.

Vesicle Type	Diameter [Å]	Colloid	Method
Injection method			
SUV	250	lecithin	injection of an ethanolic or ether solution of the
SUV	600–1200		phospholipid into water above the Krafft point; concentration by ultrafiltration [17, 18].
Evaporation method			
LUV + MLV	2000–10 000	phospho-lipids	addition of an aqueous buffer solution to a mixture of phospholipid and organic solvent; vacuum distillation of the organic solvent; uni- and oligolamellar vesicles form enclosing the aqueous solution (comprising up to 65% of the vesicle), which may contain water-soluble active substances [19].
Mechanical shearing			
MLV	>5000	egg	manual stirring (5 min)
MLV	1000–10 000	phospholipid	vortex stirring (1 h)
SUV	300–1000	emulsion	ultrasonication (15 min) [20].
Vesicle enlargement			
LUV	3000–10 000	phosphatidyl serine	addition of Ca^{2+} to SUV followed by complexation of Ca^{2+} with EDTA [21].

The diagram in Figure 3.7 represents the apparatus used for the production of a formulation of unilamellar vesicles containing the active substance "Econazole" on the 500-kg scale [22]. The procedure relies on the ethanol injection method of Batzri and Korn listed in Table 3.2 [17]. An ethanolic solution of active substance and lecithin is injected under pressure (pump characteristics: 0.1–300 bar) into an aqueous buffer solution containing electrolytes. The rate at which the aqueous solution is pumped around must be at least 50 times the rate of injection of the organic solution. The vesicle formation occurs in the homogenizers (1500–20 000 rpm). The most stable liposomes containing this particular active substance form at around pH 5–7.

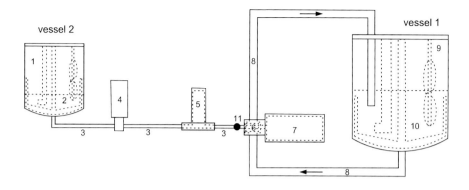

1 mixing unit
2 ethanolic solution of active substance and lecithin
3 feed to injector
4 pump
5 filter
6 injector/rotor-stator

7 homogenizer
8 circulation pipes for electrolyte
9 mixing unit
10 electrolyte solution
11 entry valve for pH regulation

Figure 3.7. Apparatus for preparation of vesicle formulations (Cilag, Schaffhausen; from [22]).

The cumulative mass curves for different batches of the vesicle formulation made under varied conditions in this apparatus are plotted in Figure 3.8. The bimodal distributions indicate the presence of multilamellar vesicles alongside the unilamellar ones. Depending on the process parameters – duration of homogenization, rotation rate, ratio of injection to circulation rate, etc. – formulations can contain 0–70% multilamellar vesicles by mass.

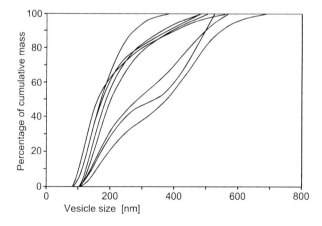

Figure 3.8. Cumulative mass plot for batches of lecithin vesicles containing "Econazole" [22].

Sterilization of the products can pose a problem. Fine formulations can be filtered through filters with pores smaller than 200 nm; larger vesicles have to be prepared from sterile solutions under aseptic conditions.

Similar considerations apply to the storage of liquid vesicle formulations as to that of emulsions. Thus, for example, flocculation and sediment caking must be prevented. Alternatively, vesicles can be lyophilized by the addition of disaccharides or spray-dried to give more stable forms.

3.2.4 Amphiphiles

In the previous section we looked at some types of amphiphiles suitable for the preparation of vesicles. The most interesting amphiphiles, though, are the naturally occuring ones, mixtures of different lipids. Phosphatidylcholine, also known as lecithin, is an important amphiphile (see Figure 3.9). It contains a variety of components with different fatty acid residues, mostly of unsaturated fatty acids, as can be seen from Table 3.3, which lists the fatty acid residues in purified soya phosphatidylcholine ("Epicuron 200").

In the broadest sense, the term "lecithin" is used to refer to soya lecithin; it consists of a mixture of different phospholipids (Figure 3.9) and 33–35% soya oil as well as 2–5% sterols and 5% carbohydrates.

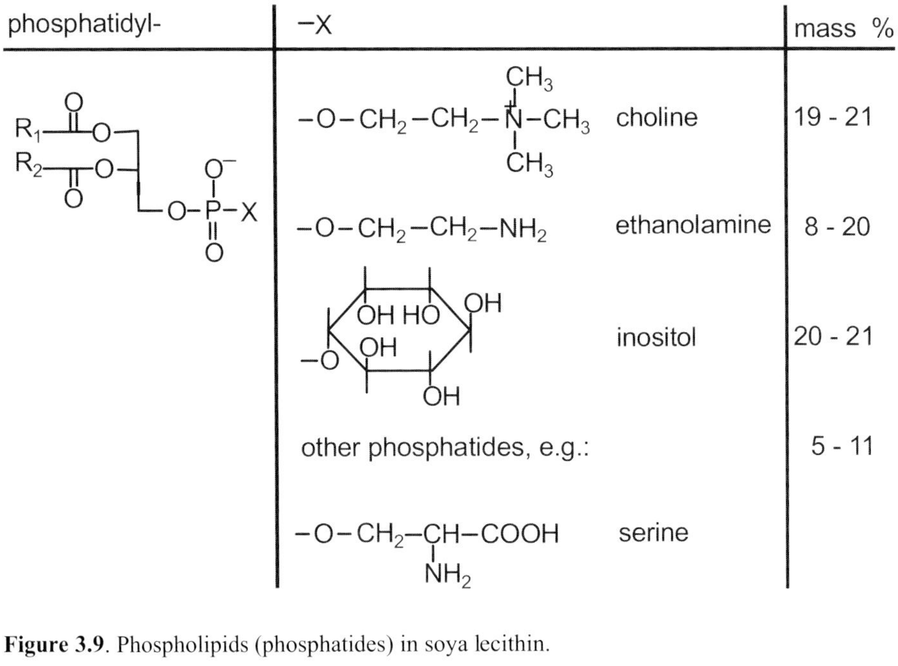

phosphatidyl-	−X		mass %
	$-O-CH_2-CH_2-\overset{CH_3}{\underset{CH_3}{N}}-CH_3$	choline	19 - 21
	$-O-CH_2-CH_2-NH_2$	ethanolamine	8 - 20
	(inositol structure)	inositol	20 - 21
	other phosphatides, e.g.:		5 - 11
	$-O-CH_2-\underset{NH_2}{\overset{}{C}}H-COOH$	serine	

Figure 3.9. Phospholipids (phosphatides) in soya lecithin.

Table 3.3. Fatty acid residues in soya phosphatidylcholine "Epicuron 200" [23].

Fatty acids	C_8	C_{16}	$C_{16:1}$	C_{18}	$C_{18:1}$	$C_{18:2}$	$C_{18:3}$
Mass fraction [%]	0.8	12.2	0.4	2.7	10.7	67.2	6.0

In order to optimize the vesicle properties, it is often necessary to blend a number of different lipids. Table 3.4 shows some mixtures of lipids in use. Such mixtures are well suited to the manufacture of long-circulating liposomes (LCL), which are administered parenterally in cancer therapy.

Polymerizable amphiphiles too can be used to increase the stability of vesicles. Because of the much lower rate of monomer exchange, polymerization in vesicles is much more successful than in micelles.

Table 3.4. Lipid mixtures used in the preparation of vesicles.

Lipids	Mole ratio	Reference
HPI/HSPC/Chol/TC	1:9:5–8:0.1	[24]
PEG–DSPE/HSPC/Chol/TC	5.5:56.1:38.2:0.2	[25]
DSPC/Chol	2:1	[26]
DPPC/Chol	55:45	[27]

Chol: Cholesterol; DSPC: distearoylphosphatidylcholine; DPPC: dipalmitoylphosphatidylcholine; HPI: hydrogenated phosphatidylinositol; HSPC: hydrogenated soya phosphatidylcholine; PEG–DSPE: polyethylene glycol (M 1900)-modified distearoylphosphatidylethanolamine; TC: α-tocopherol.

3.2.5 Properties

The kinetic stability of bilayer membranes is considerably greater than that of micelles. Micelles are in a state of dynamic equilibrium involving rapid exchange with their monomers, while vesicles, once formed, are stable over weeks or months because of the much slower diffusion of monomers out of the vesicle (Table 3.5).

Table 3.5. Comparison of micelles, microemulsions, and vesicles [28].

	Micelles	Microemulsions	Vesicles
molar mass $\overline{M}_w$	2000–6000	10^5–10^6	$>10^7$
diameter [Å]	30–60	50–1000	300–5000
solubilizate per aggregate	little	much	much
monomer exchange rate [s]	$\sim 10^{-5}$	$\sim 10^{-5}$	> 1
duration of solubilizate retention [s]	10^3–10^5	10^3–10^5	> 1
dilution with H_2O	destroyed	altered	stable

Just as for micelles, which begin to form at the critical micelle concentration, there is a minimum concentration at which vesicles can be made. For DODAC vesicles, this is $(8\pm4)\cdot10^{-6}$ mol/L [29]. Unlike micelles, though, once vesicles are formed dilution below the critical concentration does not destroy them.

Cationic and anionic vesicles are stable over a large pH range. However, they can be disrupted by electrolyte solutions with concentrations over 0.1 M or by various substances such as low molecular weight alcohols or dodecyl sulfate. Polymerized vesicles behave differently; they are stable to such influences, and although flocculation and precipitation may occur by mechanisms similar to those in emulsions and dispersions of solids, the spherical structures can still be observed in the sediment.

The stabilization of vesicle dispersions also follows principles similar to those for emulsions and solid dispersions. The coalescence of small vesicles can be prevented by addition of polymers such as polyethylene glycol (6000) and polyvinylpyrrolidone.

The sketch of the cationic DODAC (dioctadecyldimethylammonium chloride) vesicle in Figure 3.10 gives an idea of how vesicles are organized. It can be seen that the fine structure of the vesicle consists of a series of different zones: an outer aqueous phase, a lipophilic phase, an inner aqueous phase, and in addition inner and outer charged zones that can be subdivided as with polyelectrolytes. A tightly packed aqueous layer is oriented in the electrical field around the mostly ionized head groups, and is itself surrounded by a cloud of counterions.

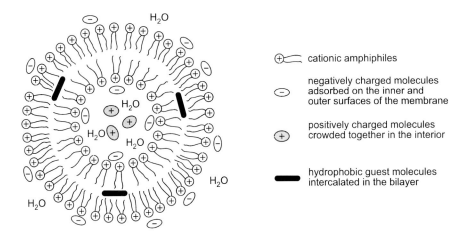

Figure 3.10. Schematic diagram of a cationic DODAC vesicle (counterions omitted); cf. [28].

As indicated in Figure 3.10, guest molecules can be incorporated into and concentrated in any of these areas, depending on their physicochemical properties. Hydroxy ions, for example, may be concentrated in DODAC vesicles: when the pH in the external aqueous phase is adjusted to 8.75, it is greater than 10 inside the vesicle [29].

Large water-soluble molecules can be enclosed in the vesicle interior; their electrostatic interactions with the membrane play an important role here. Thus DODAC vesicles incorporate zwitterionic amino acids only poorly, but enclose a large proportion 8-azaguanine in its anionic form. It can take hours or days for this uptake to occur. In contrast, lipophilic guest molecules enclosed in the lipid membrane remain there for the life of the vesicle.

Vesicles respond to osmotic pressure. They shrink when the electrolyte concentration in the external aqueous phase is higher than the internal concentration and swell when it is lower.

The permeability of the lipid membrane can nevertheless be controlled, for instance through the composition of the lipids (see Table 3.4; cholesterol acts as a "sealant"). The permeability is greatest at the phase inversion temperature.

References for Chapter 3:

[1]   J. H. Schulman, T. P. Hoar, Nature (London) *152*, 102 (1943).
[2]   G. Gillberg, H. Lethinen, S. Friberg, J. Coll. Interface Sci. *33*, 40 (1970).
[3]   K. Shinoda, S. Friberg, Adv. Coll. Interface Sci. *4*, 281 (1975).
[4]   S. Friberg, Chemtech *6*, 124 (1976).
[5]   M. Borkovec, J. Chem Phys. *91*, 6268 (1989).
[6]   J. T. Overbeek, Proceedings *889*, 87 (1986).
[7]   S. Friberg, J. Dispersion Sci. Technol. *6*, 317 (1985).
[8]   P. G. de Gennes, C. Taupin, J. Phys. Chem. *86*, 2294 (1982).
[9]   L. M. Prince, Microemulsions, Academic Press, Inc., New York, San Francisco, 1977.
[10] H.-F. Eicke, Seifen – Öle – Fette – Waschmittel *118*, 311 (1992).
[11] M. Borovec, H.-F. Eicke, H. Hammerich, B. D. Gupta, J. Phys. Chem. *92*, 206 (1988).
[12] G. Gillberg, in Emulsions and Emulsion Technology, Part III, 1 (K. J. Lissant, Ed.), Marcel Dekker, Inc., New York, Basel, 1984.
[13] A. D. Bangham, H. M. Standish, H. M. Watkins, J. Mol. Biol. *13*, 238 (1965).
[14] R. A. Mortara, F. H. Quiana, H. Chaimovich, Biochim. Biophys. Res. Comm. *81*, 1080 (1978).
[15] K. Deguchi, J. Mino, J. Colloid Interface Sci. *65*, 155 (1978).
[16] T. Kunitake, J. Macromol. Sci. Chem. *A 13*, 587 (1979).
[17] S. Batzri, E. D. Korn, Biochim. Biophys. Acta *298*, 1015 (1973).
[18] D. Deamer, A. D. Bangham, Biochim. Biophys. Acta *443*, 629 (1976).
[19] F. Szoka, D. Papahadjopoulos, Proc. Nat. Acad. Sci. USA *75*, 4194 (1978).
[20] E. Marchelli, U. Bucciarelli, DOS 29 46 601 (17.11.78).
[21] D. Papahadjopoulos, W. J. Vail, K. Jacobson, G. Poste, Biochim. Biophys. Acta *394*, 483 (1975).

[22] R. Naeff, in Liposomes, Ergänzungskurs Pharmazie, Pharmazeutisches Institut Universität Basel, 1994; Grusbach Conference: Liposome Dermatics, Springer-Verlag, Berlin Heidelberg, 1992, p. 93.

[23] B. Bergenståhl, K. Fontell, Progr. Colloid Polymer Sci. *68*, 48 (1983).

[24] A. Gabizon, D. C. Price, J. Huberty, R. S. Bresalier, D. Papahadjopoulos, Cancer Res. *50*, 6371 (1990).

[25] J. Vaage, D. Donovan, E. Mayhew, R. Abra, A. Huang, Cancer *72*, 3671 (1993).

[26] E. A. Forssen, D. M. Coulter, R. T. Profitt, Cancer Res. *52*, 3255 (1992).

[27] M. B. Bally, R. Nayar, D. Masin, M. J. Hope, P. R. Cullis, L. D. Mayer, Biochim. Biophys. Acta *1023*, 133 (1990).

[28] J. H. Fendler, J. Phys. Chem. *84*, 1485 (1980).

[29] J. H. Fendler, W. L. Hinze, J. Am. Chem. Soc. *103*, 5439 (1981).

4 Foam

4.1 General Remarks

Foam is a coarse dispersion of a gas in a liquid. The majority of the volume is occupied by the gas, and the liquid is distributed in thin films between the gas bubbles, called lamellae. Typical aqueous foams consist of 95% air and only 5% liquid. The latter is itself more than 95% water; the remainder is made up of tensides and other substances. References [1–5] are good general sources of information on foams.

Like any other material, a foam retains its structure only for as long as it is unable to convert into a state with lower energy. A foam is therefore always striving to adopt a structure in which the area of the films (bubble walls, lamellae) is minimized. A foam can never be thermodynamically stable. When a liquid lamella collapses, it breaks into drops with a smaller total surface area than the original film, thus diminishing the free energy of the system. Foams consisting of pure liquids and gases are extremely unstable, with a lifetime of less than a second.

There are two types of foams, spherical and polyhedral foams, illustrated in Figure 4.1.

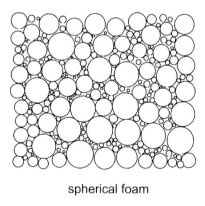 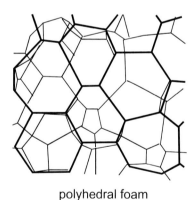

spherical foam   polyhedral foam

Figure 4.1. Types of foam.

Spherical foams consist of individual, independent bubbles. Their creation does not require the presence of surfactants, and their stability depends on the viscosity of the dispersion medium. At high viscosity, the lifetime of the foam is considerably lengthened. Polyhedral foams only develop in the presence of surface-active compounds.

One important characteristic of foams is the rapid growth of large bubbles at the expense of smaller ones. As the Young–Laplace Equation shows (Figure 4.2), the pressure in a small bubble is greater than that in one of larger radius, in accord with the greater

curvature of the surface of the small bubble. If the two bubbles are in contact with one another, the larger one grows and the smaller shrinks (Figure 4.2).

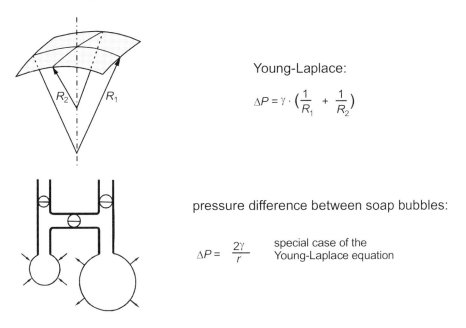

Young-Laplace:

$$\Delta P = \gamma \cdot \left(\frac{1}{R_1} + \frac{1}{R_2} \right)$$

pressure difference between soap bubbles:

$$\Delta P = \frac{2\gamma}{r}$$

special case of the Young-Laplace equation

Figure 4.2. Phenomena at curved surfaces.

The gas bubbles in a foam start off spherical, but can quickly adopt a more densely packed structure in which the bubbles are separated only by thin films of liquid. The pressure difference between these interfaces is usually very low, so that the individual foam lamellae form polyhedra joined together along straight edges.

Figure 4.3 represents the structure of a foam. The area in which three bubbles join is called the *Plateau border* after the physicist Plateau. The liquid in a foam tends to drain away because of gravity and surface tension. *The pressure at the Plateau border is smaller than in adjacent areas on account of the negative curvature.*

When a film thins owing to loss of water, it reaches a certain point at which interfacial forces counteract further thinning. These forces may be electrical or steric, and spring from the molecules that made the foam formation possible in the first place, the tensides.

When the forces of destruction and preservation are balanced, the lamellae in the foam are metastable, that is, they are stable to small changes. For example: when beer is poured, a head is formed (an ideal spherical foam), which gradually shrinks so that the liquid level drops. In reaction to the altered equilibrium of forces, the bubbles remaining on liquefaction of the foam transform from spheres into polyhedra. The wet foam becomes a dry foam and the cells begin to influence one another and to form a polyhedral foam (Figure 4.1).

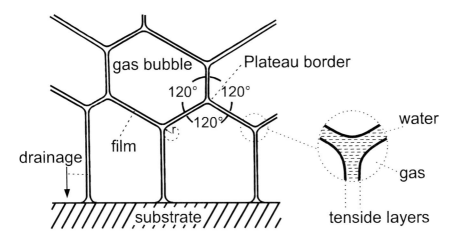

Figure 4.3. Diagrammatic representation of foam structure; see [6].

4.2 Foam Stabilization

Why doesn't a foam collapse as soon as it forms? A "two-bubble foam" with one large and one small bubble, as shown in Figure 4.2, is unstable by its very nature. The addition of tensides stabilizes it. This effect depends on two related phenomena, which were described as early as the nineteenth century by J. W. Gibbs and C. G. Marangoni. The Gibbs effect occurs when a thin film in which tenside molecules are dissolved is stretched (Figure 4.4); this increases the surface area of the film and creates room for further tenside molecules. However, once the film has been stretched, there are fewer tenside molecules at the interface than before. The interfacial tension increases, and the stretched film endeavors to contract like an elastic skin. This is known as the *Gibbs film elasticity*. Simultaneously, under the surface, liquid transport is occurring. This second phenomenon, the *Marangoni effect*, is ephemeral, since a certain time is required for the tenside molecules to migrate from the interior of the stretched film to the surface. Initially, therefore, a stretched interface has a depressed tenside concentration and the interfacial tension is still greater than would be expected from the Gibbs effect alone. As tenside molecules diffuse to the surface and the film approaches equilibrium, the surface tension declines gradually to the Gibbs value.

These two effects act to stabilize the foam against small variations in bubble size and film thickness. *Film elasticity is a necessary condition for stable foams.*

The Marangoni effect also plays an important role in preventing the coalescence of emulsions. It bears the responsibility for the technically extremely important film elasticity. Readers fond of a glass of good red wine may be amused to know that the large drops of wine that collect on the sides of the glass result from the Marangoni effect.

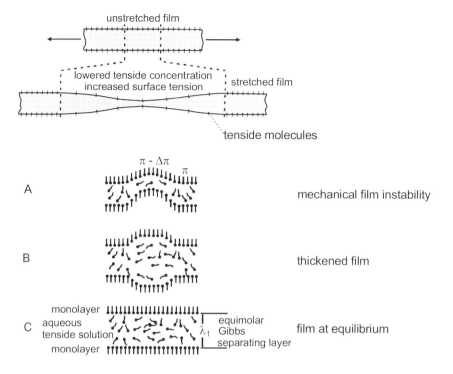

Figure 4.4. Stabilization of tenside films by the Gibbs and Marangoni effects; cf. refs. [6, 7].

4.3 Forces in Thin Films

The gradual loss of liquid from the lamellae turns a foam from a wet to a dry foam. The Plateau zone sucks liquid out of lamellae that start off thick (Figure 4.5) until eventually they become so thin that their two surfaces begin to interact with one another.

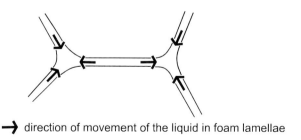

→ direction of movement of the liquid in foam lamellae

Figure 4.5. Movement of liquid at the Plateau border.

Besides this "suction" effect, the film is affected by van der Waals interactions and by electrical and steric forces, though these latter tend to stabilize it. The van der Waals forces create an attraction between the two surfaces of a thin film, pressing them together, so that more liquid is forced out of the film. In a thin film of liquid containing dissolved ions of which one variety preferentially occupies the interface and forms a layer of charge, the electrostatic forces result in repulsion as the film dries out; that is, as the interfaces approach one another, further shrinkage is counteracted (Figures 4.6 and 4.7).

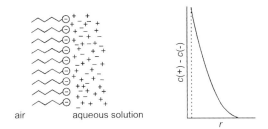

electrical double layer due to adsorption of anionic tenside at the water/air interface

excess of positive charge in the diffuse double layer as a function of distance r from the surface

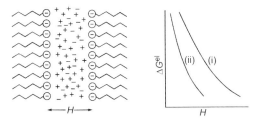

mutual approach of two charged film surfaces

free energy of interaction of the surfaces as a function of film thickness H; (i) low, (ii) high electrolyte concentration

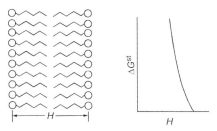

film stabilized with nonionic tenside

free energy of interaction as a function of film thickness H

Figure 4.6. Interactions between the surfaces of foam lamellae [8].

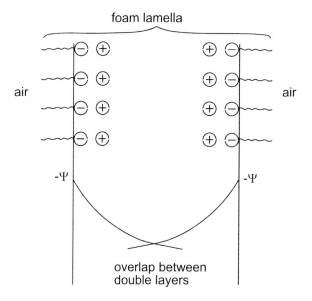

Figure 4.7. Electrostatic stabilization of a foam lamella [9].

Steric interactions too can cause repulsion when macromolecular stabilizers are present. *Sometimes, liquid crystals can form in a thin film when they would not do so in a large volume of the same liquid.* If the forces acting against thinning are large enough, the film shrinks only to the thickness at which it is metastable.

A qualitative correlation has been observed in some individual cases between foam stability and film viscosity, but the connection remains unclear. *In general, it has been found that films are unstable when the film viscosity is either very high or very low.*

4.4 Foaming Agents

In general, the lower the CMC of a tenside, the more effective a foaming agent it is. The foaming action of tensides depends on their structure. For practical purposes, the foam created must be stable to mechanical and thermal shock. This requires molecules that can be densely packed to form a firm film; for this, the hydrophobic groups must be long and straight. Since longer hydrophobic chains reduce the surface activity of the molecule, good foam formers should have a medium-length chain, around 12–14 C atoms for sodium alkyl sulfates and soaps at 20°C. For higher temperatures, longer chains are optimal: 16 C atoms for 60°C, for example, and 18-C compounds for temperatures near boiling.

In ionic media, nonionic tensides usually produce smaller amounts of considerably less stable foam than ionic tensides.

4.5 Foam Stabilizers

The foaming properties of tenside solutions can be effectively modified by the addition of suitable organic compounds. Solutions with excellent foaming properties can be converted into formulations that scarcely foam or do not foam at all, while poorly or non-foaming formulations can be made to produce large quantities of foam, simply by means of small amounts of additives. Modification of foaming properties is of considerable practical importance.

The most effective additives for stabilizing foams created with the aid of tenside solutions are long-chain, often water-insoluble polar compounds with straight-chain hydrocarbon groups of about the same length as the hydrophobic group of the tenside. For example: lauryl alcohol as an additive for sodium dodecyl sulfate, N,N-bis(hydroxy-ethyl)lauramide as an additive for dodecyl benzenesulfonate, lauric acid for sodium laurate, N,N-dimethyldodecylamine oxide for dodecyl benzenesulfonate and other anionic tensides.

It is interesting to discover that the addition of compounds that neutralize or buffer the charge of an ionic tenside lowers the surface tension further and increases foam formation. Long-chain amine oxides of the $RN(CH_3)_2O$ type, which themselves show surface activity, are used as foam stabilizers in dishwashing liquids formulated with anionic tensides. The amine oxides accept a proton from the anionic tenside to form $RN(CH_3)_2OH^+ \cdot R\text{–}SO_3^-$, which results in a much more densely packed film than would be formed by the components individually, because of the neutralization of the ionic head groups. This compound has a much stronger surface activity than does the amine oxide or the anionic tenside, and has an outstanding foaming behavior, one which surpasses that of the individual components. The optimal effect depends on the ratio of the two components and on R in the amine oxide.

4.6 Antifoaming Additives

Antifoaming additives destroy foams by replacing the foam-forming surface film by a completely different type of film. For this purpose they have to be able to replace all the foam stabilizers and tensides in the film. Therefore, in their pure state they must have a sufficiently low surface tension to be able to spread out spontaneously over the film. Their spreading coefficient $S_{LS} = \gamma_{SA} - \gamma_{SL} - \gamma_{LA}$ must therefore be positive. They must be almost insoluble in the foam solution, but only such that they are still present in the surface film.

There are two types of antifoaming agents which must be distinguished; foam breakers and foam inhibitors. The former destroy existing foams; they work by depressing the surface tension to extremely low values over small local areas, causing these areas to thin rapidly so that the pull of the surrounding areas in which the surface tension is higher makes the foam break. Diethyl ether, which has $\gamma = 17$ mN/m, and isoamyl alcohol in small quantities act in this way.

Another mode of action involves an increase in the drainage of the films in the foam and thus a shortening of their lifetime. This is how tributyl phosphate works; it effects a large decrease in the surface viscosity.

Foam inhibitors prevent the formation of foam. They act by destroying the elasticity of the film. Inhibitors create a surface which maintains constant surface tension when it expands or contracts. Some foam inhibitors achieve this effect by flooding the surface with nonfoaming, rapidly diffusing, noncohesive molecules with only weak surface activity, so that any increase in the surface tension as a result of film expansion is immediately countered. Some wetting agents and EOPO block copolymers function thus.

Other inhibitors act by replacing the elastic surface film with a fragile, densely packed one. This is the case for calcium salts of long-chain fatty acids (stearic and palmitic acids for foams made with sodium dodecyl benzenesulfonate and sodium lauryl sulfate). The calcium soap films create an unstable foam. If the calcium soap is able to form a true mixed film with the tenside, the foam is not destroyed.

Foam breakers and foam inhibitors can have additive effects, and mixtures can have both good foam-destroying and good foam-inhibiting characteristics.

Other foam inhibitors include:

– Octanol, used in the paper industry and in electroplating baths. Silicones find wide application at concentrations of 10 ppm. These additives work by preferential, strong adsorption.
– Perfluoroalcohols act as both foam breakers and foam inhibitors.
– 4-Methyl-2-pentanol and 2-ethylhexanol are effective foam inhibitors for detergents.

References for Chapter 4:

[1]   A. W. Adamson, Physical Chemistry of Surfaces, 4th ed., Wiley-Interscience, New York, 1982.
[2]   J. J. Bikerman, Foams and Emulsions, Ind. Eng. Chem. *57*, 57 (1965).
[3]   N. Pilpel, Foams in Pharmacy, Endeavour, New Series Vol. 9, No. 2, 87 (1985).
[4]   S. Ross, Ind. Eng. Chem. *61*, 48 (1969).
[5]   A. J. Wilson (Ed.), Foams: Physics, Chemistry and Structure, Springer-Verlag, Berlin, 1989.
[6]   T. C. Patton, Paint Flow and Pigment Dispersion, Wiley-Interscience, N.Y., 1979.
[7]   J. A. Wingrave, T. P. Matson, J. Am. Oil Chem. Soc., 1981, 349 A.
[8]   D. H. Everett, Basic Principles of Colloid Science, Royal Society of Chemistry, London, 1988.
[9]   D. Eklund, T. Lindström, Paper Chemistry, DT Paper Sci. Publ., Grankulla, Finland, 1991.

5 Manufacture and Properties of Colloidal Suspensions and Dispersions

Distinction between colloidal dispersions and suspensions on the basis of size range:

colloidal dispersions:	1 nm (10 Å) to 1 μm
suspensions:	≥ 1 μm

Despite its somewhat arbitrary nature, this division is used by the principal sources in the field. Our chief interest will be in hydrophobic dispersions and suspensions. Often no distinction is made between dispersions and suspensions and the two words are used interchangeably, with no stipulations about particle size range (see Section 1.2.1). In contrast, the term sol is used to refer to very fine dispersions.

The technical significance of lyophobic disperse systems lies in the fact that sparingly soluble substances can be dispersed in very finely divided form in a liquid phase to give a system that behaves like a concentrated, low-viscosity solution. Examples have already been cited in Chapter 1.

Other examples of lyophobic suspensions of industrial importance:

- sparingly soluble pharmaceutical and agricultural agents;
- finely divided pigments in H_2O or organic liquids:
  dispersion colors, printers' inks, inks, and paints;
- sparingly soluble textile dyes (disperse vat dyes);
- magnetic particles, e.g. Fe_3O_4 in aqueous or organic suspension for the manufacture of magnetic tape;
- aqueous and nonaqueous polymer dispersions for surface coatings;
- monodisperse polymer dispersions with controlled particle size, e.g. of latex;
- monodisperse suspensions of ceramics such as Al_2O_3 for use in electronics (high-tech ceramics);
- fine dispersions of AgBr in gelatin for photographic films;
- pigment dispersions for the coating of TV screens;
- dispersions of soot in polymers (toner) for photocopying and laser printing.

5.1 The Dispersion Procedure; Definition

The powder is dispersed in a liquid such that every particle is completely surrounded by the liquid. *The solid/air interfaces are exchanged for solid/liquid interfaces.* In principle there are two types of methods for the manufacture of lyophobic dispersions:

1. Condensation methods

The disperse phase is formed from a soluble compound by condensation and aggregation. For example: sulfur sols are made by mixing a concentrated solution of sulfur in alcohol with water. Seed crystals form from the supersaturated solution and grow rapidly to result in a colloidal suspension. Colloids can also be obtained from chemical reactions resulting in precipitation, but generally only in low concentrations. In the case of gold sols the colloid particles are prepared by the reduction of gold (III) chloride with formaldehyde, with hydrazine, or with other reducing agents. Sols of inorganic oxides are often formed by simple hydrolysis. A solution of iron acetate, for example, gives a sol of $Fe(OH)_3$ on boiling. Except in a few special cases, such techniques are not economic or suited to production on an industrial scale. On the other hand, undesirable colloidal dispersions are sometimes formed in organic synthesis. They have to be destroyed by boiling or by addition of suitable non-solvents or reagents. The formation of such colloids may also cause problems in the preparation of clear solutions, or in the case of waste water.

2. Dispersion and comminution methods

In contrast to condensation and aggregation, these methods involve division of the solid to obtain the particles, particularly by *grinding*. Although *over 99% of the energy used in grinding is lost to friction*, nearly all industrial production of colloidal dispersions and suspensions relies on comminution, usually wet grinding, since concentrated colloidal products can be manufactured in high volumes this way. Dry grinding procedures have a lower attainable particle size limit of about 10 μm (jet mills). Modern agitator ball mills have large throughputs (up to ~5 t/h) at a power of up to ~200 kW and can produce dispersions in the submicron size range. The fineness of the particles attainable depends chiefly on the optimization of the many operating parameters (see below) and of the suspension to be ground in terms of wetting agents and so on.

As an example, the energy transfer in wet grinding of particles of size <1 μm occurs mainly hydrodynamically and depends on the surface of the particles to be ground. Without grinding auxiliaries, it is impossible to produce particles smaller than about 0.5 μm from crystalline material. These additives coat the surfaces of the particles thinly and prevent the comminuted but still agglomerated crystals from aggregating ("growing back together"). A common grinding additive for use with organic pigments is Stabelite resin (tetrahydroabietic acid), which must be used in quantities sufficient to coat the surface created to a depth of about two molecular layers. No matter how long the pigments are milled, if less resin is used then the particles are not ground any smaller. It should be noted that consecutive recrystallizations (Chapter 9) can affect the achievable particle size. A further point is that the material of the mill may be abraded and thus enter the product, which may be undesirable.

The procedure of dispersion can be split into three steps, though these generally overlap:

- wetting of the powder
- comminution and distribution of the particles (deagglomeration)
- stabilization of the particles

The dispersion process causes the solid/air interfaces to turn into solid/liquid interfaces.

5.2 The First Step in the Dispersion Procedure: Wetting of the Powder

For optimal dispersion, the powder from which the dispersion is to be manufactured must be completely wetted. Often this requirement is not taken fully into account. The choice of wetting agent is therefore not trivial. The criteria have already been discussed in Section 1.7. Good wetting agents must cause complete wetting ($\theta \sim 0°$) with rapid equilibrium and in low concentrations, without depressing γ_L greatly (see Section 1.7.3). In addition, they must adsorb onto the particle surface if possible; this lowers γ_L and thus also θ (Equation 1.26). The concentrations in which wetting agents are used lie in the range of the *CMC*, for example ~0.2% for sodium lauryl sulfate. The *CMC* values for common tensides can be found in any of the relevant textbooks. Wetting agents are dealt with in Section 5.3. One compound which fulfils the requirements for wetting agents in dispersions very nearly perfectly and also acts as a deflocculant (next section) is the condensation product of β-naphthalenesulfonic acid and formaldehyde, sodium dinaphthyl-methanedisulfonate (trade names: Tamol varieties), which has a relative molecular mass of ~500–2000, depending on its degree of polymerization (Figure 5.17). These compounds only lower the surface tension of water slightly (~60 mN/m), but they adsorb onto a very wide range of solids. They are *polyelectrolytes*. Therefore, they fulfil the conditions stipulated in Section 1.7.3 for good wetting of agglomerated powders or filter cakes almost perfectly. As an example, we describe the wetting effect of Tamol when it is added to a hard filter cake of an apolar organic dye. When the powdered wetting agent is stirred into the dry-looking filter cake, the cake becomes a liquid slurry. It contains about 60% sequestered water that is liberated once the surface is wetted.

5.3 The Second Step in the Dispersion Procedure: Comminution and Distribution of the Particles in the Liquid

When a substance is ground, an equilibrium is achieved: the fine particles produced gather together into agglomerates or flocs because of their increased surface energy (Figure 5.1).

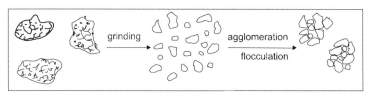

Figure 5.1. Grinding equilibria.

Deflocculants or *dispersants* must be added to stabilize the ground powder. A distinction must be made between crystalline and agglomerated materials: crystalline materials are usually harder to pulverize and require *grinding aids*, such as polyacrylates, acrylamides, or polystyrenesulfonates. Agglomerated powders are easier to pulverize; the process is usually not really one of grinding but one of separation of the agglomerated particles. The primary particles or fragments thus formed must be prevented from reagglomerating or flocculating; this is the purpose of *deflocculants*.

Deflocculants of higher molecular mass can act simultaneously as dispersants, as do, for example, Tamol and lignin sulfonates.

Most deflocculants are charged tensides that adsorb onto the particles and cause repulsion between them. Their charge must be of the same sign as the charge on the particles. When the charges have opposite signs, flocculation occurs (Figure 5.2a). On adsorption, the hydrophobic tails of the tensides are located in the area known as the Stern layer (see Section 5.6.1). At higher concentrations, the tenside can form a second adsorption layer, giving the particle an electric charge once more (Figure 5.2b).

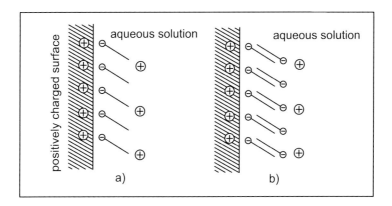

Figure 5.2. Adsorption of an anionic tenside onto a positively charged surface: a) monolayer → charge neutralization, flocculation; b) double layer, charge reversal → overall negative charge [1].

Typical deflocculants:
Anionic type: Dodecylbenzenesulfonate, isopropylnaphthalenesulfonate, diamylsulfosuccinate.
Cationic type: Dodecyltrimethylammonium bromide, cetyltrimethylammonium bromide. Deflocculants that are also effective dispersants are the Tamol group of compounds and lignin sulfonates, preferably of high molecular mass.

The deflocculant concentration must be greater than that required to cover the fresh surfaces, because the deflocculant must be immediately available wherever the new surfaces are formed in sufficient concentration to prevent reagglomeration, which occurs very quickly. Depending on the fineness of the dispersion, the requisite concentrations are 10–20% of the mass of the particles.

5.3.1 Comminution

Wet grinding relies on the mechanisms shown schematically in Figure 5.3. The shearing forces in a viscous medium are often sufficient to break up agglomerates (for example, pigments in molten polymers). Pulverization by compression, though, is hindered by viscosity.

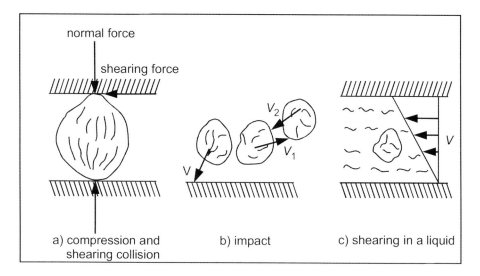

Figure 5.3. Stress mechanisms (according to Rumpf): a) compression and shearing between two solid surfaces; b) parallel stresses at a solid surface; c) shearing forces in a liquid environment [2].

The division of particles in shearing flows is proportional to the energy liberated. This energy is given by the product of shear stress τ and shear rate D, Figure 5.4. From the definition of viscosity η (Equation 5.1) we can thus derive the important Equation 5.2.

$$\eta = \frac{\tau}{D} \tag{5.1}$$

(η: viscosity $[\text{N·s·m}^{-2}]$; τ: shear stress $[\text{N·m}^{-2}]$; D: shear rate $[\text{s}^{-1}]$)

$$E = \tau \cdot D = \eta \cdot D^2 \quad [\text{W/m}^3] \tag{5.2}$$

E has the units of energy liberated in volume (m^3) per unit time (s), or *power per unit volume*.

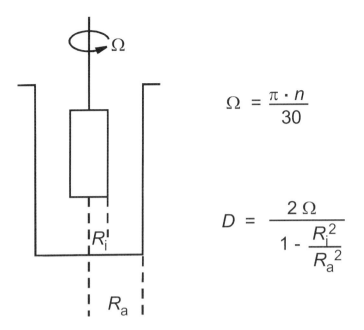

$$\Omega = \frac{\pi \cdot n}{30}$$

$$D = \frac{2\,\Omega}{1 - \dfrac{R_i^2}{R_a^2}}$$

Figure 5.4. Energy conversion in the grinding process, analogous to the rotational viscometer. Division of a particle in a shear flow (viscous system). Ω: angular velocity [s^{-1}]; *n*: rotation rate [rpm].

For a given shear rate *D* (which depends on the rotation rate of the machine and the design of its components) the viscosity is variable, as is the case, for instance, in extruders and roll mills. Therefore, for such apparatus dispersion is most successful at high viscosities. For machines in which the rotation rate can vary widely, like agitator ball mills, effective comminution is possible even at low velocities.

The energy required to divide a solid increases disproportionately with the fineness of the powder produced. For example, it can rise to over ~1 kWh/kg for the wet grinding of pigments to less than 1 µm in size. As has already been mentioned, over 99% of the mechanical energy used in this process is converted to heat by friction.

Nevertheless wet grinding is the most effective method for the manufacture of liquid dispersions. The costs of dry grinding to the finest level (by jet mills using blasts of air, to about 10 µm) are around twice as high as for wet grinding.

5.3.2 Survey of Wet-Grinding Mills

Figure 5.5 classifies dispersion apparatus according to the viscosity of the material to be ground. The apparatus is divided according to mechanism of action into "smashers" (compression, shearing, and impact in Figure 5.3) and "smearers" (shear stressing), and the intermediate "hybrids" (according to T. C. Patton).

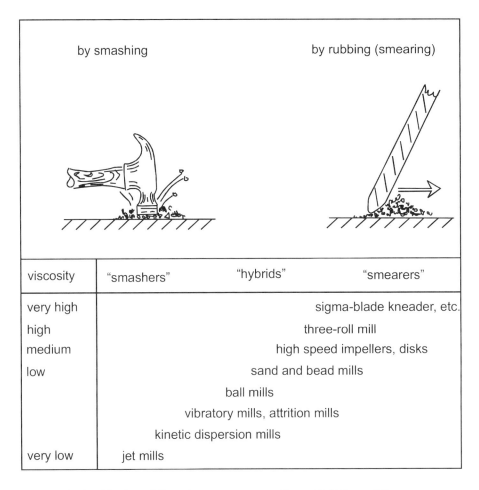

Figure 5.5. Classification of dispersion apparatus according to T. C. Patton [3].

The usual classification of wet-grinding mills is based on whether they use grinding media. Figures 5.6 to 5.11 survey the types. It should be noted that the apparatus shown in Figures 5.6–5.8 is only suitable for producing relatively coarse suspensions. It is used to work the material to be ground into a paste or for "predispersion", and achieves a particle size over 1 μm, generally 10, 20 μm or more. The expression "colloid mill" is not justified for such apparatus (except when an easily dispersible product is ground).

In contrast, vibratory mills (Figure 5.9) are able to produce particles in the colloidal range; still better are agitator ball mills, which operate on liquids, Figure 5.11. Besides their effective dispersing action in the 1 μm range and, depending on the material to be ground, down to about 0.01 μm, these machines have an outstanding throughput, operating by a continuous process.

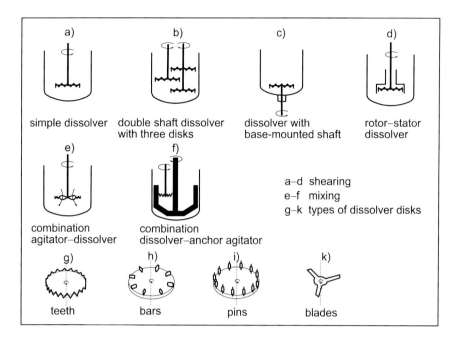

Figure 5.6. Types of dissolvers, from reference [4].

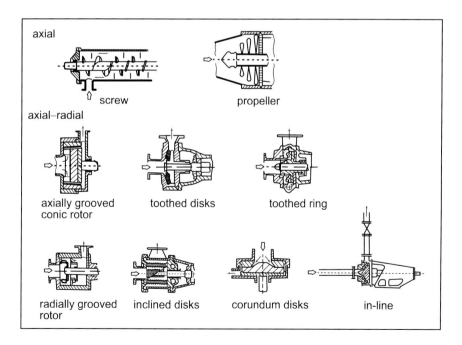

Figure 5.7. Types of colloid mills, from reference [4].

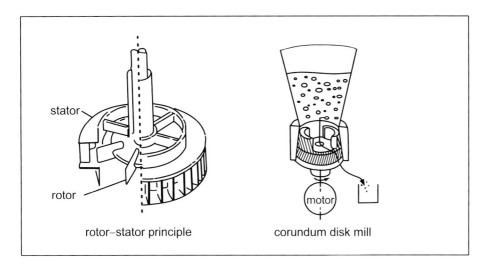

Figure 5.8. Principles of colloid mills, according to reference [4].

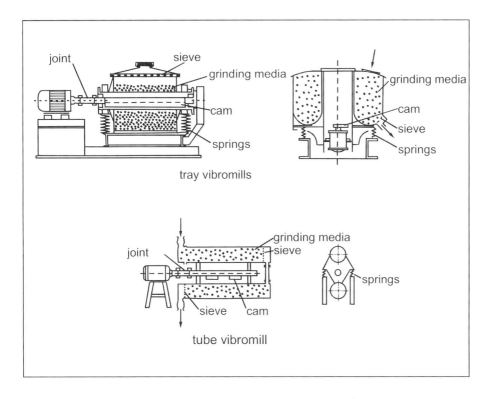

Figure 5.9. Types of vibratory mill (oscillatory mill), according to reference [4].

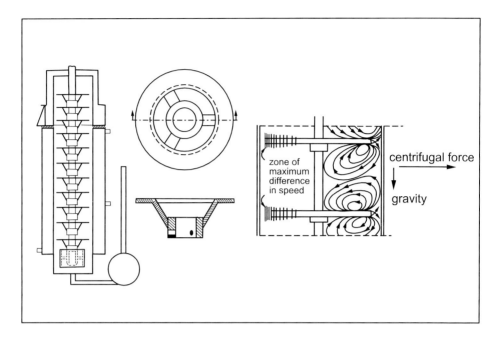

Figure 5.10. Design of a sand mill, according to reference [5].

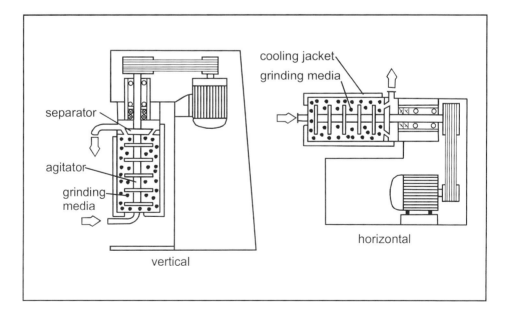

Figure 5.11. Modern types of agitator ball mills, according to reference [4].

For good dispersion with a narrow size range (as few oversize grains as possible), the flow profile through the machine of the material to be ground is critical. The product must pass through all the grinding areas, so a plug flow is best. To optimize the passage, numerous apparatus and operating parameters must be taken into account; about 30 such parameters for optimum dispersion have been identified. The ten most important of these are summarized below. Only when they are set correctly will the product have a narrow grain size range.

The most important parameters for agitator ball mills:
– rate of rotation of the agitator
– diameter of the agitator
– shape of the agitating components
– distance between the agitating components
– fill level of the grinding media
– solid concentration
– throughput
– diameter of the grinding media
– tenside concentration
– viscosity

The ideal viscosity range for a grinding suspension is 0.1–1 Pa·s, and the upper limit to the viscosity is about 10 Pa·s. On average, the material being ground takes about 10 min to pass through the machine.

There are dozens of makes of agitator ball mills on the market; in principle, they all adopt one or other of the forms depicted in Figure 5.11, even though the details of their construction differ. Du Pont's sand mill was the first high-speed agitator ball mill to be widely used in industry. Two modern varieties are shown in Figure 5.12.

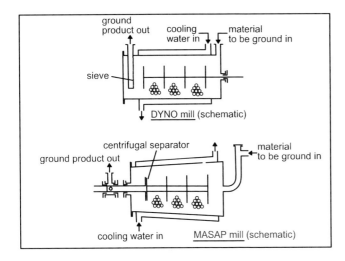

Figure 5.12. Modern dispersion apparatus.

Recently, annular- and split-gap mills have been developed. The energy density and thus the efficiency of these new designs is increased, heat exchange is improved, and so on. The CoBall Mill (Fryma), amongst others, should be mentioned. These types of mills are of particular interest for special cases in which division or dispersion is otherwise problematic.

Example of a typical grinding mixture: Given the numerous parameters that must be considered in order to achieve optimal dispersion, it will be useful to look at the example of pigment dispersion in some detail. Pigments are usually agglomerated or aggregated powders and will serve here as a model for many other substances. This laboratory procedure can be scaled up without difficulties, if the pilot or production plant corresponds to the laboratory apparatus. Suitable makes include the Drais bead mill and the Bachofen Dyna-Mill.

Effective capacity of the apparatus:	1 L
Grinding mixture:	300 mL
Concentration of solid:	25–35 weight %, depending on substance.

The solid is used in the form of a powder (containing ~0.1% wetting agent) or as a filter cake, and 10–20 weight % dispersant and water are added to obtain a final volume of 300 mL. This mixture is homogenized with a disperser–agitator or a dissolver and reduced to a maximum particle size of 50 μm.

This predispersed mixture is then combined with ~800 mL glass beads of 2 mm in diameter (grinding media filling degree: 0.8).

The grinding effect depends chiefly on the duration of the grinding process, but also on the rotation speed of the stirring apparatus. The grain size distribution must be measured at intervals. The temperature rises by an amount that depends on the nature of the material being ground, and this rise is accompanied by a change in viscosity, which is particularly marked in paint products.

The above remarks are only general notes concerning the most important factors; these can vary widely from product to product.

5.4 Special Methods of Dispersion

High-pressure homogenization:

The suspension is compressed to 50–700 bar with a piston and then allowed to expand rapidly through a homogenizer valve (Figure 5.13). Cavitation, turbulence, torque, and shearing forces create a large pulverizing effect. Individual crystals cannot usually be broken up, but hard agglomerates and aggregates can. The nozzles are heavily abraded in use, so are made of hard materials such as tungsten carbide. A model commonly used in industry is the Manton Gaulin submicron disperser, which produces fine dispersions (~1 μm) at high volume. At low pressure, this machine can also be used to manufacture emulsions.

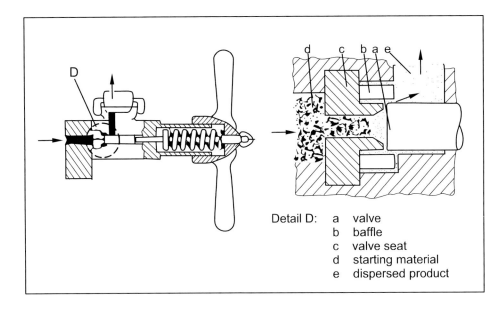

Detail D: a   valve
b   baffle
c   valve seat
d   starting material
e   dispersed product

Figure 5.13. Manton Gaulin dispersion apparatus for emulsions.

Dispersion by ultrasonication:

Ultrasound (US) waves with a frequency greater than 20 kHz are used; the source is a mechanical sound generator such as a sonolator or a US whistle as described by Janovski and Pohlmann (Figure 5.14).

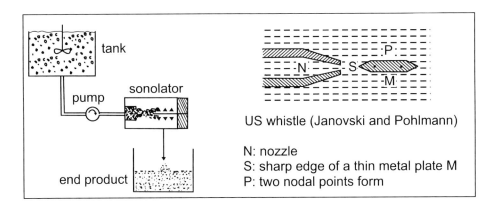

US whistle (Janovski and Pohlmann)

N: nozzle
S: sharp edge of a thin metal plate M
P: two nodal points form

Figure 5.14. Ultrasound dispersion apparatus.

These instruments function in the lower frequency range. The actual US generators use piezoelectric (quartz, tourmaline, zinc blende) and magnetostrictive sound sources.

Frequencies of 200 kHz and considerably more are possible, as are sound intensities of up to 50 W/cm^2. The effect is highly localized, and is caused by cavitation. Large volumes such as a suspension in a container cannot be evenly sonicated; in contrast, this is easily achieved in continuous-flow apparatus (in-line operation). The effectiveness of pulverization is heavily dependent on frequency, but the equipment works only with one or a few frequencies. Homogeneous colloidal dispersions can only be produced on a large scale with US in a few very special cases (depending on the product). Not everything marketed as US equipment deserves the term! Sonolators are used in predispersion, but cannot produce colloidal dispersions.

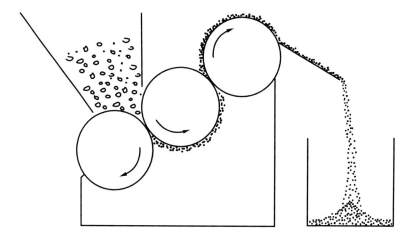

Figure 5.15. Three-roll mill.

Roll mills and dispersion kneaders (Figures 5.15 and 5.16):
These machines can produce very fine dispersions (submicron range) from high-viscosity mixtures on an industrial scale. They are the most common type of dispersion equipment in the production of printing inks, in the dispersion of pigments in polymers, for example in polyolefins. Their products are free from oversize grains (>1 μm) and therefore they are used, for example, in nozzle-spinning of polymer melts, since they create no problems with spinneret blockage; in addition, the absence of oversize grains means that the spun fibers are less brittle. In extruders, calcium carbonate and other inorganic fillers are mixed into various polymers colloidally. The resultant composites have improved mechanical strength. At present there are no alternatives of comparable efficiency for dispersing additives in polymers. The high viscosity of the polymer medium surrounding the colloidal particles protects them from agglomeration and stabilizes the dispersion.

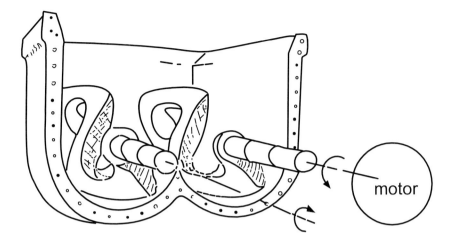

Figure 5.16. Dispersion kneader.

5.5 The Third Step in the Dispersion Procedure: Stabilization of the Dispersion

A "stable dispersion" is one in which the total number and size of the particles in the dispersion does not change over time. Dispersants are used to stabilize the dispersion. In viscous media the particles are prevented from agglomerating (see Section 5.3).

The particles can be stabilized in three possible ways:

1. By electrostatic repulsion (DLVO theory, Deryagin–Landau–Verwey–Overbeek)
2. By steric repulsion (HVO theory, Hesselink–Vrij–Overbeek)
3. By a combination of electrostatic and steric repulsion

There is a large number of different dispersants available for this purpose. Selection amongst them is based chiefly on the type and electrical charge of the particles and the nature of the liquid medium (water, polar or nonpolar organic liquid), Figures 5.17 and 5.18.

Even nonpolar solids usually have a surface electric charge due to ionization or adsorption in contact with a polar medium such as water. The ionization depends on the pH value of the suspension: because of their carboxy and amino groups, protein molecules are positively charged at low pH and negatively charged at high pH. Multiply charged or surface-active ions adsorbed at the interface determine the charge of the interface (*potential-determining ions*).

Dispersed particles are more often negatively than positively charged because of the preferential adsorption of anions, especially OH^-. Even oil drops and air bubbles in water carry a negative charge.

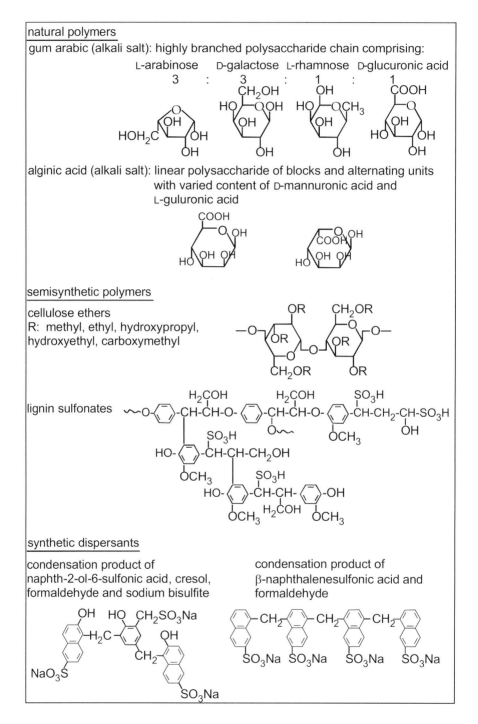

Figure 5.17. Dispersants for aqueous suspensions.

In order to decide which dispersant is suitable for a particular task, we need to know or to determine the charge on the particles. For this we can use the zeta potential; see below.

The dispersant must be adsorbed on the surface of the solid. The mechanisms of adsorption of tensides are illustrated in Figure 5.19. These create an energy barrier to attraction. In general, the barriers arising from the use of short-chain tensides (wetting agents, deflocculants) are insufficient for effective, long-term stabilization. For electrical stabilization, protective layers creating steric repulsion, such as layers of hydrocarbon chains, must also be present. Good dispersants are therefore molecules with both an electrical charge and various hydrophobic groups, for example high molecular mass hydrocarbon chains known as polyelectrolytes. The hydrophobic groups are preferentially adsorbed at the solid surface in aqueous suspensions. In dispersions with organic dispersion media, special conditions exist, as will be discussed later.

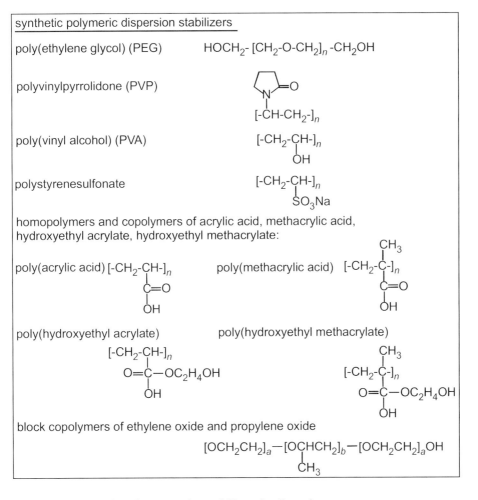

Figure 5.18. Synthetic polymers used as stabilizers for dispersions.

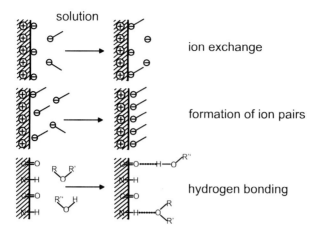

Figure 5.19. Mechanisms of tenside adsorption onto solids: ion exchange, formation of ion pairs, hydrogen bonding, polarization of π electrons, dispersion (London, van der Waals) forces; according to reference [6].

5.5.1 Choice of Dispersants for Use in Aqueous Media

In general, ionic tensides are used to create electrostatic barriers to flocculation.

The nonpolar parts of the tensides are adsorbed onto the surfaces of uncharged particles; the polar groups project into the water. Thus, all particles have the same charge sign, and therefore repel each other. Adsorption of the hydrophobic groups on the particle surface lowers the interfacial tension γ_{SL}. *Longer chain tensides adsorb more strongly and are more effective dispersants.*

The usual tensides are not suitable for stabilization of electrically charged particles. If the particles and the tenside have opposing charges, flocculation occurs. On the surface, now neutralized, a second layer can be adsorbed (Figure 5.2). However, this results in ineffective stabilization, since the second layer is easily removed. If the particles and the tenside have the same charge sign, adsorption of the polar head of the tenside onto the particle surface is hindered by electrostatic repulsion, and the nonpolar groups of the tenside are oriented towards the water. Stabilization is possible only at high tenside concentrations.

As a consequence, ionic dispersants for the stabilization of polar and nonpolar solids in water are constructed as follows: *they have several ionic groups (multiple charges), which are distributed over the entire molecule, and hydrophobic groups with polarizable structures such as aromatic rings and ether groups instead of hydrocarbon chains. The ionic groups have a number of functions* [1]:

1. They prevent adsorption of the dispersant in such a manner that the hydrophobic groups face the water (this would result in flocculation). One of the many ionic groups with a charge opposite to that on the surface is adsorbed, while another ionic group is

oriented towards the aqueous phase. Thus the tenside cannot adsorb in such a way that a hydrophobic group is oriented towards the aqueous phase.

2. The presence of more than one ionic group strengthens the repulsive effect of the electrostatic barrier. The more ionic groups with the same charge sign per molecule, the greater the electrostatic barrier per molecule adsorbed on a surface with like charge, and the greater the neutralization of the electrical charge that leads to formation of an electrostatic barrier of the same charge sign, like a dispersant molecule at a surface with opposing charge.

3. They permit the dispersant molecule to expand in the aqueous phase, forming a steric barrier to flocculation, without increasing the free energy of the system. The decrease in free energy due to hydration of the hydrophilic ionic groups is compensated for by its growth due to the greater degree of contact between the hydrophobic phases and the aqueous medium.

The polarizable hydrophobic groups enable greater adsorption onto charged particles, the hydrophobic groups oriented towards the particle surface.

Commonly used dispersants with multiple ionic groups and aromatic hydrophobic groups are exemplified by the condensation products of β-naphthalenesulfonic acid with formaldehyde (the Tamol compounds already mentioned) and by lignin sulfonates (Figure 5.20). Figure 5.20 also shows an idealized model of a lignin sulfonate with multiple charge. Other dispersants of this type in common use are copolymers of maleic anhydride or acrylic acid neutralized with alkali. These are *polyelectrolytes*.

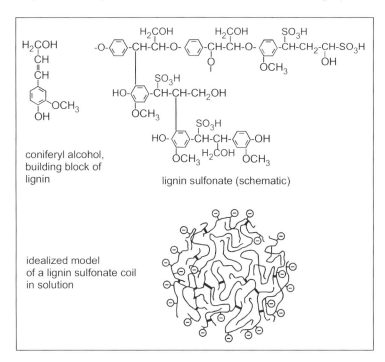

Figure 5.20. Lignin sulfonates.

5.5.2 Steric Stabilization

Particles in aqueous media can be stabilized by nonelectrostatic barriers too. Both ionic and nonionic tensides can serve as dispersants in this case. Steric barriers to flocculation are created by the projection of adsorbed tenside molecules out into the aqueous phase; this prevents particles coming into close contact. Longer molecules are more effective than short ones, always provided that their increased solubility in water does not interfere with their adsorption on the particle surfaces. For example: nonionic tensides of the poly-oxyethylene type are excellent multipurpose dispersants. The highly hydrated POE chains spread out as coils in the aqueous phase to become an excellent steric barrier to flocculation. They also depress the van der Waals attraction between the particles.

Block copolymers of propylene oxide and ethylene oxide (Figure 5.18) are particularly effective nonionic dispersants for use in aqueous media.

The propylene groups whose ether residues cause the adsorption at the particle surface are hydrophobic. The POE groups are oriented towards the aqueous phase and are hydrated. There are different types of compound in which the POE and POP groups have varying lengths; commercially, these are available under the trade name Pluronic. The most effective dispersants of this type are those with long POE *and* long POP chains.

The oldest known dispersants for aqueous systems are *protective colloids*, natural macromolecular compounds with hydrophilic natures, such as gum arabic, alginic acid, casein, lecithin, and gelatin. Their stabilizing action is steric (Figure 5.17).

Synthetic polymers suitable for stabilizing aqueous suspensions include poly(vinyl alcohol), polyvinylpyrrolidone, polyacrylamide, poly(styrene–oxyethylene), and poly(vinyl alcohol–vinyl acetate), Figure 5.18.

5.5.3 Stabilization of Nonaqueous Suspensions

The stabilization of dispersions in organic media is a subject mentioned scarcely or not at all in textbooks, despite the industrial importance of these dispersions. They include, for example,

- paints, particularly in the large field of automobile lacquers
- coatings based on film-forming systems
- inks for reprographic use
- dispersions of magnetic particles in polymeric binders
- dispersions of carbon black
- emulsion polymerization in organic media
- dispersion resins
- organosols and plastisols

In nonaqueous dispersions with low dielectric constants (DC), the diffuse electrical double layer (Section 5.6.1) is well spread out, having a thickness of 2000 Å, for example, in the case of a (1–1) electrolyte dissolved at a concentration of 10^{-9} mol/L in a solvent with a DC of 4. This is about 20 times broader than for a 10^{-3} molar aqueous KNO_3 solution. The drop in potential with distance from the particle is thus very small, so *the zeta potential can be taken as equal to the potential at the particle surface. A small surface charge density suffices to give a high surface potential.*

The suspensions in nonpolar liquids are stabilized by electrostatic repulsion but, more importantly, by steric repulsion.

Even in nonionic solvents, suspended particles may carry a charge as a result of dissociation of surface groups and adsorption of ionic tensides. The clearest proof for the existence of electrical charges in nonaqueous systems is the danger of explosion when petroleum becomes electrostatically charged. High zeta potentials can arise in oils when certain dispersants are used; these dispersants are acids or bases. The acidity or basicity of the particle surface is an important factor in determining whether electrical repulsion energy contributes to the stabilization. Protons are transferred from acidic areas of the particle surface to basic, adsorbed dispersant molecules, and the dispersant cation then desorbs. *Electrical stabilization should only be possible at zeta potentials of at least 100 mV.*

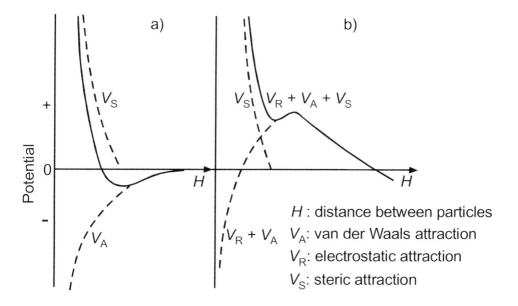

Figure 5.21. Plots of the potential for the steric stabilization of spherical particles: a) without an electrical double layer; b) with an electrical double layer [7].

If electrical charge does not create an energy barrier high enough to prevent flocculation of a dispersion in a nonpolar medium, the effect may be achieved by steric hindrance, that is, by an adsorption layer of a suitable polymer. Steric repulsion is characterized by a steep rise in the potential at small distances between particles (Figure 5.21).

It can be seen from these plots that, for a sterically stabilized dispersion, descent into the deep primary energy minimum can be fairly well excluded as a possibility, whether an electrical double layer exists or not. The combination of steric and electrostatic repulsion offers numerous possibilities for the stabilization of formulations using organic media.

In dispersions in organic media, particle size has an interesting influence on stability, in contrast to the case with aqueous systems, in which particle size is of only secondary importance. In water, near the flocculation point the range of electrical forces is generally small in comparison with the particle size, so the decisive forces and energies can come into play in the region of the closest approach of the two particles. The average particle size is therefore only of secondary importance.

In apolar media, we can expect electrostatic repulsion to extend over a large volume and steric repulsion to rise sharply at small distances between particles. The electrostatic repulsion is approximately proportional to the square of the charge on the particle. For a constant surface potential, it decreases with particle size, and for small particles may not be able to overcome the van der Waals force. Steric repulsion, on the other hand, is only sufficient to keep the particles a certain distance apart from one another. The van der Waals force increases with the size of the particle (Equation 1.11). It follows that, for the stabilization of coarse particles (micron-sized), electrostatic repulsion is more effective than steric, whereas for very small particles (submicron range) the relationship is reversed, a result of great importance for formulation chemists.

Most dispersants for nonpolar media are specific to a particular system. *There are no generally effective dispersants suitable for stabilization of a wide variety of solids in organic media* like, say, lignin sulfonates or dinaphthylmethanedisulfonates in aqueous systems. On the other hand, many nonaqueous media such as paints, latices, and printers' inks themselves possess dispersant properties.

Effective dispersants for nonaqueous dispersions must meet the following criteria:

They must interact strongly with both the particles and the dispersion medium, that is, adsorb strongly onto the particle surfaces and be readily soluble in the solvent. Therefore, they must have "anchor groups" that adsorb onto the solid and solvated groups that, in the form of loops and trailing tails, create a sufficiently thick layer to prevent attraction between the particles. There should also be a mechanism for charge transfer between the dispersant and the particle surface so that, on top of the steric repulsion, electrostatic repulsion is active.

A wide variety of tensides and polymers are suitable for use as dispersants for dispersions in organic media, depending on the solid, as shown in Figure 5.22. Alkyl-modified polyvinylpyrrolidones with relative molecular masses of 7300–20 000 and varying hydrophobic–hydrophilic properties should come under consideration when a broader range of application is necessary. The same is true of block copolymers of propylene oxide and ethylene oxide (Pluronics). The relatively new Avecia Solsperse "hyperdispersants" (Figure 5.23) should also be mentioned; "Disperbyk" dispersants from BYK-Chemie are similar.

"comb" surfactants

poly(12-hydroxystearic acid)

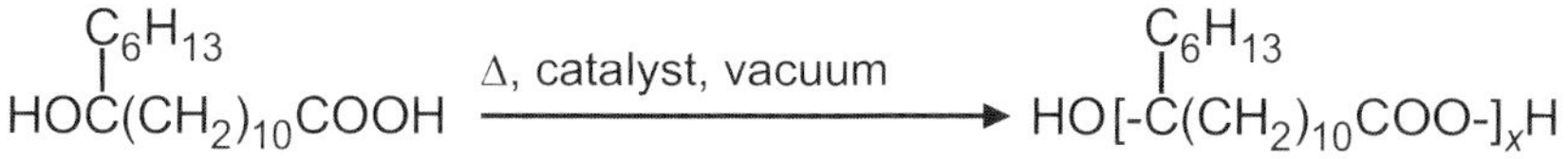

$$\underset{\text{HO}\overset{|}{\text{C}}(\text{CH}_2)_{10}\text{COOH}}{\overset{\text{C}_6\text{H}_{13}}{|}} \xrightarrow{\Delta,\ \text{catalyst, vacuum}} \underset{\text{HO}[\text{-}\overset{|}{\text{C}}(\text{CH}_2)_{10}\text{COO-}]_x\text{H}}{\overset{\text{C}_6\text{H}_{13}}{|}}$$

condensation with glycidyl methacrylate yields a macromonomer which can be copolymerized with methyl acrylate and methyl methacrylate to give a comb surfactant

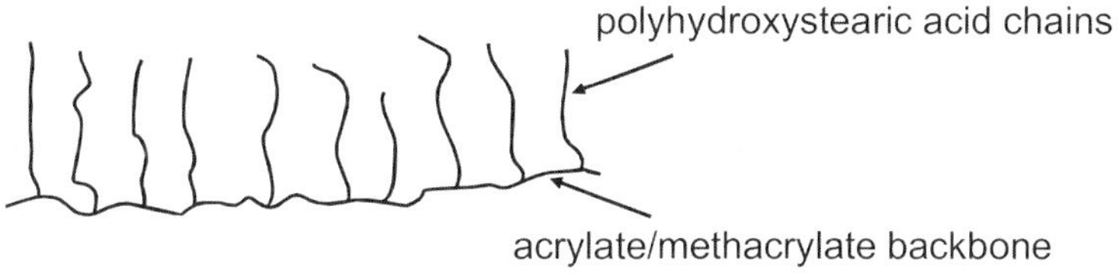

polyhydroxystearic acid chains

acrylate/methacrylate backbone

alkyl-modified polyvinylpyrrolidone 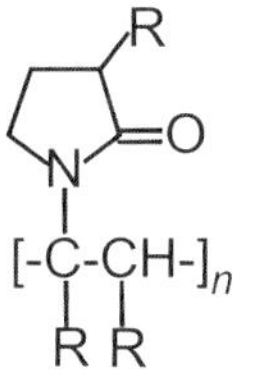   R = H or alkyl

$[\text{-}\overset{\text{R}}{\underset{\text{R}}{\text{C}}}\text{-}\overset{}{\underset{\text{R}}{\text{CH}}}\text{-}]_n$

block copolymers

e.g. poly(styrene-butylstyrene), poly(styrene-acrylonitrile);
anchor group = polystyrene

polyureas synthesized by addition of fatty amines to polyisocyanates

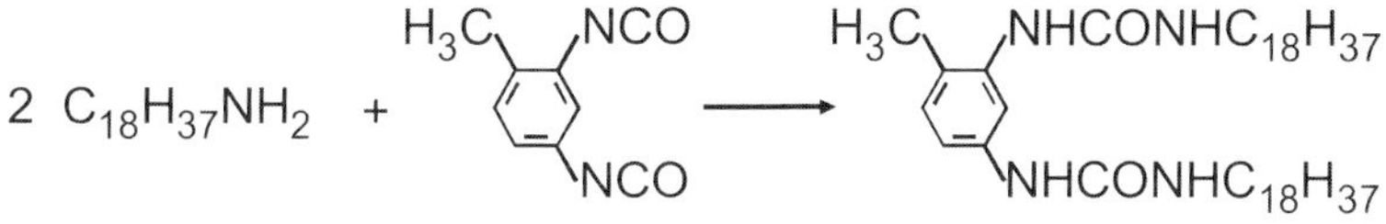

$2\ \text{C}_{18}\text{H}_{37}\text{NH}_2\ +$ (H$_3$C, NCO, NCO toluene diisocyanate) $\longrightarrow$ (H$_3$C, NHCONHC$_{18}$H$_{37}$, NHCONHC$_{18}$H$_{37}$)

Figure 5.22. Dispersants for organic suspensions.

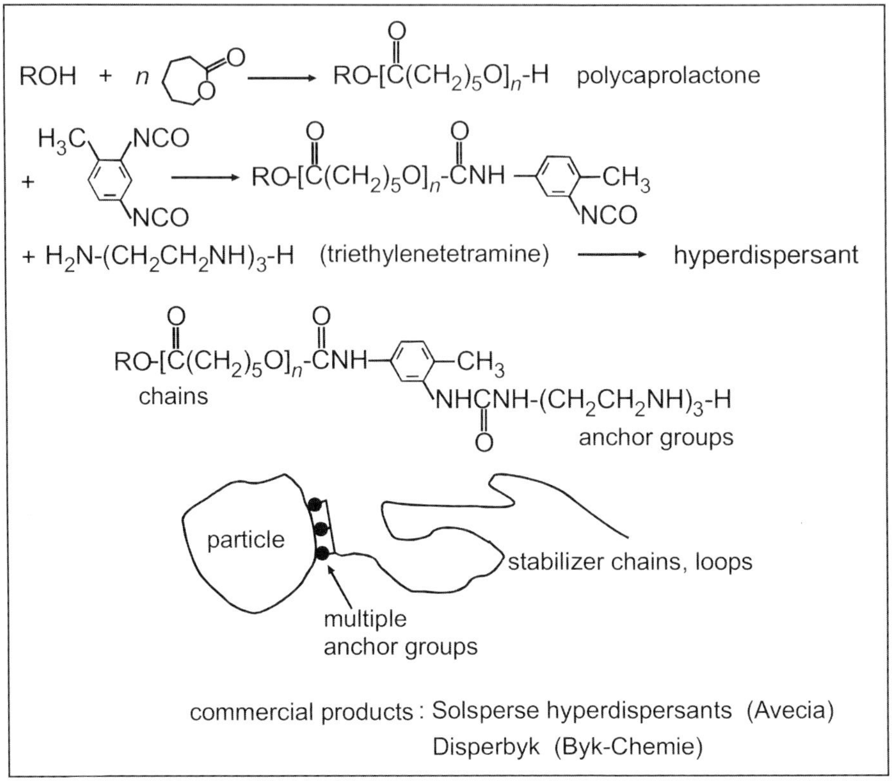

Figure 5.23. Hyperdispersants.

5.6 The Most Important Points for Formulation Chemists from the Theory of Colloid Stability

All colloidal dispersions are thermodynamically unstable, since they have greater free energy (owing to their greater surface area) than the undivided starting material. The system has a tendency to revert spontaneously to a state of lower free energy through flocculation of the particles, unless this is prevented by an energy barrier. If such a barrier exists, the system is metastable and can remain in this state for a long time; Faraday's original gold sols (~1850), for example, are still stable today (they may be seen at the Royal Institution in London). Brownian motion of the particles can impart the necessary energy to permit colloids to overcome the barrier. The average translation energy of colloids in Brownian motion is $3/2\, kT$ per particle (see Section 1.4). At a temperature of 300 K, two particles have a collision energy of 10^{-20} J. It is unlikely that collisions of energy equivalent to a multiple of kT will occur. An energy barrier of ~10 kT is therefore sufficient to preserve the dispersion in its metastable state; it is colloidally stable.

This energy barrier is the result of the combination of two forces, van der Waals attraction and a repulsive force due to surface charges and adsorption layers.

5.6.1 The Electrical Double Layer

An electrical double layer forms at any interface. The opposing charges of the ions on each side of the interface cause an electrical potential difference. Amongst other factors, the charges can result from the presence of ionizable groups on the particle surface (-COOH, -SO$_3$H, -SiOH, etc.), or from the adsorption of ions at the interface. When cellulose fibers are dispersed in water, for example, the polysaccharide carboxy groups are ionized. In the case of $CaCO_3$, the particle surface is charged because of the adsorption of ions such as Ca^{2+} and CO_3^{2-} at the interface.

To preserve electrical neutrality, the charged colloidal particles must be surrounded by ions of the opposite charge, counterions. In addition, ions of the same charge are present in the vicinity of the interface, coions. The electrical double layer theory describes the distribution of the co- and counterions near a charged interface in contact with a polar medium in terms of the size of the electrical potentials in the region.

In the absence of thermal motion, the counterions would be firmly in contact with the surface of the colloidal particles and would neutralize their charge. In this *Helmholz model*, the potential drops to zero at a small distance from the charged surface, Figure 5.24.

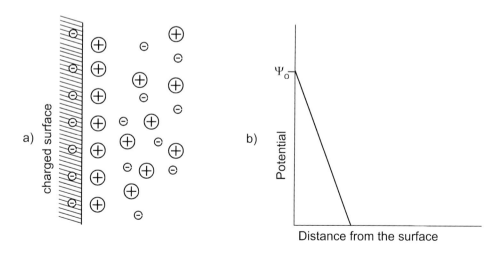

Figure 5.24. Helmholz model of the electrical double layer: a) ion distribution in the vicinity of the charged surface; b) change in potential with distance from the charged surface.

However, this model is unrealistic, since thermal motion prevents the formation of a compact double layer. Electrical forces and thermal motion combine to create a diffuse double layer, described by the *Gouy–Chapman* model. In this model, the counterion concentration – and thus the potential – decreases rapidly at first with distance from the charged surface, but the decline becomes slower and slower as this distance increases, Figure 5.25.

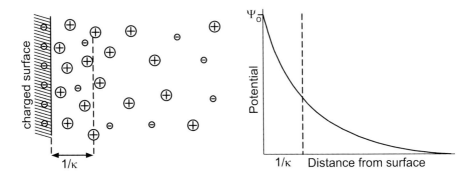

Figure 5.25. Gouy–Chapman model of the electrical double layer.

In Figure 5.25, the dashed vertical line corresponds to the distance at which the potential has fallen to an *e*-th fraction of its original value ($1/2.718 = 0.37$). The distance $1/\kappa$ is called the Debye length. κ can be calculated from known quantities (see below).

The Gouy–Chapman model works with the unrealistic assumption of point charges. Stern introduced a correction in the form of the effective ionic radius (*Stern model*, Figure 5.26).

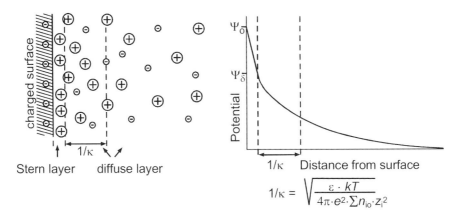

$$1/\kappa = \sqrt{\frac{\varepsilon \cdot kT}{4\pi \cdot e^2 \cdot \sum n_{io} \cdot z_i^2}}$$

Figure 5.26. Stern model of the electrical double layer.

This model also takes into account the possibilities of the specific adsorption of ions, which leads to the formation at the interface of a complete layer of counterions bound to the surface strongly enough to withstand thermal motion. Thus, *the electrical double layer is split into two levels, a compact Stern layer, in which the potential drops from ψ_0 to ψ_δ, and a diffuse layer, in which the potential drops from ψ_δ to zero.* ψ_δ is the potential on the outer side of the Stern layer.

Debye and Hückel's mathematical treatment of the diffuse layer yielded the concept of the effective thickness $1/\kappa$ (known as the Debye length of the layer). This can be calculated with the expression given in Figure 5.26, in which the symbols have the following meanings:

ε:	dielectric constant
e:	elementary unit of charge
n_{io}:	concentration of the i-th type of ion
z_i:	charge on the i-th type of ion

The presence of electrolytes curtails the range of coulombic interactions and *compresses the electrical double layer; this compression by electrolytes eventually results in flocculation of the dispersion.* The effect is much more pronounced for multiply charged ions than for those with a single charge (see the Schulze–Hardy rule for the behavior of dispersions in the presence of electrolytes, below).

5.6.2 The Zeta Potential

Various electrokinetic phenomena are related to the relative motion of charged surfaces. This is the case for disperse solid particles.

In order to characterize the electrical properties of disperse particles, we must know the electrical potential; we use the *electrokinetic* or *zeta potential* ζ (Figure 5.28). The ζ *potential* can be measured by means of *the electrokinetic effects*: *electrophoresis*, *electroosmosis*, *streaming potential*, and *sedimentation potential* (Figure 5.27). *Microelectrophoresis* is the most commonly used effect.

When charged particles move in a dispersion medium, or when the dispersion medium moves relative to a fixed barrier, a zeta potential is created. The zeta potential can therefore be measured by forced tangential displacement of the diffuse part of the double layer. The magnitude of the ζ potential is determined by the distance between the slip plane of this motion and the particle surface. It is usually assumed that this slip or shear plane is immediately adjacent to the Stern layer. This is seldom actually the case. Usually, a monomolecular layer of water molecules is bound to the surface of the charged particle, and, in addition, the surface is usually rough. These factors displace the slip plane beyond the outside edge of the Stern layer. The ζ potential is therefore usually smaller than ψ_δ. The exact position of the slip plane and the structure of the liquid within it remain unknown, so the determination of the value of the ζ potential is problematic.

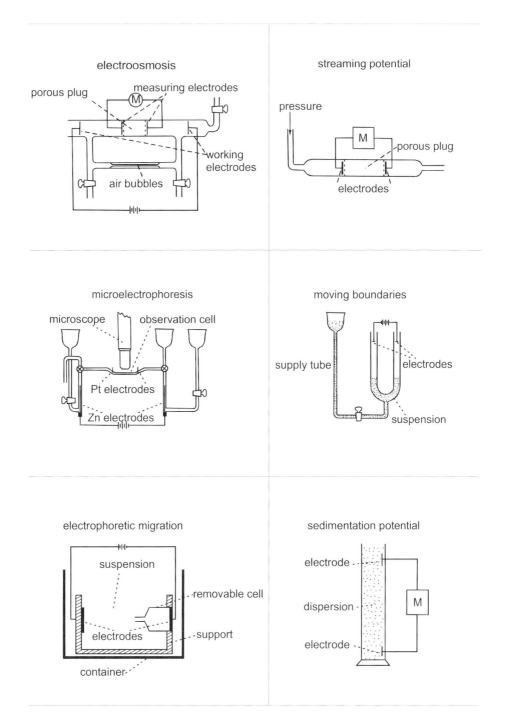

Figure 5.27. Measurement of electrokinetic effects [8].

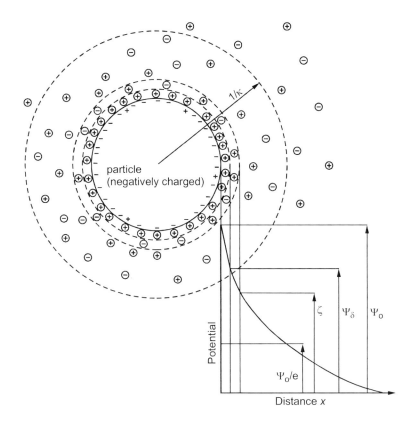

Figure 5.28. Potentials in the electrical double layer

For the evaluation of the zeta potential measurements, the dispersed particles are considered as spheres of charge q in an electrical field E (Figure 5.29).

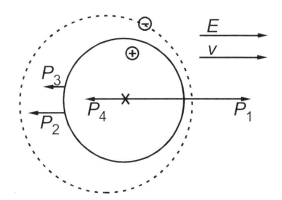

Figure 5.29. Forces to be taken into account in measurements of zeta potentials [9].

The following forces are acting:

P_1: force of E on the particle $P_1 = E \cdot q$ (5.3)

P_2: force of friction $P_2 = -6\eta r v$ (5.4)

P_3: hydrodynamic force resulting from the effect of E on the counterions and thus on the solvent molecules (electrophoretic effect)

P_4: relaxation effect due to asymmetry of the charge distribution in the movable fraction of the double layer

In the stationary state, Equation 5.5 holds:

$$P_1 + P_2 + P_3 + P_4 = 0$$ (5.5)

For dispersions with $\kappa a \geq 10^3$, the Smoluchowski Equation (Equation 5.6) is valid:

$$v = \frac{\varsigma \cdot E \cdot \varepsilon}{4\pi\eta}$$ (5.6)

For the other extreme, the Hückel Equation (Equation 5.7) applies when $\kappa a \leq 10^3$:

$$v = \frac{\varsigma \cdot E \cdot \varepsilon}{6\pi\eta}$$ (5.7)

These equations are only valid provided that:

– the particles are nonconducting and spherical
– the concentration is low; double layers do not interpenetrate
– κa is unambiguously defined
– η around the particle is independent of distance from the slip plane
– Brownian motion is negligible.

In principle, zeta potentials can be calculated from the equations given here. However, the viscosity and the dielectric constant must be known, and these values in the electrical double layer differ from those in the bulk phase, and are generally unknown. *It is therefore very difficult to measure the numerical value of the ς potential experimentally to a reasonable degree of accuracy. Instead, the easily measurable quantity electrophoretic mobility v/E is determined and the results obtained under varied conditions are compared.*

In the paper industry an instrument known as a "zeta meter" is used, but this makes comparative rather than absolute measurements.

The electrophoretic mobility u ($= v/E$) can serve as a basis for estimation of the zeta potential in aqueous dispersions according to Equation 5.8:

$$\varsigma\,(mV) = 12.86 \cdot u \quad (25°C), \quad u \text{ measured in } \left[\frac{\mu m \cdot cm}{S \cdot Volt}\right]$$ (5.8)

5.6.3 Potential Energy Curves

A potential energy curve shows the potential energy of two particles as a function of their distance apart. It tells us whether a colloidal system will remain stable (no flocculation) or is unstable and will flocculate.

In order to obtain the total energy of interaction between two particles, we must add the individual components. These are:

1. electrostatic and steric repulsion energy
2. van der Waals attraction energy.

As far as electrostatic repulsion is concerned, a complicated formula for the potential energy V_R can be derived from the theory of the electrical double layer. When greatly simplified, this formula becomes Equation 5.9.

$$V_R = \exp{(\kappa \cdot d)} \qquad (5.9)$$

d: distance between the centers of gravity of the two particles

The van der Waals attraction between two spherical particles is approximately

$$V_A \approx -\frac{A \cdot a}{12H} \qquad (5.10)$$

a: radius of the particles
A: Hamaker constant
H: distance between the surfaces

The total interaction is:

$$V_T = V_R + V_A \qquad (5.11)$$

This summation can also be carried out graphically; see Figure 5.30.

The theory behind dispersion stability was developed by Deryagin, Landau, Verwey, and Overbeek, and is therefore known familiarly as the DLVO theory. The calculation of dispersion stability by the DLVO theory is extremely difficult. Using approximations, however, we can now carry out calculations for practical examples, for instance PVA–TiO_2 or copper phthalocyanine dispersions in water. The calculations are based on fourteen equations [10]. For qualitative discussions, it is sufficient to consider potential energy curves and their physical meaning.

Except when the particles are compressed, attraction is the dominant force at very small and very great distances. In between these ranges, on the other hand, repulsion dominates, depending on the current values of the two forces. In our example

(Figure 5.30), there is a clear secondary energy minimum, when the particles are well separated ($\sim 4/\kappa$), as well as the primary minimum when they are close to each other.

When the energy barrier V_{max} is high ($\sim{>}25\ kT$) compared with the thermal energy kT, the colloidal system ought to be stable; otherwise, flocculation occurs.

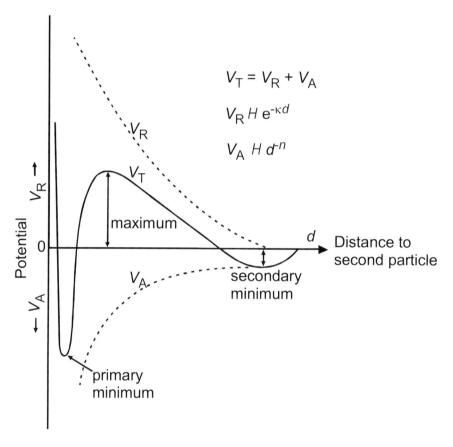

$$V_T = V_R + V_A$$

$$V_R\ H\ e^{-\kappa d}$$

$$V_A\ H\ d^{-n}$$

Figure 5.30. Potential as a function of the distance between particles. V_T: overall potential; V_R: repulsion; V_A: attraction.

The presence of electrolytes influences the stability of the dispersion heavily (Figure 5.31). *Electrolytes compress the electrical double layer and the potential declines faster with distance.* The flocculation concentrations of mono-, di-, and trivalent ions have the ratio 100:1.56:0.137 (cf. Schulze–Hardy rule). Thus Al^{3+} has a flocculating effect $\sim$64 times greater than that of Ca^{2+} and $\sim$729 times greater than that of Na^{+}. Flocculation is determined in mmol/L electrolyte over the flocculation threshold. *For formulation chemists, this is one of the most important corollaries of the theory of dispersion stability;* it follows from DLVO theory.

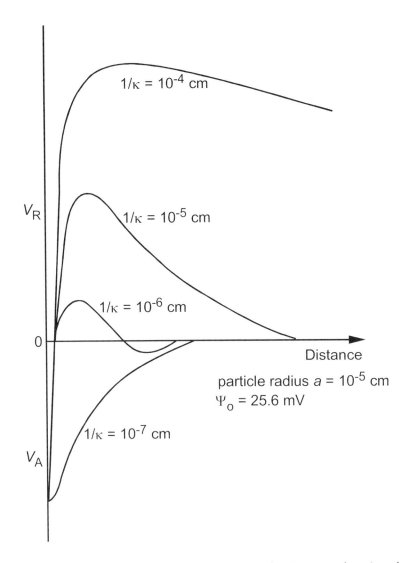

Figure 5.31. Influence of the electrolyte concentration (represented as κ) on the total potential of two spherical particles.

Schulze–Hardy rule:

- Flocculation concentrations for singly, doubly, and trebly charged ions are related by the term $1/z^6$ [11]:
  $1 : 1/2^6 : 1/3^6 = 100 : 1.56 : 0.137$
- Flocculation at the primary minimum, for $a \cdot \kappa \ll 1$
  (a: particle radius, $1/\kappa$ Debye length)
- Flocculation at the secondary minimum, for $a \cdot \kappa \gg 1$

Secondary minimum:

If the ratio of particle size to thickness of the electrical double layer is much greater than unity, a *secondary energy minimum* occurs at some distance from the particles. Unlike flocculation at the primary minimum, *this flocculation is weak and can be reversed by stirring*. Flocculation occurs only when the particle size is about 1 μm or greater, or for asymmetric particles such as needles or flakes. The resulting structure is thixotropic. The secondary minimum may be used for the deliberate production of thixotropic structures in dispersions by *controlled flocculation*; these do not cake and can be returned to their original, highly fluid state simply by gentle shaking (Figure 5.32). Sedimenting dispersions with no pronounced secondary energy minimum flocculate at the primary minimum, and cannot then be redispersed by stirring.

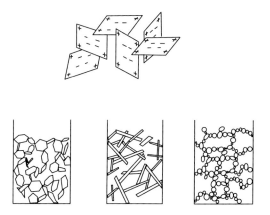

Figure 5.32. Thixotropic structures.

5.6.4 Steric Stabilization

Steric repulsion occurs when a particle surface adsorbs or chemically reacts with molecules with a large volume, usually with long chains, so that it has a "hairy" covering. Only part of the polymer is anchored to the solid surface; the remainder extends into the dispersion medium in the form of loops, trains, and tails (Figure 5.33).

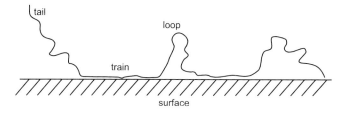

Figure 5.33. Adsorption of linear polymers onto surfaces.

When two particles approach one another and each has a layer of thickness d adsorbed on its surface, they begin to interact at a distance of $2d$ or less. The adsorbed layers either interpenetrate or become compressed, or both (Figure 5.34). This interpenetration or compression results in repulsion between the particles (Figure 5.35) for thermodynamic reasons (change in free energy or entropy) which cannot be described in detail here.

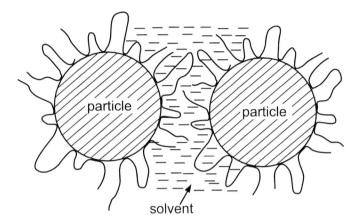

Figure 5.34. Particles with polymer coating.

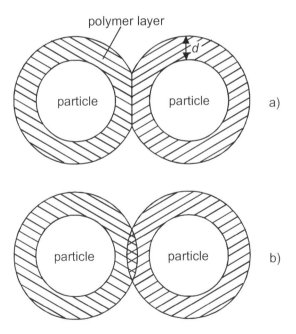

Figure 5.35. Interaction of particles coated with polymers: a) compression; b) interpenetration.

Figure 5.36 shows a plot of the interaction energy for two particles that repel each other sterically. *The solubility of the polymer chains in the dispersion medium is a crucial factor in the success of the steric stabilization.*

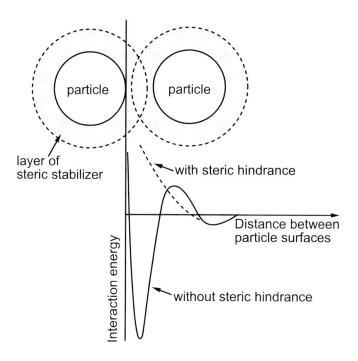

Figure 5.36. Steric stabilization.

Sterically stabilized dispersions cannot be flocculated by electrolytes; up to 10 mol/L of electrolyte can be added to an aqueous dispersion of latex stabilized with polyethylene oxide, for example, without causing it to coagulate. *In addition, more concentrated aqueous dispersions that nevertheless retain good flow properties can be manufactured by means of steric stabilization.* The reason for this is that, in contrast to the situation in electrostatically stabilized dispersions, no electroviscous effect occurs when the dispersion flows.

5.7 Flocculation or Coagulation of Suspensions

The electrostatic interactions between particles finely dispersed in an aqueous suspension are most strongly influenced by the electrolyte concentration over the Debye length $1/\kappa$. Addition of electrolyte to an aqueous suspension causes the thickness of the electrical double layer, and thus the height of the energy barrier, to decrease until flocculation or coagulation occurs. According to the DLVO theory, the energy barrier of electrostatic repulsion should have a minimum size of 25 kT. From Figure 5.26, the κ value of a salt of ions of the same numerical charge in aqueous solution at 25°C is defined as in Equation 5.12 (z: charge; c: molar concentration):

$$\kappa = 0.329 \cdot 10^8 \cdot z \cdot \sqrt{c} \qquad\qquad (5.12)$$

The Debye length $1/\kappa$ is therefore inversely proportional to the numerical charge of the ions in the solution and to the square root of their concentration. For divalent ions, the thickness of the diffuse layer $1/\kappa$ is reduced to about half of the value for monovalent ions, and for trivalent ions to one third. Since the Debye length is proportional to the dielectric constant of the solution (Figure 5.26), we expect that in solutions with a high ε, such as aqueous solutions, the electrical interactions will extend considerably further into the solution than they do in those with low values of ε, such as solutions in hydrocarbons. An important practical consequence for us is that coulombic repulsion is usually ineffective for the stabilization of dispersions in organic media with low ε, and a layer of steric protection is necessary instead. Another important effect on dispersion stability is the curtailment of the range of coulombic interactions by electrolytes. In other words, addition of electrolytes compresses the extent of the interactions.

Table 5.1 lists some values for the thickness of the diffuse layer as a function of the electrolyte concentration of monovalent salts at 25°C. It can be seen that this layer, which is responsible for the electrostatic stabilization of dispersions, is very thin even in a 0.1 molar salt solution.

The influence of electrolyte concentration on the potential energy curve for the interaction of two spherical particles in aqueous solution can be seen in Figure 5.31.

Table 5.1. Debye lengths for aqueous electrolyte solutions of monovalent salts at 25°C.

Concentration (mol/L)	κ (cm^{-1})	$1/\kappa$ (Å)
10^{-6}	$3.29 \cdot 10^4$	3040
10^{-5}	$1.04 \cdot 10^5$	962
10^{-4}	$3.29 \cdot 10^5$	304
10^{-3}	$1.04 \cdot 10^6$	96
10^{-2}	$3.29 \cdot 10^6$	30
0.1	$1.04 \cdot 10^7$	10
0.5	$2.32 \cdot 10^7$	4.3
1.0	$3.29 \cdot 10^7$	3.0
10	$1.04 \cdot 10^8$	1.0

As has already been mentioned, addition of electrolytes to electrostatically stabilized dispersions can result in flocculation. The charge of the ions is critical in this process:

Doubly charged ions are about 60 times more effective and trebly charged ions about 770 times more effective than singly charged ions as flocculants. This law, referred to as the Schulze–Hardy rule, can be derived from the DLVO theory [11]. It follows from the theory that the flocculation concentration of counterions of charge 1, 2, and 3 are related to each other as $1/z^6$, that is, $1:2^{-6}:3^{-6}$. However, this rule only applies when there are no counterions in the Stern layer, or, in other words, when ψ_δ is unaltered by the various electrolytes added (cf. Figure 5.28).

5.7.1 Determination of the Coagulation Concentration

The transition between stability and flocculation usually occurs in a narrow electrolyte concentration range; the critical flocculation concentration can therefore be determined with accuracy. This concentration is also known as the flocculation threshold or the *critical coagulation concentration* (c.c.c.). The c.c.c. can be defined as the minimum electrolyte concentration necessary to make the colloid under examination flocculate.

The process can be followed closely by means of optical measurement of opacity. An acceptable and useful practical method that gives semiquantitative results is the direct observation at specified intervals, without optical equipment, of flocculation experiments carried out in test tubes.

The experiment measures the concentration of electrolyte that causes a clearly visible opacity in the colloidal sample ten minutes after its addition. The electrolyte solutions might be, say, 0.5 N KCl, CaCl$_2$, AlCl$_3$, and Na$_2$SO$_4$. The solutions are first diluted: for each solution, five test tubes are used. Into the first of these, 10 mL of the original electrolyte solution is placed, and 9 mL of water goes into each of the remaining four. Then 1 mL from the first is transferred to the second, which is shaken; 1 mL from the second is transferred to the third, etc., to give five solutions for each salt with a geometric series of concentrations, related by the factor 1/10. The colloidal solution (1 mL) to be tested is added to each of these solutions, which are shaken and then checked after ten minutes for flocculation. The rough value for flocculation threshold thus obtained, correct to one order of magnitude, is refined to a more exact value by use of a further series of electrolyte solutions with concentrations in the critical range and related e.g. by a factor of 1:2 (5 mL of the 10 mL solution in the first tube transferred to the next and mixed with 5 mL water, and so on). Once more, 1 mL colloid solution is added to each and the mixture in which a just-visible flocculation occurs is noted. The corresponding electrolyte concentration is the coagulation value. If the test tubes are observed over a longer period, a quantitative picture of the rate of flocculation by different electrolyte concentrations is obtained.

In this procedure it is important to note that the first clearly perceptible signs of coagulation, rather than complete flocculation, are the best sign of the flocculating effect of an electrolyte. For comparative observations in a series of experiments, the decision must always be taken after the same period, be it 10 minutes or 2, 18, or 24 hours.

The coagulation values depend on the concentration of the colloid, and the nature of the dependence is controlled in turn by the charge of the precipitating counterion. In many systems, this dependence is described by Burton's rule [12]: the flocculation threshold declines with increasing sol (dispersion) concentration when the precipitating ions are singly charged, but it increases when the ions are multiply charged; see also reference [13].

For nonabsorbing spherical particles, such as polystyrene–latex beads, the c.c.c. can be determined by means of opacity measurements with a spectrophotometer [14].

Dispersions stabilized with nonionic dispersants like poly(vinyl alcohol) or block co-polymers of poly(propylene oxide) and poly(ethylene oxide) usually begin to flocculate when the temperature reaches a critical point (the CFT). Formulation chemists need to know this temperature, since suspensions should not be exposed to temperatures greater than the CFT (for example, when in storage). A dispersant must therefore be chosen to give the formulation a CFT greater than the maximum storage temperature.

The flocculation process plays an important role in practical solid–liquid separation, such as in water treatment or mineral extraction. The flocculation properties of the relevant suspension must therefore be known when plants for the treatment of water or minerals are designed.

5.7.2 Controlled Flocculation

In colloid chemistry, the word "stable", when used to refer to a dispersion, is synonymous with "deflocculated". A flocculated dispersion which sediments fast is termed "unstable". Chemists in industry therefore usually assume that dispersions and suspensions should be kept in the deflocculated state. This viewpoint ignores the fact that hard sediments usually settle out of deflocculated systems on storage, a phenomenon known as "caking". However, if the sediment thus formed can easily be redispersed, the dispersion can be considered to have qualified stability to storage, since it must be homogenized before use.

The potential curves of electrostatically stabilized dispersions have a secondary minimum when the particles are long distances apart, say 1000–2000 Å (Figure 5.30). This minimum results from the facts that the curve for the attraction V_A falls off according to a hyperbolic power law whereas the curve for the repulsion V_R declines exponentially. The existence of these secondary minima was first described by E. J. Verwey and J. T. G. Overbeek [15] (Section 5.6.3). If the secondary minimum is sufficiently low in comparison to the thermal energy kT of the system, light flocculation occurs. The depth of such secondary minima is generally only a small multiple of kT. The structures formed are therefore easily destroyed once more by shaking or pumping.

It must be remembered that, on settling out, electrostatically stabilized dispersions that lack a deep secondary minimum give a hard sediment that cannot be redispersed. This is because thermal energy, combined with gravity, suffices to overcome the potential maximum as shown in Figure 5.30. The particles then coagulate at the deep primary minimum. This statement holds when the ratio of particle size to the electric double layer thickness $a/(1/\kappa) \ll 1$. For the case $a/(1/\kappa) \gg 1$, a secondary energy minimum occurs at some distance (Figure 5.30). However, for particles of $\ll 1$ μm this secondary minimum is not deep enough for light coagulation. Such dispersions yield hard sediments through flocculation at the primary minimum; these sediments cannot be redispersed. In contrast, for particles over about 1 µm in size the secondary minimum is adequate for the loose flocs to form and remain. This is particularly true for asymmetric particles such as needles or flakes. Coagulation at the secondary minimum creates a thixotropic structure that can easily be returned to a highly fluid state. Figure 5.32 illustrates some thixotropic structures.

The depth of the secondary minimum depends not only on the particle size but also on the Hamaker constant, the zeta potential, and the electrolyte concentration (Section 5.6.3). It is theoretically possible to optimize these parameters such that the secondary minimum has the desired depth. This is known as "controlled flocculation". In practice, the principle is approximated by addition of electrolyte. The conditions for controlled flocculation are actually more often found by accident than by strict optimization of the parameters.

For sterically stabilized dispersions too the plot for potential against distance has a minimum and controlled flocculation is possible. However, it is strongly temperature-dependent and thus its application is limited [16].

The principle of the controlled flocculation of electrostatically stabilized dispersions has chiefly found application in the preparation of pharmaceuticals. One example is the controlled flocculation of sulfamerazine by $AlCl_3$ [17]. A 2% aqueous suspension of sulfamerazine also containing 0.1% Aerosol OT as a wetting agent and dispersant was flocculated with a solution of $AlCl_3(H_2O)$. Addition of 0.6 to 10 mmol/L flocculant causes light flocculation of loose sediment, more than that results in a hard sediment that cannot be redispersed.

Other examples include the pharmaceutical active substances sulfaguanidine, griseofulvin, and hydrocortisone [18].

Application of the principle of controlled flocculation to the formulation of stable dispersions is nevertheless tricky, especially if the flocculation should occur exclusively at the secondary minimum. Formulation chemists therefore make use of additional means to influence the sedimentation characteristics of dispersions when they wish to manufacture a dispersion that will be stable to storage. A suitable method is the combination of flocculation and thickening. When flocculation is utilized to manufacture a stable dispersion, the sediment volume of the flocculated dispersion should not alter over time. In order to delay the sedimentation of the flocs, thickeners are added to the suspension; this technique is particularly common in agrochemicals. "Flowables" to be diluted with water may be exposed to high temperatures in storage and must not form hard sediments; the concentrates must be easily homogenized once more by stirring before dilution.

5.7.3 Flocculation by Bridging and Sensitization

In suspensions stabilized by adsorption of polymers it is possible that flocculation will occur when, for example, a polymer chain is adsorbed on two or more adjacent particles at the same time. This can happen when the particle is only partially covered by polymer, which is possible at low polymer concentration. This sort of flocculation is not limited to systems in which the colloidal particles and the polymer have opposite charges. A silver iodide sol, for example, can be flocculated by poly(vinyl alcohol). This *bridging flocculation* is important in many modern processes, especially in water treatment and ore processing. Kitchener [19] gives a good overview of the field; La Mer developed a quantitative theory to cover it [20]. Modern polymeric flocculants have relative molecular masses of one to five million. They are considerably more expensive than inorganic coagulants but the required dosage is much lower, about 0.1 mg/L for polymers compared with 50–400 mg/L for inorganic flocculants. A comprehensive review may be found in reference [21].

5.7.4 Depletion Flocculation

A sterically stabilized dispersion can also be flocculated by addition of free, nonadsorbing polymer. In this case, flocculation depends on the relative molecular mass and the concentration of the polymer added. The phenomenon occurs when the colloidal particles are so close to one another that the polymer chains are unable to fit into the spaces between the particles on account of their length, and those already present are forced out of the solution between the particles to create a polymer-free zone. The decrease in osmotic pressure in the intervening region creates an attractive force (Figure 5.37). Vincent has shown that systems stabilized with low molecular mass polyoxyethylene can be flocculated by addition of free, higher molecular mass POE [22].

The effect of added free, nonadsorbing polymer on the stability of suspensions has long been applied in numerous industrial processes, but nobody realized that the added polymer was not adsorbed onto the particles. The phenomenon has a thermodynamic explanation, but we must direct the reader to references [23, 24] for details.

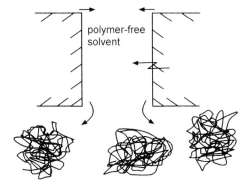

polymer-free solvent

Figure 5.37. Flocculation by nonadsorbing polymers.

5.8 Formulation of Stable Dispersions

5.8.1 Criteria

The formulation chemist's work is by no means finished with the production of a stable (colloidally speaking) dispersion that does not coagulate or flocculate irreversibly, even after long storage.

Normal commercially available dispersions or suspensions usually have to fulfil extremely high specifications regarding stability. Storage times of two or more years, in a temperature range of –10 to +50°C, are common for paint dispersions and agro-chemicals. As well as colloidal stability, discussed in Section 5.6, the following criteria must be fulfilled by commercial products:

- no claying or caking (formation of hard sediments that cannot be redispersed)
- no crystallization (Ostwald ripening, Section 9.2.5)
- no microbial or chemical decomposition.

In addition the formulations created in the laboratory should possess a certain "robustness" so that they survive the little variations which can occur from batch to batch in production.

The most difficult criterion to fulfil is probably the formulation of dispersions or suspensions that contain no sediment or only sediment that is easily redispersed and consists of primary particles even after the maximum suggested storage period has passed. The following possibilities are the best for meeting this challenge:

- division of the solid to a particle size that will not result in sedimentation even after long storage, that is, no particles with diameters greater than about 0.1–0.5 μm, depending on the density of the solid and the viscosity of the suspension medium
- adjustment of the liquid density to match that of the solid. This brings the sedimentation rate to zero (Stokes)
- addition of thickeners to increase the viscosity of the suspension medium, and formation of three-dimensional gel structures by addition of fine, inert particles
- controlled flocculation of the dispersion/suspension.

Since the sedimentation properties of polydisperse suspensions decide the long-term behavior of these liquid formulations, they will be dealt with in more detail in the following section.

5.8.2 Sedimentation of Suspensions

The sedimentation behavior of polydisperse suspensions is complex, especially at high concentrations and for flocculated systems. The sediment characteristics must be included in any assessment of suspensions, or, in other words, the entire process of settling. This has two stages, depending on whether the particles are inert or they attract one another and the system flocculates.

First stage: free sedimentation of the particles, flocs, or agglomerates due to gravity. For inert particles, the Stokes Equation for free fall holds (r = radius of the spherical particle, v = sedimentation rate, ρ = density of the particle, ρ_0 = density of the dispersion medium, η = viscosity; Equation 5.13):

$$v = \frac{2r^2 \cdot (\rho - \rho_0)}{g \cdot \eta} \tag{5.13}$$

The equation is valid for spherical particles only; otherwise, correction factors must be included [25]. Also, the equation only holds strictly true for particle concentrations of less than 0.5 vol% and to a first approximation up to about 2 vol%. The lower size limit for the sedimenting particles is set by Brownian motion, which tends to counteract sedimentation. According to Burton [26], there is a critical radius r_k of 0.2–2 μm, depending on the system. Particles with a smaller radius are kept suspended by the molecules of the suspension medium. In suspensions of inert particles with con-centrations >5% settling is hindered, so Stokes' Law becomes invalid. The particles are not yet actually touching, but the flow around them is hindered by other particles in the system, so the fast particles are forced to sink more slowly. Ideally, all particles end up sinking at the same rate. A clear boundary forms between the sediment and the liquid. All the dispersed material sinks in a cloud, the speed of which is controlled by its porosity and depends on the concentration of the particles. The rate of sinking Q is described by Steinour's Equation [27], Equation 5.14 (ε: porosity, A: constant = 1.82, V_s: Stokes fall rate):

$$Q = V_S \cdot \varepsilon^2 \cdot 10^{-A(1-\varepsilon)} \tag{5.14}$$

The sediment is easily redispersible; no compression zone forms.

In flocculating suspensions, during sedimentation flocs form and are destroyed by friction with other flocs. The size and shape of the flocs affects the sedimentation. At low concentrations, the flocs settle according to Stokes' Law, at higher concentrations they form structures and all settle out at the same rate (*zone sedimentation*).

At high concentrations, the floc structure becomes compressed and a sludge forms. There is a clear boundary between the sediment and the liquid. Even the smallest particles are associated into flocs. This is not the case in a deflocculated system, in which the liquid over the sediment remains cloudy for a long time. Only in monodisperse systems does a sharp boundary form between the sedimenting zone and the liquid above it, which contains no particles. The clarity of the liquid over the sedimenting zone is a good pointer to the nature of the suspension, whether it is flocculated (clear) or deflocculated with particles of different sizes (cloudy).

Second stage: Compression of the sediment. At very high concentrations, the floc structure is compressed by the weight of particles resting on it. The effect of gravity on the sedimentation is reduced, since it must act against the increasing force of compression in the sediment. For inert particles, no compression of the sediment occurs; the particles touch only when they adopt a close-packed structure. It is almost impossible to redisperse such a sediment. Flocculated and deflocculated suspensions can be distinguished by the condition of their sediments (Figure 5.38):

   flocculated suspension → large sediment volume, loosely packed
                   sediment easily redispersed
   deflocculated suspension → low-volume, densely packed sediment
                   redispersed only with difficulty or not at all

The various types of sediment can be represented as in the diagram according to Fitch (Figure 5.39), and the sedimentation characteristics of a concentrated suspension are depicted schematically in Figure 5.40.

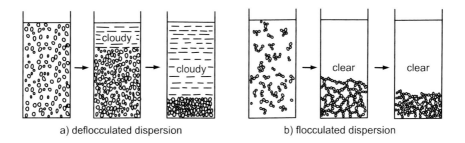

a) deflocculated dispersion   b) flocculated dispersion

Figure 5.38. Sedimentation of deflocculated and flocculated dispersions.

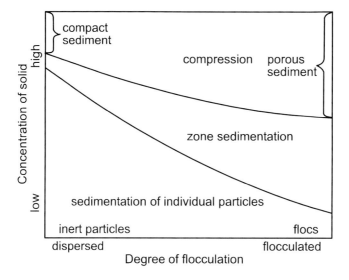

Figure 5.39. Sedimentation types according to Fitch [28].

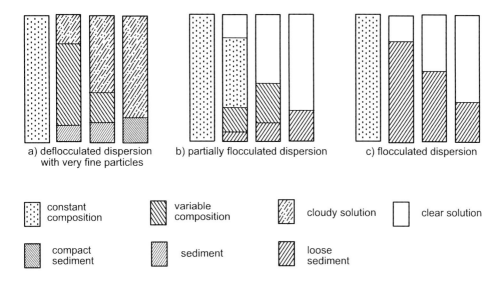

a) deflocculated dispersion
with very fine particles

b) partially flocculated dispersion

c) flocculated dispersion

constant composition

variable composition

cloudy solution

clear solution

compact sediment

sediment

loose sediment

Figure 5.40. Sedimentation characteristics of concentrated dispersions [29].

The time course of a sedimentation is easily observed in a sedimentation cylinder. Provided that the suspension does not contain a large proportion of very fine particles, a clear interface forms between the sedimentation zone and the zone of clear liquid above it soon after sedimentation starts. The height of this interface is measured at intervals and plotted in a settling curve (Figure 5.41).

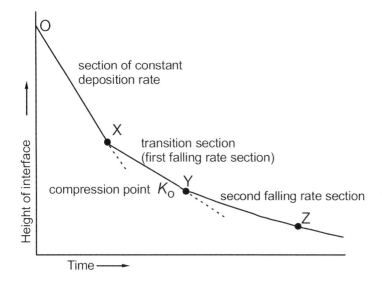

Figure 5.41. Sedimentation ranges in a settling curve.

A settling curve usually comprises three parts: an initial linear section with a constant settling rate (O—X), a transition section in which the sedimentation rate declines (X—Y), and, from the compression point K_0, a section in which the sediment is collapsing on itself. At Z the sediment is so solidly compressed that it can no longer be easily redispersed. The three parts of the sedimentation process are not always all represented; it depends on the concentration and whether the particles are inert or flocculated. Some typical settling curves are shown in Figure 5.42.

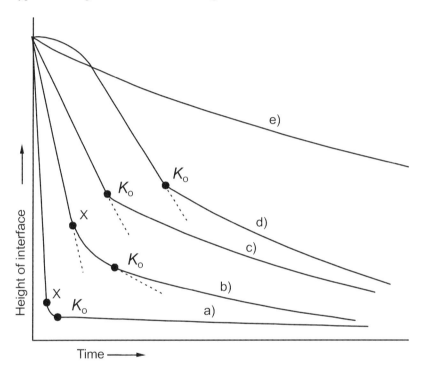

Figure 5.42. Typical settling curves: a) low concentration, deflocculated; b) low concentration, flocculated; c) medium concentration; d) medium concentration with initial delay; e) high concentration; K_0: compression point [30].

Ideally, the particles in a commercially available suspension should remain homogeneously distributed for a long period, perhaps even for years. The settling curve should therefore be horizontal. A flocculated suspension that behaves thus can therefore be considered to be a stable form. However, in colloid science, flocculated suspensions are considered to be unstable dispersions. Chwala calls this situation the "sedimentation paradox". The defining characteristic for a suspension that is stable in this sense is that the sediment formed can easily be redispersed into primary particles. The difficulty in testing stability therefore lies in the location of point Z on the settling curve (Figure 5.41). Direct measurement would require a very long wait for the storage period to elapse. It is possible to speed matters up by measuring sedimentation under centrifugation. The extrapolation from centrifuge tests to deposition under normal gravity is, however, somewhat problematic and is only possible for inert particles and liquids with Newtonian flow characteristics. At high accelerations, flocs can be destroyed, giving incorrect results. It is therefore better not to exceed an acceleration of $4g$ [31]. The best technique is to carry out centrifuge tests at various g values, as low as possible, and then to extrapolate the results to $1g$.

Sediment quality cannot often be assessed visually. A variety of equipment is available for this purpose, such as penetrometers and gel test apparatus. The Brookfield "Helipath" and the "Rheoprobe" should be mentioned (for further details, see reference [3]). The sedimentation volume is an important parameter in sedimentation tests. The height H_0 or the volume V_0 of the suspension is compared with the height H_t or the volume V_t after time t. The ratio of these variables H_t/H_0 or V_t/V_0 is the degree of sedimentation F, which provides qualitative information about flocculation. For $F = 1$, the suspension is in equilibrium as regards flocculation. F can also be greater than 1. This is the case when the mesh built by the flocs is very loose, so that the volume that they occupy is greater than the original volume of the suspension. The concept of sedimentation volume must nevertheless be applied with caution, since an unstable (in the colloidal sense) suspension which flocculates irreversibly may also yield a large volume of sediment. An attempt to redisperse the sediment is necessary in order to distinguish between weak and strong flocculation; this is simply achieved by mechanical rotation (inversion) of the measuring cylinder and enumeration of the revolutions necessary for redispersion. To determine the sedimentation behavior over a very long storage period, the sedimentation volume is measured repeatedly as a few weeks pass and these values are extrapolated to the total storage time.

5.8.3 Prevention of Caking

It is possible to influence settling in various ways to avoid caking (the formation of hard, nonredispersible sediments):

1. By balancing the densities of the disperse phase and the liquid medium. However, this is only possible when the difference between the original densities is small, and only over a small temperature range.

2. By the use of thickeners: thickeners are employed to increase the viscosity; even small concentrations have a noticeable effect. Suitable additives are macromolecules such as modified cellulose and modified starch, natural rubbers, poly(vinyl alcohol), and poly(ethylene oxide), amongst many others (see the chapters on cosmetics and food formulation). Such polymers form structures through the continuous medium, with the effect that they usually have an initial yield value. A disadvantage is that solutions of polymeric thickeners can only be diluted by the application of large shearing forces. In addition, the temperature dependence of the viscosity can negatively effect the storage stability if the storage temperature varies. Furthermore, many polymer solutions decompose over time, so their viscosity decreases.

Common thickeners include: "Carbopols" (high molecular mass acrylic acid polymers), carboxymethylcellulose, hydroxypropylmethylcellulose ("Klucel"), methylcellulose, alginates, poly(ethylene glycol), poly(vinyl alcohol), polyvinylpyrrolidone, and xanthan gum ("Kelzan"). Besides polymeric thickeners, those based on colloidally dispersed inorganic substances, such as clays or oxides, are also important. Clays like montmorillonite and oxides like silicic acid join up into three-dimensional networks in the liquid dispersion medium under certain conditions (concentration, pH, electrolyte content, etc.). These gel networks prevent the formation of hard sediment. Another clay worthy of note is bentonite, different types of which are suitable for aqueous and for organic suspensions. Pyrogenic silicic acid ("aerosil") is well suited to use as a thickener for both polar and apolar media. Inorganic colloidal thickeners have recently been joined by organic products with microcrystalline structures, most prominently microcrystalline cellulose ("Avicel" varieties).

When inorganic and polymeric thickeners are mixed, the effects of the individual components in hindering suspension settling are magnified. Three-dimensional structures form in which the particles form nodes joined together by polymers. The thickening effect is heavily dependent on the mixing ratio of the thickeners.

3. By controlled flocculation: this topic has already been described in Section 5.7.2. In contrast to the formation by polymeric and microcrystalline additives of gels in which the disperse phase is trapped, in this case the dispersed material itself forms a three-dimensional network.

4. If a suspension is to remain stable over long periods of storage, crystallization should not occur, or only to an insignificant degree. Smaller particles are more soluble than large ones. As a result, larger particles grow at the expense of smaller ones (see Chapter 9):

$$\frac{RT}{M} \cdot \ln\frac{S_1}{S_2} = \frac{2\sigma}{\rho} \cdot \left(\frac{1}{r_1} - \frac{1}{r_2}\right) \tag{5.15}$$

(R: gas constant; M: relative molecular mass; S_1, S_2: solubility of a particle of radius r_1, r_2; σ: solid/liquid interfacial tension; ρ: particle density)

The criteria for retardation of crystal growth are low solubility and a narrow particle size distribution.

5.8.4 Preservation of Suspensions

Suspensions containing organic substances are often prone to decomposition by microorganisms. They must therefore be protected against a large variety of possible microorganisms; if the types of bacteria likely to be present are unknown, formulation chemists will choose preservatives with a broad spectrum of activity (Table 5.2). Since the activity of most preservatives is optimal only in a particular pH range, the pH value of the suspension must be taken into account when choosing a preservative. Wallhäusser and Fink have produced a publication that is very helpful in the evaluation of preservatives for use in suspensions [32]. It must be remembered that some substances may no longer be considered suitable for use in commercial products, for toxicological and environmental reasons. Sometimes, other preservatives are used in pharmaceuticals and cosmetics; more details will be given in Chapter 12.

Table 5.2 Preservatives with a broad spectrum of activity [32].

Compound	pH range	Required concn. [%]	Commercial products
formaldehyde	3–10	0.05–0.2	Formalin (40 %)
Na pentachlorophenol	4–10	0.1–0.3	Preventol PN, Witophen N, Cryptogil Na
N-(3-chloroallyl)-hexaminium chloride	4–10	0.05–0.2	Dowicil 200, Quaternum15, Preventol D1
1,3-dimethylol-4,4-dimethylhydantoin	3.5–10	0.05–0.2	Glydant 55
1,2-dibromo-2,4-dicyanobutane	broad	0.01–0.025	Tektamer
1,2-benzisothiazol-3-one	3.5–11	0.02–0.5	Mergal K10, Proxel AB, Proxel BD and other varieties of Proxel
sorbic acid	2–4.5	0.5	(as K sorbate)
benzoic acid	2–4.5	0.5	(as Na benzoate)
o-phenylphenol Na	8–12	0.1–0.2	Dowicide A, Mergal KM, Preventol ON
m-cresol	8–9	0.1–0.2	

References for Chapter 5:

[1]   M. J. Rosen, Surfactants and Interfacial Phenomena, John Wiley & Sons, Inc., New York, 1978.
[2]   H. Rumpf, Chem.-Ing.-Tech. *37*, 187 (1965).
[3]   T. C. Patton, Paint flow and pigment dispersion, 2nd ed., Wiley-Interscience, New York, 1979.
[4]   B. Heinrich, L. Kreitner, Aufbereitungstechnik *10* (1981), *11* (1981).
[5]   US Patent 2,855,156.
[6]   M. J. Rosen, J. Am. Oil Chem. Soc. *49*, 431 (1972).
[7]   R. J. Pugh, T. Matsunaga, F. M. Fowkes, Colloids and Surfaces *7*, 183 (1983).
[8]   P. Sennett, J. P. Olivier, Ind. Eng. Chem. *8*, 45 (1965).
[9]   H. Sonntag, Lehrbuch der Kolloidwissenschaft, VEB Deutscher Verlag der Wissenschaften, Berlin, 1977.
[10] L. Dulog, M. Hilt, Farbe + Lack *96*, 180 (1990).
[11] T. G. Overbeek, Pure Appl. Chem. *52*, 1151 (1980).
[12] E. F. Burton, E. Bishop, J. Phys. Chem. *24*, 701 (1920).

[13]   W. Ostwald, Kleines Praktikum der Kolloidchemie, 9th ed., Leipzig, 1943.
[14]   W. Heller, W. J. Panagonis, Chem. Phys. *29*, 498 (1957).
[15]   E. J. Verwey, J. T. G. Overbeek, The Theory of the Stability of Lyophobic Colloids, Elsevier, Amsterdam, 1948.
[16]   T. F. Tadros, Adv. Colloid Interface Sci. *72*, 505 (1979).
[17]   B. A. Haines, A. N. Martin, J. Pharm. Sci. *50*, 288, 753 (1961).
[18]   R. D. Jones, B. A. Mathews, C. T. Rhodes, J. Pharm. Sci. *57*, 569 (1968).
[19]   J. A. Kitchener, Brit. Polymer J. *4*, 217 (1972).
[20]   V. K. La Mer, Discuss. Faraday Soc. *42*, 246 (1966).
[21]   J. Vostrcil, F. Juracka, Commercial Organic Flocculants, Noyes Data Corp.
[22]   F. K. R. Li-in-on, B. Vincent, F. A. Waite, A.C.S. Symp. Ser. *9*, 165 (1975).
[23]   R. J. Feigin, D. H. Napper, J. Colloid Interface Sci. *74*, 567 (1980).
[24]   S. Asakura, F. Oosawa, J. Polym. Sci. *33*, 183 (1958).
[25]   H. Heywood, Symposium on the Interaction between Fluids and Particles, Inst. Chem. Eng. *1* (1962).
[26]   E. F. Burton, in Colloid Chemistry, Vol. I, 165 (A. E. Alexander, Ed.), Reinhold Publ., New York, 1926.
[27]   H. H. Steinour, Ind. Eng. Chem. *36*, 618 (1944).
[28]   E. B. Fitch, Filtration Separation *12*, 40 (1975).
[29]   G. D. Parfitt, Dispersions of Powders in Liquids, Applied Science Publishers Ltd., London, 1981.
[30]   M. J. Pearse, Gravity Thickening Theories, Report No. LR 261 (MP), Warren Spring Laboratory, Stevenage (UK), 1977.
[31]   W. Jones, J. Grimshaw, Pharm. J. *19*, 459 (1963).
[32]   K. Wallhäußer, W. Fink, Farbe + Lack *91*, 277 (1985).

6 Solid Forms

Solid formulations are a huge topic; they range from simple powders through agglomerates to state-of-the-art formulations such as liposomes and preparations of encapsulated or coated active substances. They reveal a spectrum of new and extremely interesting possibilities to the formulation chemist: for example, slow- or controlled-release (retard) preparations, instant products, and not least, nondusting, easily poured solid formulations. The technologies used in the preparation of such "high-tech" products are correspondingly sophisticated. Recently, nanoparticles and their formulations have joined this list. These advanced types of formulation are often still protected by current patents, and vigorous research and development activities are continually releasing new formulations onto the market. We can only cover the most important and proven members of this group of advanced formulations here.

6.1 Powders and Powder Mixtures

British Standard 2955 defines a dry material as a powder if it consists of particles of a maximum dimension of less than 1000 μm. Powder that can pass through a test sieve of 76 μm is known as sub-sieve powder. This limit serves as a rough criterion for the distinction between grit and dust. Fine powder is an inexact expression for a material with a grain size of less than 1 μm.

Like agglomerates and granules, powders are classified as loose bulk goods. In nearly all branches of industry, solids are processed as bulk goods and must be moved and stored accordingly. These two processes are amongst the basic operations of mechanical process technology. Unlike a liquid, a solid in bulk can adopt a surface of any shape up to its maximum angle of slope. It can transmit static thrust, although the pressure it exerts on the walls and floor of its container does not increase linearly with depth but rather quickly reaches a maximum, since the walls take part of the weight of the material by friction. Furthermore, the pressure depends on the direction in which it is measured and differs between filling, emptying, and storage. Loose bulk goods cannot transmit tension, or only in a very limited fashion, and do not obey the rules that apply to undivided solids. Thus, they are neither solids nor liquids.

A powder can be regarded as a two-phase disperse system in which solid particles of varying diameters are distributed throughout a gaseous continuous medium. The solid particles form a network on the basis of their interactions. This network can be expanded, under specific conditions, but tends to return to the "packed" state under the influence of gravity. It therefore possesses a certain mechanical strength and a certain elasticity. There is a strong interaction between the gas phase and the dispersed particles; the stronger this interaction (increased by higher gas viscosity or by adsorption of gas), the more the powder can be expanded. A greater degree of expansion means the powder has better flow properties. A curious result is that the greater the viscosity of the gas, the lower the viscosity of the powder.

These properties have a role to play in the case of all more or less cohesive powders. They must be taken into account whenever a powder is involved in a process, for instance mixing, agitating, or fluidizing powder. Cohesive and noncohesive powders will be treated in more detail later.

Geldart's classification of powders into the groups depicted in Figure 6.1 should also be mentioned here [1].

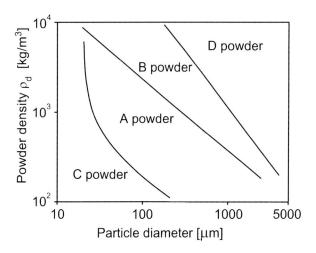

Figure 6.1. Geldart's classification of powders.

An *A powder* is recognizable by the fact that on fluidization with a gas velocity near the critical fluidization velocity, it is distributed homogeneously over the bed. *B powders* generally comprise larger particles and cannot be homogeneously fluidized. *C powders* are finer than A powders and are distinguished by relatively strong cohesion. This means that when the minimum fluidization velocity is exceeded, horizontal fracture surfaces appear in the powder bed, joined by irregular vertical channels through which the excess gas escapes upwards. The bed barely moves. *D powders* are the coarsest and require high gas velocities for fluidization. Gas bubbles coalesce and carry particles upwards, a phenomenon known as spouting.

Geldart's powder classification is suitable for the description not only of fluidization, but also of most other powder-handling operations such as transport, mixing, grinding, agglomeration, and separation. In all these operations, powders are kept in an expanded state, be it through the piping in of gas or simply as a consequence of continuous agitation and mixing, which keeps introducing gas. As a result, the powder is fluidized and its flow properties improved. D powders are not included in this description, because of their particle size. Interparticle forces have no effect here, so their behavior is due purely to hydrodynamics.

Dust particles are usually electrically charged; the charge can be determined shortly after their creation. Charges between 134 and 172 e (elementary unit of electronic charge) have been measured for quartz particles in the 4–5 µm range. Particles of less than 1 µm carry up to 100 e. The increase in charge is roughly proportional to the surface area; this has been confirmed for sawdust, flour, icing sugar, soot, and other substances. The charge influences the flow characteristics and the cohesion of the powder. Charging is mostly connected with powder-handling operations like filling and emptying, sieving, mixing, dust separation, pneumatic transport, etc. As an example of the charging that occurs between the particles and the walls of the apparatus, for instance in pneumatic transport, charges of 10^{-6} to 10^{-4} Coulomb/kg have been measured. Slower movement such as pouring produces values between 10^{-9} and 10^{-7} C/kg. Even a small charge of 10^{-7} C/kg can, in practice, be critical; the danger of ignition or explosion in powder handling can be greatly reduced if all parts of the apparatus are made of electrical conductors and grounded [2].

6.1.1 Flow Characteristics of Powders

The mechanics and flow properties of loose bulk goods and their foundations cannot be treated in detail here; we recommend references [3, 4] to the reader. The flow characteristics of powders are of great technical importance and, accordingly, so are those of powder formulations. The terms trickling ability, crumbliness, pulverulence, flowability, and dispersibility are also used. Loose goods can be divided into two groups according to their flow characteristics:

- noncohesive, free-flowing loose bulk goods
- cohesive, loose bulk goods.

The grain size above which powders do not cohere (as long as they are not too damp) is 100–200 µm. Cohesion is therefore significant under 100 µm. It is based on the cohesive forces between individual particles, which, it is true, still exist for particles larger than 100 µm, but are negligible compared with the weight of the particle. For a particle of 1 µm in size, the cohesive force can be 10^6 times that of gravity.

The flow behavior of powders is often determined by their angle of incline. However, the results for cohesive powders are unusable, because these powders can adopt any angle when piled up, sometimes even over 90°. The slope depends on the degree of compression of the powder. The many tests in which powder flows from a funnel onto a surface and the slope of the resulting pile is measured are equally useless in assessing the flowing ability of a cohesive powder. The Pfrengle method for determination of powder flowing ability (DIN standard 53916), originally developed for testing laundry detergents, should be mentioned here. No limitation to noncohesive powders is stipulated. There is also a standard, DIN 53492, for determination of the flowing ability of grains of synthetic plastics. In this it is stated that unambiguous and reproducible measurement of the angle of inclination is only possible for those loose bulk goods which flow with little difficulty anyway.

For the determination of the cohesion of powders and of loose goods generally, measurement with a shearing cell is now considered to be the best on theoretical grounds and for reasons of reproducibility and reliability. This relies on the principles of the mechanics of loose powders, for which we must refer the reader to the literature [3, 4]. Measurement with this apparatus requires a good knowledge of these principles. About five different parameters must be measured to describe the flow properties; the most significant properties that must be known in order to quantify the processes involved in storage and flow are:

– the pile-up stability of the powder
– the internal friction angle
– the angle of internal friction in stationary flow
– the angle of friction with the wall
– the apparent density of the bulk material.

As Jenike found on the basis of theoretical work, a shearing cell is suitable for the measurement of these variables. A cell comprising base, shearing ring, and lid (Figure 6.2) is filled with powder; the powder is preconsolidated with a standard weight N and sheared with a force S, permitting the measurement of these parameters of importance for powder flow. Nowadays there is a wide variety of shearing equipment available for this task. For details of the principle behind the measurements, the construction of the cells, and their use, see Kulicke in Chapter 10 of Wilms and Schwedes [5]. Nevertheless, the values obtained from all these instruments are only useful, that is, only give a quantitative and sufficiently accurate picture of the flow properties and permit their interpretation, if the user has the necessary understanding of the principles of powder mechanics. This is also the case for the simple shear apparatus designed by Schwedes, reference [5], p. 460.

Peschl's shearing apparatus, the RO-200 Automatic, is an automatic device for industrial use relying on the principle of rotational shearing that furnishes a so-called "flowability index" from the various, automatically measured mechanical properties of the powder. This index is suitable for process control of powder handling, and can be measured without knowledge of powder technology. (Institute of Powder Technology/Institut für Schüttguttechnologie, Schaan-Vaduz, Liechtenstein.)

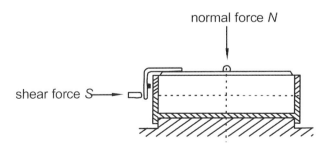

Figure 6.2. Jenike shear cell.

The ability of a powder to flow can be assessed quantitatively by a single parameter, the shear index n. The shear indices defined by N. Pilpel (Table 6.1) range from 1 (free-flowing powder) to 2 (caking powder) [6]. They are measured in a special shearing cell.

Table 6.1. Ability of powders to flow [6].

Substance	Flowing ability	Shear index n
sand	free-flowing	1.2
Portland cement	free-flowing	1.4
powdered lime	free-flowing	1.3
dried egg	cohesive	1.6
titanium dioxide	cohesive	1.8
Griseofulvin	cohesive	1.9

The shear index of a powder increases as its particle size decreases, as does its cohesion; for example, for precipitated chalk with a mean particle diameter of 1 μm, $n = 1.86$; with a mean particle diameter of 20 μm, $n = 1.29$.

6.1.2 Lowering Cohesion and Improving Flow Properties

Additives that improve the flow behavior of powders, lubricants, are widely used to decrease the cohesion of fine powders. A coating of a few layers of stearic acid molecules on the surface of particles of agrochemicals to be sprayed onto fruit or other produce to protect against pests and diseases prevents clumping and facilitates application as a fine, effective dust. The atomization of pharmaceuticals into a cloud of free-flowing powdered medication amplifies their effect on inhalation.

Besides the fatty acids, other water-repellent additives that can be applied as a spray or a vapor to reduce the adsorption of water onto powders are dimethyldichlorosilane and quaternary ammonium compounds. Adsorption of such additives renders the powder hydrophobic.

To improve the flow properties of powdered foods, starches and lactose may be used. Magnesium carbonate mixed into table salt at about 1% makes it more easy to sprinkle, even when the atmosphere is humid.

To reduce the cohesion of pharmaceutical powders, aerosil is added in carefully controlled quantities so that capsule-filling machines or fast tablet presses can handle the powder better. Talc, calcium phosphate, and various metal stearates serve the same purpose.

The effect of a lubricant on the cohesion and the flowing ability of a powder depends on many factors: on its physical and chemical affinity to the powder, on its average particle size and shape in relation to those of the powder, on its concentration, and on its moisture content.

6.1.3 Dust

In the seventeenth century, the English doctor Sir Thomas Brown wrote "Time, which antiquates antiquities, hath an art to make dust of all things". Particles in the atmosphere have diameters between 0.002 and 100 μm. The particles at the lower edge of this range are usually unstable and coagulate into larger particles; fine drops of liquid tend to evaporate quickly. The largest particles, on the other hand, settle out fast; they are usually irregular in shape and are often aggregates. Their settling speed can be calculated from the Stokes Equation (see the chapter on colloidal suspensions). Dust particles with a diameter greater than 2 μm cannot remain in the air for long, since gravity will pull them down, whereas smaller particles can remain suspended for lengthy periods. Very small dust particles, condensation nuclei, as they are called, are responsible for the formation of haze, fog, and raindrops. They have a concentration of about 1000 particles per cubic centimeter in rural areas and more than 100 000 per cubic centimeter in towns. The problems caused by dust from various sources, especially human and industrial activity, are well known. About 20 g of dust is created per kWh of electricity generated in a coal-fired power plant; a steel converter throws about a tonne of dust into the air every hour.

The dust-producing properties of products in the chemical industry are of interest to formulation chemists too, especially in the case of toxic or highly physiologically active substances, and also of dyes. We therefore need to be able to measure the dusting of powders and, thence, to bring them under control. Here we summarize some of the possibilities.

6.1.4 Measurement of the Dust-Producing Properties of Powders

The dust-producing properties of a powder are related to its slope when heaped: dusting powders have a small angle of inclination, while those that produce little dust can be piled up at 90°. This is only a rough qualitative indication of dusting; methods have been developed in the trade for semiquantitative tests for dyes in which a weighed amount of powder is tipped from a glass cylinder into a 500 cm^3 measuring cylinder through a funnel that has a moistened circular filter mounted around its elongated stem. The impact of the dust on the base of the measuring cylinder creates a dust cloud that, depending on the intensity, leaves powder adhering to the perforated filter attached to the stem of the funnel at a height of the measuring cylinder corresponding to a volume of about 400 cm^3. The discoloration of the moist filter is used to determine the intensity of dusting. The "Casella" dust-measuring apparatus provides a quantitative method, in which a known amount of dust is allowed to fall in a chamber and the relative change in the intensity of light from a source is recorded over time with a photocell (Figure 14.4).

We should mention one of the first scientific examinations of dusting and its measurement, the work of Andreasen [7]. This covers the essential aspects of the dusting behavior of a wide variety of inorganic and organic substances.

6.1.5 Dust Reduction Methods

Humans have been fighting dust in their communities and industries for a long time; as an instance, let us cite the removal of dust from the streets. People tried to control the problem with water or aqueous salt solutions to wet the dust. Here, the principles of the wetting of solids by liquids are critical: adhesion of the droplets of spray to a rough surface, spreading on the surface, capillary penetration into the dust, escape of air bubbles from the capillaries through the aqueous medium. The tensides used as wetting agents must, most importantly, be able to lower the advancing contact angle, but at the same time they should only decrease the surface tension of the solution slightly.

The interface-active substances that meet these demands include lignin sulfonates. As an otherwise useless by-product of cellulose manufacture, these are very cheap. Aqueous solutions of calcium sulfate and other deliquescent salts (hydrates) are also used. Of the tensides used to bind dust, most importantly Lissapol N (a condensation product of alkylphenol and ethylene oxide) and Teepol (sulfated olefin) have been employed for a long time. The condensation products of naphthalenedisulfonic acid with formaldehyde are also worthy of note; see Chapter 1. An undesirable side-effect of the use of wetting agents for dust suppression in powders is foaming. Wetting and foam formation are opposing properties in certain wetting agents, a fact that should be taken into consideration when a tenside-based dust binder is selected.

Agglomeration of powders is a generally applicable method for dust binding. The various ways of achieving this will be described below. Individual dust particles must be bound sufficiently well to resist fragmentation on handling. On the other hand, the agglomerates must be easily separated into their individual particles once more in dispersion or dissolution processes. Wet spherical agglomeration is a process with many possibilities of application; its principles are described by Capes [8], and the patents assigned to Mollet, Rabassa, et al. give numerous examples ranging from dyes and pharmaceuticals to coffee, powdered milk, laundry detergent, textile auxiliaries, and so on (CH Patent 602 276 and other patents). In addition, the *Harshaw dustless process* should be described as a special process in which Teflon K is added to the powder to be treated, after which the mixture is mixed intensively in a ball mill, hammer mill, extruder, or other apparatus. This disperses the Teflon K Type 20 into fibrils of about 0.2 µm in diameter, which bind the powder particles together in a fine network to create a dust-free product in which the dust is captured by the fine fibrils. Since Teflon K is insoluble in solvents such as water, however, the method is not suitable for use with products which must pass a filtration test on application. The maximum quantity of Teflon K added is 0.5–1%, but the optimum dosage is usually lower, although it requires a longer mixing period. Users of the process must pay a license fee (Harshaw Dustless, Cleveland, Ohio).

6.1.6 Powder Mixing

The mixing of solid particles is a complex process-technological problem. Nearly all processes in the chemical industry require a mixing operation at some point. The theoretical foundations of basic mixing procedures cannot be discussed here; the reader is directed to the literature (references [9, 10]). Here we shall simply summarize the principles and the practical knowledge most important for formulation chemists.

The powder characteristics of each component in a mixture of powders influence the mixing behavior. The differences in relevant properties such as particle size, size distribution, shape, density, and surface characteristics dominate the mixing process and render it difficult.

A few cases of typical mixing problems will be mentioned here to demonstrate the challenges that can be posed regarding the homogeneity of a mixture.

1. In the manufacture of a ceramic mass, 1% pigment must be blended with 99% sand, TiO_2, and other components; the fraction of pigment in the mixture may not vary by more than 0.1%.

2. Powdered cork for linoleum production must contain a mixture of grain sizes to yield the desired color; 15% of the grains have a diameter of 80 µm and 85% 330 µm. The percentage variation of particle sizes in the mixture may not be more than 2.2%.

Any homogeneous mixture tends to separate. The quality of a powder mixture is improved if the degree and rate of demixing can be minimized. In contrast to the case for gaseous and miscible liquid systems (including suspensions), this reduction does not result from molecular diffusion; instead, the system must be manipulated so that the particles move relative to one another. The mass may be kneaded, rolled, cut, or scooped, as long as the individual particles have the opportunity to settle back into the mixture. The ability of the particles to move freely is the basis of a general classification of powder mixtures into:

– free-flowing mixtures
– cohesive or non-free-flowing mixtures.

Particle size is probably the dominant influence on the flow behavior. For large particles, gravitational force is much stronger than the inter-particle forces which act to retard flow, so individual particles retain their freedom of movement. For smaller particles, various inter-particle forces may dominate and the particles tend to retain an ordered structure. To a reasonable approximation, particles of diameter greater than 50 µm tend to be free-flowing, whereas those under 50 µm tend to cohere. The properties and the effect of the powder flow behavior on the mixing process vary for the two types of powder.

a) Free-flowing mixtures

These create less dust owing to their larger particle size. After transport, they still flow smoothly out of a container. The disadvantage of such mixtures is that they demix easily precisely because the ability of the individual particles to move freely and independently of their neighbors also enables them to move preferentially in a particular direction. Thus, small particles will trickle downwards through a bed of larger ones when the latter are moving. Even when the difference in particle size is extremely small, this results in demixing. When a large quantity of free-flowing powder is perfectly mixed, subsequent handling or storage can destroy the quality of the mixture. Only when the mixture is "frozen" in its final use, that is, when the particles lose their mobility, is its quality certain. The final application could be the making of a tablet or the filling of a package.

b) Cohesive masses

In contrast to free-flowing mixtures, for cohesive mixtures the problem lies in the need to disrupt their natural structure repeatedly in order to give the individual particles the chance of moving to a different point within this structure. It is relatively simple to produce a good cohesive mixture with most industrial mixing equipment.

Both the nature and the resistance of the forces between particles acting within a cohesive system determine how easy or difficult it is to move the particles within the mixture. If it is certain that the particles are not solidly joined to one another, then linking mechanisms must be due to:

– moisture
– electrostatic charge
– van der Waals forces.

Figure 6.12, taken from Rumpf, gives an overview of these forces. If a tensile strength of 0.01 kg/cm^2 is chosen arbitrarily as the limit of "insignificant" powder resistance, then it can be seen from the diagram that the van der Waals force only has a cohesive effect for particles under 1 μm in diameter. If a film of adsorbed moisture or chemical bonding forces are present at the points of contact, then the threshold particle size is about 80 μm.

For wet powders or those with very high moisture contents, "liquid bridges" form between the particles. In this case, particles in the 500 μm size range can demonstrate cohesion. Under most atmospheric conditions, a "dry" powder mixture contains no liquid bridges. The cohesive forces stem from van der Waals forces, electrostatic forces, or adsorbed moisture. According to Rumpf [9], a powder tends to be cohesive if its particles have a diameter of less than 50 μm.

6.1.7 Quantitative Determination of the Quality of the Mixture

The variance of one constituent in a number of samples is a measure of the quality or consistency of a product. The smaller the variance, the better the mixture. Nevertheless, the value of the variance alone does not provide an absolute estimate of the quality of the mixture. To obtain this, we need to compare the value obtained from experiments with the limiting mixture variance values for complete demixing and for random mixing.

For most industrial mixtures, the limiting random variance s^2 can be calculated. The simplest case will be discussed here, the mixing of two components with particles of equal sizes (Lacey, reference [11]; Equation 6.1).

$$s^2 = \frac{a \cdot b}{B}$$

(6.1)

s: standard deviation
s^2:   variance
a,b: fractions of the two components
B: number of particles in sample

The mixing procedure is an attempt to reduce the variance between the samples to a minimum. Ideally, s^2 should be as small as possible. It follows from the equation that the random variance can be reduced either if one component is present as a very small percentage, or if each sample contains a large number of particles. If the sample weight is fixed, then the only possible freedom remaining in the process is the reduction of the particle size. The importance of size reduction and the control of particle size in order to achieve a consistent, high-quality mixture is confirmed by inspection of expressions for the random variance in more complex systems. Likewise, for systems containing two or several components with particles of different sizes, relationships exist [12, 13]. However, these do not take into account inadequacies in the mixer or actual demixing processes.

6.1.8 Commonly Used Mixing Equipment

There are many types of mixers for the efficient mixing of powders, ranging in size from those suitable for the laboratory through pilot plants to industrial size. Ries [14] gives a somewhat dated review of mixers and their classification; Perry reviews mixers for powders and agglomerates [15]. The journal *Chemie-Ingenieur-Technik* surveys the market for agitators, mixers, and kneaders periodically.

Most mixtures rely on the mechanisms of *diffusion*, *convection*, and *shearing*. Processes that encourage the mobility of individual particles support mixing by diffusion. These movements are small in themselves. If there are no forces favoring the opposing process of separation, mixing by diffusion is quick and thorough. It occurs if particles are distributed over a fresh surface and if they have a raised individual mobility. For rapid mixing, an additional mixing technique must be used alongside diffusion in order to mix in groups of particles. This second effect may be convection or shearing. Most mixers employ all of these mechanisms, but one is always predominant. In tumbler mixers, it is diffusion, in ribbon mixers convection, in universal mixers it is shearing. Peschl's universal mixer offers an original solution to the problem of mixing [16]. It relies on an up-to-date knowledge of the properties of loose goods, specifically, of flow conditions. The principle behind the mixing is the even distribution of the product onto a special mixing floor from a vertical column, the conversion of horizontal layers into vertical ones, and repetition of the process. This results in a perfect mix even in the smallest samples, whether for free-flowing or cohesive substances (manufacturer IPT, Vaduz).

6.2 Agglomerates, Granules

Size reduction of particles has been discussed in the chapters on the manufacture of stable liquid dispersions and emulsions. The production of agglomerates is the conceptual opposite of size reduction, as illustrated in Figure 6.3.

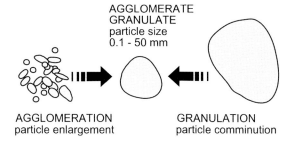

AGGLOMERATE
GRANULATE
particle size
0.1 - 50 mm

AGGLOMERATION
particle enlargement

GRANULATION
particle comminution

Figure 6.3. The concepts of agglomeration and granulation.

Rumpf [17] defines agglomeration as the desirable or undesirable coagulation of finely dispersed material into coarser. The joined particles thus created are known as agglomerate, regardless of their shape and strength. In older literature and in technical usage, the term granules is also used. This refers to a loose solid with individual particles called grains, granules, or pellets; the grains are of roughly the same size, in the range 0.1–50 mm. "To granulate" literally means "to make into grains", and as such can apply not only to comminution of a solid but also to aggregation of tiny particles. The term agglomeration, used to mean particle enlargement, has been in common use in the literature for more than 20 years, while the term granulation has been restricted to particle size reduction and special cases of particle enlargement. The word granule itself refers only to the shape and size of the final product and not to the method of manufacture.

Since the principal business of formulation chemists is solid formulations that can easily be reconverted to dispersions or solutions for use (instant products, in the extreme case), this chapter will concentrate on the manufacture of agglomerates with good reconstitution properties. *Reconstitution* is used to mean the entire process of dispersion or dissolution of the agglomerate. We will not deal with those agglomeration or granulation methods which aim to produce the hardest possible granules, such as the manufacture of pellets in mineral refining or of coal briquettes; nor will we touch on the methods of particle enlargement typically found in the pharmaceutical industry such as pill-making and coating, since this is covered in Chapter 12 (see also the review in reference [18]).

The purpose of the technology applied to agglomeration is the production of agglomerates systematically and with properties defined as well as possible. However, agglomerates often form by chance and are unwanted. Since as particle size decreases, weight declines with the cube of the particle diameter but cohesive forces, to a first approximation, are directly proportional to the particle size, cohesive forces are greater than particle weight for small particles. For a 1 µm particle the cohesive force may be a million times greater than the force of gravity, depending on the type of cohesive force and the density and shape of the particle. Cohesive forces can make a nuisance of themselves when fine-grained material is being worked, such as on grinding, mixing, and sieving. The growth of kidney stones is surely a good example of undesirable agglomeration, whereas the spherical lumps of poplar seeds or pine needles found in the woods are perhaps one of nature's jokes. Agglomeration as intended by humans has a variety of different goals, including:

– products with good or defined flowing ability
– products that produce little dust
– products with a low volume when packed or settled
– products that are rapidly redispersed or dissolved in liquid media; instant products
– products with depot activity due to suitable additives, for example slow-release pharmaceutical formulations.

Agglomerated particles measure between 0.1 and 5 mm (Figure 6.4), depending on the manufacturing process. Smaller grains are known as microagglomerates, and have a greater tendency to emit dust. The best shape for agglomerate grains is spherical, since spheres have no edges and corners to be rubbed off.

The mechanical stability is a measure of the suitability of an agglomerate for its application. It is connected to all other properties of the agglomerate, especially porosity, the size of the particles making up the grains of the agglomerate, the proportion of liquid present, and the wetting by the liquid. These relationships will be discussed in more detail below. The mechanical divisibility of the agglomerate depends on its mechanical strength; if necessary, its divisibility in liquid media can be improved by the addition of so-called disintegrating agents, that is, readily soluble or swelling additives.

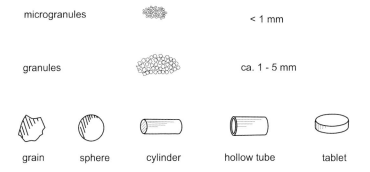

Figure 6.4. Sizes of agglomerates and microagglomerates, and shapes of agglomerates and granules, both porous and nonporous.

6.2.1 Granule Formation

In principle, granules can be formed by the following basic methods:
– agglomeration by molding, compaction
– agglomeration by build-up
– agglomeration by drying.

6.2.1.1 Agglomeration by Molding, Compaction

The powder is molded under pressure into agglomerates (Figure 6.5). This not only creates more points of contact between the particles, but also deformations which can increase the cohesion. This increase in cohesion makes the most important contribution to the increase in strength of the agglomerate during the molding process. The most common pressure agglomeration processes are roller compaction, briquetting, tablet production, aperture molding, and vibratory compaction.

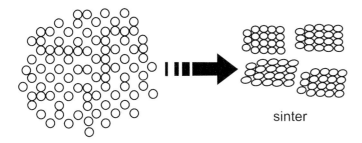

sinter

Figure 6.5. Particle deformation.

6.2.1.2 Agglomeration by Build-Up

A distinction must be made here between agglomeration by the actual build-up of particles and agglomeration by accumulation of particles (Figure 6.6). Particle build-up occurs when suspended solids are atomized and gather on the surfaces of existing particles so that the masses and volumes of the agglomerates grow but the number of associated grains remains unaltered.

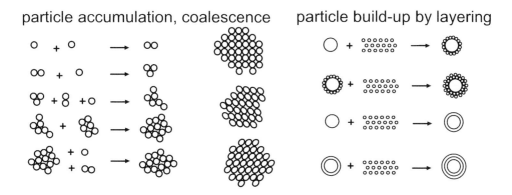

Figure 6.6. Particle accumulation and particle build-up.

Particle accumulation occurs when addition of liquid results in the joining of primary particles, fragments of agglomerates, or agglomerates onto existing agglomerates. Again, the masses and volumes of the agglomerates increase, but the number of particles present decreases over the course of the agglomeration until it reaches equilibrium. This is called coalescence.

The starting point of the build-up process is the formation of nuclei. These kernels are created by the addition of starting material, that is, moist powder or solution containing solids, because of the capillary attraction between the particles. The further build-up of the agglomerate can now occur by various mechanisms, namely, fracture of the agglomerates and accumulation of the fragments, or accumulation of starting material (known as layering or snowballing). A vivid example of agglomeration by build-up is that of an avalanche. Pressure and motion cause ice crystals to join onto the agglomerates already formed, which thus grow ever larger.

6.2.1.3 Agglomeration by Drying

Agglomerates can also be manufactured from solutions, suspensions, or moist loose solids by suitably controlled drying processes. In this method, crystallization is usually the important process. When a solution is spray-dried, crystals or amorphous particles connected together by solid bridges are formed. Similarly, when moist loose solids are dried, solid bridges form, since the liquids always contain dissolved substances. As the material dries out, these are concentrated chiefly in the liquid bridges between the particles, where they are eventually precipitated. The drying conditions can be controlled to affect the strength, porosity, and "instant" properties (characteristics relevant to use in instant products) of the agglomerates.

The most common methods of agglomeration by drying are:
- spray drying
- fluidized-bed drying
- roll drying
- freeze drying

6.2.2 Cohesive Forces in Agglomerates

For agglomerates to come into being, cohesive or binding forces must exist between the individual particles. The binding may be due to solid bridges, to binders, or to liquids, or to van der Waals forces or electrostatic attraction, which act without material links. In the case of ferromagnetic materials, magnetic attraction can be added to this list. Rumpf has classified the cohesive forces that may act in agglomeration [19]; see Table 6.2. The possible types of binding that hold agglomerates together will be summarized briefly.

Table 6.2. Schematic survey of the principles of agglomeration.

Binding forces between primary particles	Process
solid bridges	sintering, melting, moistening with solvents or adhesive solutions
interfacial forces:	
– surface tension	moistening sparingly with liquid
– lowered capillary pressure	moistening with liquid
– surface tension of droplets	moistening with ample liquid
adhesive and cohesive forces	moistening with binder
attractive forces	electrostatic charge

6.2.2.1 Solid Bridges

These can be constructed in the following manners:

a) *By sintering processes*: A rise in temperature increases the mobility of atoms, so that sinter bridges form between particles by diffusion. The agglomerates formed are full of pores.

b) *By melt adhesion and cold welding*: Temperatures near the melting point can be reached locally by friction or plastic deformation at points of contact and rough projections, especially for materials with low melting points. This leads to enlargement of the points of contact and bridges created by melting. Such bridges resemble sinter bridges, but in this case the required heat is generated by friction and plastic deformation.

c) *By binders that set (agglutination effect)*: An adhesive solution is used as the agglomerating liquid. After the solvent has evaporated, the individual particles remain attached to one another by the glue. Bridges of mortar, which form when lime is present and the atmosphere is humid, are of this type.

d) *By crystallization of dissolved substances (encrustation)*: If the liquid present in wet or damp agglomeration processes contains dissolved substances or dissolves the solid, the solute crystallizes out at the points of contact as the material dries and forms solid bridges between the particles.

6.2.2.2 Adhesive and Cohesive Forces in Liquid and Binder Bridges with Restricted Mobility

a) *Highly viscous binders*: Instead of forming liquid bridges, the drops of the binder retain their shape, because the deformation energy necessary to alter it would be greater than the energy gain at the surface. On the other hand, the adhesive forces at the interface between liquid and solid and the cohesive forces in the binder can be exploited to the full. Interfacial forces are unimportant in comparison.

b) *Adsorbed layers*: Nearly all solids adsorb water vapor from the atmosphere, so that a few layers of water molecules are present and held firmly with limited mobility on the solid surface. These layers, which are less than 30 Å thick, cause full-strength molecular attraction between particles in contact with one another, that is, in a powder.

6.2.2.3 Forces of Attraction between Solid Particles

Molecular forces: Valence forces cannot be involved in adhesive mechanisms because of their short range. For distances between grains of up to 100 Å, however, van der Waals forces are active and can hold particles together.

Electrostatic forces: Particles of solid can become electrostatically charged by contact, friction, and size reduction. The degree of charge is dependent on the material and the load. The attraction caused by the contact potential is greater for conductors than for insulators, since the charge concentrates at the surface in the former. For smooth spheres less than 100 μm in size, electrostatic cohesive forces are negligible compared with van der Waals attraction, whereas for larger particles in dry systems they can have a critical effect on cohesion and agglomeration. The influence of electrostatic forces becomes still more significant if attractive forces, rather than cohesion, are considered for larger distances between particles. For distances greater than 1 μm only electrostatic forces are still effective, and these bring about the accumulation of particles before cohesion comes into play.

6.2.2.4 Cohesive Forces at Freely Mobile Liquid Surfaces

In the presence of liquid that wets the loose goods, as is the case in agglomeration by moisture, by build-up, and by accumulation, forces holding the particles together come into being owing to the lowered capillary pressure in the liquid and the boundary force at the solid–liquid–gas tangent. This boundary force is caused by the surface tension of the liquid.

At a freely mobile liquid surface, there must be an equilibrium for each liquid element between the components surface tension γ and the pressure difference ΔP between the concave and convex sides of the interface. From this we can derive the Laplace formula for the dependence of the capillary pressure p_c on the radii of curvature R_1 and R_2 of a curved surface, Figure 6.7.

$$p_c = \gamma \cdot \left(\frac{1}{R_1} + \frac{1}{R_2} \right)$$

(6.2)

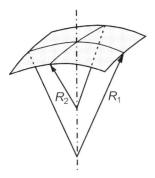

Figure 6.7. Illustration of the Laplace formula.

When two spherical particles are joined by a liquid bridge, we can see from Figure 6.8 that:

$$p_c = \pi \cdot R_2^2 \cdot \gamma \cdot \left(\frac{1}{R_1} - \frac{1}{R_2} \right)$$

(6.3)

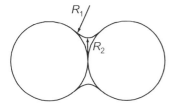

Figure 6.8. Capillary pressure between two particles linked by a liquid bridge.

For capillary pressure in a cylindrical capillary (Figure 6.9), Equation 6.4 holds:

$$p_c = \frac{2\gamma}{r} \cdot \cos\theta$$

(6.4)

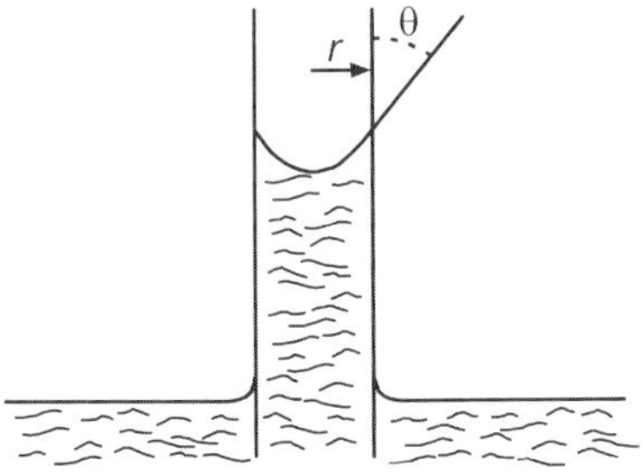

Figure 6.9. Capillary pressure in a cylindrical capillary.

Depending on the quantity of liquid present in the loose solid, and assuming that wetting is complete, the following possibilities for the distribution of liquid and solid particles exist (Figure 6.10):

 a) bridging: pendular state

 b) transitional: funicular state

 c) capillary state

 d) droplet filled with particles

Figure 6.10. Adhesion of mobile liquid to powders.

a) *Pendular state (bridge formation)*: There is only enough liquid present to form discrete liquid bridges at the points of contact where the liquid is drawn by capillary forces. The surface tension acting at the solid–liquid–gas point of contact pulls in the direction of the liquid surface. In the interior of the liquid bridge, a capillary underpressure exists. Together, these two forces act to attract the particles to one another (Figure 6.10a).

b) *Funicular (transitional) state*: When an increased proportion of liquid is present, some liquid bridges coalesce to form small volumes filled with liquid. Elsewhere individual liquid bridges still exist (Figure 6.10b).

c) *Capillary state*: If the liquid completely fills the spaces between particles and has concave menisci at the outer edges of the agglomerate granules, a capillary underpressure exists throughout the liquid volume; this lends a certain tensile strength to the agglomerate (Figure 6.10c). Agglomerates practically full of liquid are formed by build-up agglomeration in a granulation pan.

d) *Liquid droplet filled with particles*: As soon as the liquid surrounds the solid particles completely, the concave meniscus in each pore of the agglomerate is replaced by the convex surface of a drop of liquid. The capillary cohesion between the particles has vanished (Figure 6.10d). The liquid experiences a small overpressure; the particles in the droplet are held together by its surface tension. Coalescence of adjacent droplets is favorable. The division of a suspension into drops and the tendency of the drops to unite can be used for agglomeration, for example in spray drying.

6.2.2.5 Tensile Strength of Agglomerates

Rumpf suggested tensile strength as a suitable measure of the mechanical strength of agglomerates for comparative experiments [19]. To be precise, the force required to break the agglomerate apart related to its cross-sectional area is intended. This can be determined unambiguously by a rending test. If we consider an agglomerate as a continuum through which can be transmitted a pure tensile strain that remains constant over the fracture cross-section (that is, without plastic deformation, so without shear stress), then its strength can be determined in terms of the maximum transmissible tensile strain σ_{max}. In a perfect tensile test, the measured tensile strength σ_k can reach the value of σ_{max}.

In estimating the tensile strength of agglomerates, we must make a distinction between systems in which the bonds between particles are localized at the points of contact, and those in which the space between the particles is partially or completely filled with liquid that transmits the tensile strain.

For *a collection of particles with local areas of contact*, the mean tensile strength of an agglomerate is given by Equation 6.5 (Rumpf, reference [20]):

$$\sigma_z = \frac{9}{8} \cdot \frac{1-\varepsilon}{\varepsilon} \cdot \frac{H}{d^2} \qquad (6.5)$$

where:

$$\varepsilon = 1 - \frac{\rho_a}{\rho_s} \qquad (6.6)$$

(H: cohesive force; d: particle diameter; ε: porosity; ρ_a, ρ_s: density of agglomerate and particles)

For a mean porosity of $\varepsilon = 0.35$, one obtains Equation 6.7:

$$\sigma_z \cong \frac{2H}{d^2} \qquad (6.7)$$

For cohesion mechanisms based on mobile liquid bridges, van der Waals forces, and electrostatic attraction, the value of H can be calculated for simple models. In contrast, the stronger forces involved in solid bridges and bridges of very viscous liquids only lend themselves to calculations in the most simple cases [21].

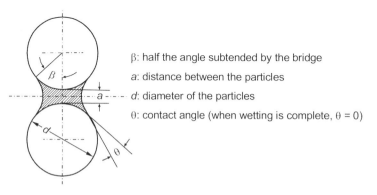

β: half the angle subtended by the bridge

a: distance between the particles

d: diameter of the particles

θ: contact angle (when wetting is complete, θ = 0)

Figure 6.11. Liquid bridge between two spherical particles.

For the *cohesive forces transmitted through mobile liquid bridges*, we can base our calculations on the work of Newitt and Conway-Jones [22]. According to these authors, the cohesive force between two fully wetted spherical particles joined by a liquid bridge (Figure 6.11) is given by Equation 6.8, below:

$$H = \gamma \cdot d \cdot F\left(\beta, \frac{a}{d}, \theta\right) \tag{6.8}$$

Pietsch and Rumpf [23] represented the function F (= related cohesive force) graphically as a function of the degree of saturation, assuming complete wetting of the solid by the liquid. The degree of saturation S describes the relationship of the liquid volume with the pore volume. For $S = 1$, the entire pore volume is filled with liquid, for $S = 0$ there is no liquid present. In the range in which bridges exist, the tensile strength remains approximately constant at a mean value of about 2 for constant γ and d. If we substitute this value into Equation 6.8 and then combine it with Equation 6.5, for $\theta = 0$ we obtain Equation 6.9:

$$\sigma_z = \frac{9}{4} \cdot \left(\frac{1-\varepsilon}{\varepsilon}\right) \cdot \frac{\gamma}{d} \tag{6.9}$$

When the liquid content is sufficiently high for the coalescence of some of the liquid rings, an intermediate, funicular state is reached. As soon as all pores are filled, the system enters the capillary state. Here, the contribution of the underpressure in the liquid to the tensile strength is much greater than the contribution resulting from the interfacial tension at the surface of the agglomerate. This underpressure is defined by the Laplace Equation, Equation 6.4. For an agglomerate, the capillary radius can be characterized in terms of a hydraulic radius, which is obtained from the specific surface area of the particles and the porosity. This yields Equation 6.10 for the tensile strength of an agglomerate in the capillary state with spherical particles of equal sizes:

$$\sigma_z \approx p_k \approx 6 \cdot \left(\frac{1-\varepsilon}{\varepsilon}\right) \cdot \frac{\gamma}{d} \cdot \cos\theta \tag{6.10}$$

In the case of nonspherical particles and complete wetting by the liquid, the expression is:

$$\sigma_z = C \cdot \left(\frac{1-\varepsilon}{\varepsilon}\right) \cdot \frac{\gamma}{d} \tag{6.11}$$

The constant C relates to the specific surface area of the agglomerates and the deviation from spherical geometry. For sand, C has a value between 6.5 and 8.

A comparison of Equations 6.9 and 6.11 shows that *the tensile strength in the capillary state is about three times larger than it is in the pendular state*. In the transitional, funicular state, the values are intermediate.

When the particles are completely surrounded by liquid, they are held together only by the surface tension of the convex liquid surface.

To summarize: the linking of particles by bridges is the active mechanism for values of S up to about 0.25. In the range $0.25 < S < 0.8$, the capillary pressure mechanism supersedes it. For $S > 0.8$, there are no longer any liquid bridges present, because most of the space is filled with liquid at a capillary pressure of p_k. Since the volume that is not filled with liquid does not transmit any tensile strain, the tensile strength for $S \geq 0.8$ is given by Equation 6.12:

$$\sigma_z = S \cdot p_k \tag{6.12}$$

6.2.2.6 Van der Waals Attraction

For two spherical particles at a distance $a < 1000$ Å from each other, the van der Waals attraction is:

$$H = \frac{A}{24a^2} \cdot d = 4.2 \cdot 10^{-14} \frac{d}{a^2} \qquad \left[\frac{\text{dyn}}{\text{cm}^2}\right] \tag{6.13}$$

(d: particle diameter [cm]; A: Hamaker constant [10^{-12} dyn·cm]; a: distance between particles [cm])

For the tensile strength of an agglomerate of medium porosity ($\varepsilon = 0.35$), in accordance with Equation 6.7, this yields Equation 6.14:

$$\sigma_z = 8.4 \cdot 10^{-20} \cdot \frac{1}{a^2 \cdot d} \qquad \left[\frac{\text{kg}}{\text{cm}^2}\right] \tag{6.14}$$

For an agglomerate grain of diameter 1 μm in which the individual spherical particles are 20 Å apart, the tensile strength is 21 g/cm².

Roughnesses in the particles decrease the force of their cohesion. For the calculation of the van der Waals attraction, monolayers of adsorbed water molecules have to be added to the solid to a first approximation. Thus the adsorption of moisture results in a diminution of the distance a between particle surfaces and therefore an increase in strength. The van der Waals forces are approximately one order of magnitude smaller than the cohesive forces resulting from the presence of liquid bridges.

6.2.2.7 Electrostatic Cohesive Forces

Particles can become electrically charged by many routes during the production and treatment of agglomerates. Simple movement of dustlike products creates charge by friction and impact. Likewise, when particles are in contact an electrical attraction is created by the contact potential. For smooth spheres of less than 100 μm in diameter, these electrostatic cohesive forces are negligible compared with van der Waals forces and, especially, the force produced by a liquid bridge. For large particles in dry systems, in contrast, the electrostatic attraction due to excess charge can exert a deciding effect on the cohesion of agglomerates. The influence of electrostatic forces becomes significant if, rather than the cohesion, the attraction at greater distances between particle surfaces *a* is considered. For distances greater than 1 μm, or *a/d* > 0.2, only electrostatic forces are still effective. They affect accumulation processes before cohesion can take effect [24].

6.2.2.8 Summary of the Strength of Agglomerates

The various contributions to the strength of agglomerates are depicted in Figure 6.12. In this diagram, taken from Rumpf, reference [20], the maximum transmissible tensile strains σ_{max} are plotted against particle size. However, only the figures for the van der Waals and capillary forces can be regarded as certain. The values for the tensile strengths of the other types of cohesion can merely be estimated within an order of magnitude.

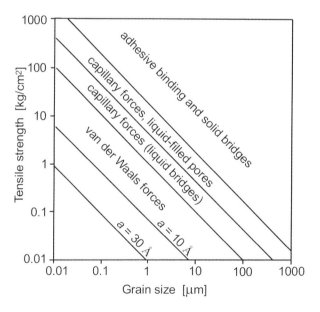

Figure 6.12. Contributions to agglomerate strength as a function of particle size for the various types of binding (from Rumpf, ref. [20]).

The following conclusions concerning the possible binding forces at work in agglomerates can be drawn from this diagram and the previous discussions. The strength of agglomerates increases on going from cohesion of submicroscopic particles through capillary forces in the presence of liquids and finally to the various possibilities of binding by solid bridges. In the range in which solid bridges occur, the strength of the agglomerate depends less on the particle size than it does in the range of van der Waals and capillary forces.

6.2.3 Testing Agglomerate Strength

Mechanical strength is one of the most important parameters in almost every use of agglomerates. Since the stress conditions encountered in use can be of many very different types, the test methods are designed to simulate their variety.

A series of precise methods and practical tests exists for the measurement of agglomerate strength. The measuring methods principally applied in the laboratory permit direct measurement of the mechanical strength. A prerequisite is a knowledge of the fracture process from which the tensile strength can be calculated. We only mention the existence of these methods here; for more information see, for example, Rumpf [25] and Schubert [26].

The tests comprise methods based on compression, impact, abrasion, dissolution, and so on. One should not seek to understand the results on the basis of theoretical principles, but in practice they are common because of the simplicity of their execution, and they allow conclusions to be drawn about the handling and the processing of agglomerates as well as about their suitability for different applications.

On transport, packaging, and storage, agglomerates must be able to withstand pressure and falls well and possess adequate resistance to abrasion. There are therefore three particular process-monitoring methods for the determination of agglomerate strength which simulate these types of wear in the laboratory: the drop test, the pressure-resistance test, and the abrasion test (Figure 6.13).

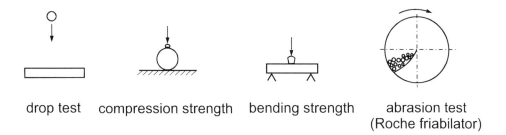

drop test   compression strength   bending strength   abrasion test
(Roche friabilator)

Figure 6.13. Technical tests for agglomerate strength.

To determine their pressure resistance, the granules are crushed between two parallel plates in a pressure test machine or in a simple process measuring apparatus. The strength is characterized in terms of the force required to initiate the fracture. In the drop test, the agglomerates fall from a known height onto a steel plate and the proportion of broken agglomerates is counted, or sieving is used to measure the proportion of fine powder created by the breakage.

Agglomerates do not often need a high resistance to pressure and to drops, but they do require adequate resistance to abrasion. A knowledge of abrasion properties is particularly important in the case of agglomerates that will pass through dosing or packaging machines. Abrasion resistance is measured by placing a sample free of dust and fragments into a cylindrical drum with an exterior wall of mesh of a known size. The sample is tumbled at a constant revolution rate and the amount of powder produced is measured after a set period. To increase the wear, plastic spheres are often added, for example of polyethylene with a diameter of 1–2 cm. The use of friability as a measure of the strength of agglomerates has been discussed by Hunter [27]. The Roche Friabilator is a common, commercially available apparatus for this test.

6.2.4 Binders and Auxiliaries

Various different additives are used for agglomeration, including binders, wetting agents, lubricants, and disintegrating agents. Binders cause a considerable increase in the agglomerate strength. Lubricants are especially useful in agglomeration by molding, where they lower friction between particles and thus improve contact between them. Disintegrating agents are used when the agglomerate must crumble rapidly when mixed with the medium for application, that is, in instant products. Hundreds of compounds are used and have been patented as binders. They can be divided into groups as follows:

– according to chemical type (organic or inorganic)
– according to physical state (liquid, semisolid, solid)
– according to mode of operation (matrix, film, chemical reaction).

In the case of matrix binders, the particles are embedded in a matrix of the binder. Film binders are used as solutions or dispersions which form films on drying. The last type of binders achieves its effect by a chemical reaction between the binder components or between the binder and the agglomerate.

Depending on the type of binder, the agglomerate strength can be seriously affected by the nature of the binder addition, that is, whether it is added dry to the powder mixture or dissolved or swelled in a liquid. Use of a binder solution always results in a so-called "agglutinate", in contrast to the crust agglomerates manufactured from liquids containing no binders. Some typical binders are listed below:

– Starch (principally maize and wheat starch), as a 10–25% paste or an 8% mixture for spray agglomeration. Special types: Prejel (heat-treated potato starch) as a 10–15% aqueous colloidal solution; STA-Rx (mechanically treated maize starch with a higher than usual content of cold-water-soluble constituents) as a 10–20% paste.
– Gelatin, type B, in 2–20% aqueous solution.
– Hydroxypropylmethylcellulose (HPMC): Methocel E and K, as a 1–5 % solution; suitable for binding hydrophobic substances.

– Methylcellulose: Methocel MC, Tylose MH and MB, as a 1–5 % aqueous solution; suitable for agglomeration of hydrophobic substances.
– Sodium carboxymethylcellulose (Na-CMC): Tylose C and CB, Hercules CMC, as a 1–6 % aqueous solution.
– Poly(ethylene glycol) (PEG): Carbowax 4000 and 6000, 2–15% as a dry binder, may also be used in organic solvents.

Further binders are:
– dextrin, dextrose, glucose
– gum arabic
– lignin sulfonates (sulfite waste)
– alginates
– poly(vinyl acetate)
– poly(vinyl alcohol).

Selected binders and lubricants are listed in reference [28].

The starches also possess the properties of disintegrating agents, especially starches soluble in cold water like Prejel and STA-Rx. Sodium carboxymethylcellulose with small substituent groups (varieties of Nymcel) and microcrystalline cellulose (Avicel) are suitable disintegrating agents. The most effective disintegrating agent is a mixture of sodium hydrogencarbonate and citric or tartaric acid; on contact with water, these two components react to release CO_2, and the granules fall apart.

6.2.5 Basic Methods of Agglomeration

A great variety of procedures and equipment is used for the manufacture of agglomerates from powders, suspensions, solutions, and melts. Ries [29] attempted to classify the procedures. In this work, no conceptual distinction was made between agglomerates and granules. The classification covers:

– rolling granulation
– granulating dryers
– solidification and melting granulation
– grinding granulation
– compaction and briquetting presses
– aperture presses
– drop-formation procedures
– vacuum densification and agglomeration procedures
– crystallization
– granule production in the liquid phase.

We shall concentrate on those procedures which permit the manufacture of agglomerates with good solubility, that is, build-up and drying agglomeration processes.

For information about agglomeration by molding, the reader should consult, for example, reference [30].

6.2.5.1 Agglomeration by Build-Up (Accumulation)

Among build-up agglomeration procedures, rolling agglomeration and mixing agglomeration are of the greatest interest to us.

6.2.5.1.1 Rolling Agglomeration

Dry or moistened powder is built up into agglomerates by being sprayed with water or another liquid while subjected to a rolling motion in a drum or a pan. Liquid bridges form between the particles of starting material, and gradually agglomeration centers develop. The rolling motion of these primary agglomerate particles and the accumulation of fresh material and atomized binder yield spherical grains. For larger apparatus, the greater distances through which the granules fall create denser and thus stronger agglomerates. Starting materials with a large proportion of fine particles produce stronger agglomerates, so the starting material should be finely ground. The moisture content also has a considerable effect on the agglomerate strength: too low and the shaping of the agglomerate is hindered, too high and the agglomerate is soft, or its formation is hindered. Binders can be used to produce a marked improvement in the strength of the agglomerate. Competition between the forces of cohesion and disintegration favors the production of evenly structured and strong agglomerates. The duration of stay in the apparatus has a strong influence on the size and the strength of the agglomerates, since the continually changing forces acting on the agglomerates as they form strengthen the bonds until all the particles have settled in the optimum position and are stabilized in a potential well.

In an *agglomerating drum*, the material is added at one end and works its way down through the inclined, rotating drum while being formed into spheres by the rolling motion. Since the agglomeration always produces granules of different sizes, some sort of classification is necessary at the end of the process. A sieve attached to the bottom of the drum is sufficient. This broad distribution of grain sizes is a disadvantage of drum agglomeration. It is difficult to produce a fine agglomerate in this fashion; the material must pass through the drum very quickly if this aim is to be attained. In the Wiklund process [31], the classification is actually performed in the agglomeration area: pellets of the desired size are removed, while smaller ones remain there. This separation is achieved by means of a conical drum. If two drums are used in series, the quantities produced can be increased. A schematic representation of drum agglomeration is shown in Figure 6.14.

An *agglomerating pan* consists of a rotating pan which can be inclined at a variable angle to the horizontal, into which the starting material is poured at a particular place. A powder bed forms, over which a liquid is sprayed through an atomizer. The agglomerate granules that form describe reniform trajectories. As they grow in size, the pellets travel from the base of the pan to the surface, so that the larger ones roll over the top of the bed and off the edge of the pan, while the smaller ones are carried higher and repeatedly brought back to the point at which the liquid is added, and thus grow larger. The size of the so-called "green pellets" that drop over the edge is astonishingly consistent. Figure 6.15 depicts the equipment and the course of motion in pan agglomeration.

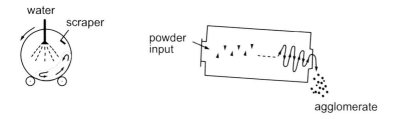

Figure 6.14. Agglomerating drum.

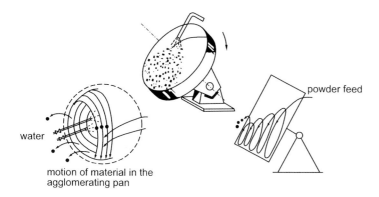

Figure 6.15. Agglomerating pan.

The great advantage of the pan over the drum is its classificatory activity. Another advantage is the transparency of the progress of the agglomeration. The effect of each factor, whether pan inclination, rotation rate, or type, quantity, and distribution of the added liquid, is immediately visible and can be regulated without delay. The quality of the green pellets and the performance of the pan are simply represented by the following parameters:

– position of raw material intake
– position of liquid nozzle
– pan inclination and rotation rate
– position of scraper.

Agglomerating pans have diameters of up to 8 m and can process over 100 t/h. The green pellets obtained from pipe or pan agglomeration must subsequently be dried in a suitable manner.

6.2.5.1.2 Mixing Agglomeration

Most solid–liquid mixers are capable of forming agglomerates; indeed, they often do so when it is not intended. In principle, nearly every powder mixer is also suitable for use in agglomeration, as long as provision is made for adequate particle cohesion. Capes [32] divided agglomeration mixers into two classes defined by the size, density, and degree of wetting of the agglomerates they produce. The first class contains horizontal pan or trough mixers and plowshare mixers, which furnish dense agglomerates in the capillary state, like those produced by rolling agglomeration. These mixers are equipped with internal agitators that create frictional and shearing effects. This kneading results in harder and stronger agglomerates than rolling agglomeration, and can be used on tough and plastic materials. Less liquid is needed than in rolling agglomeration. A disadvantage is the irregular form of the agglomerate granules, which require subsequent shaping.

In the second class of techniques for mixing agglomeration, the powders are moistened less than would be necessary for the capillary state. Loose clusters of powder particles accumulate. An example is the manufacture of instant food products by moistening, drying, and cooling.

There are specific apparatus for both of these methods. Equipment for instantizing agglomerates (rendering them suitable for instant use) is discussed in Section 6.3. The horizontal pan mixer (Figure 6.16) has been in long use to agglomerate artificial fertilizers. The rotation of the pan and the eccentric stirrers, which turn in the opposite direction, create a constant mixing effect that forms compact agglomerates when liquid is added. If the agglomerate granules must be spherical, a subsequent forming step in a pan agglomerator is necessary. A typical apparatus is the Eirich countercurrent mixing granulator.

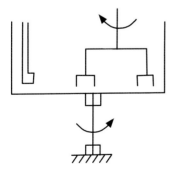

Figure 6.16. Mixing agglomerator with planetary gearing.

Agglomerating mixers are most useful when several components must be simultaneously mixed and agglomerated. Worm mixers are continuous or batch mixers with one or two spindles bearing paddles or plowshare-type mixing elements in a trough. Machines in which the spindles have a high rate of rotation, such as the Lödige mixer, have a more intensive action.

A more recent development in the field of fast mixing agglomerators is the Schugi mixer (Figure 6.17).

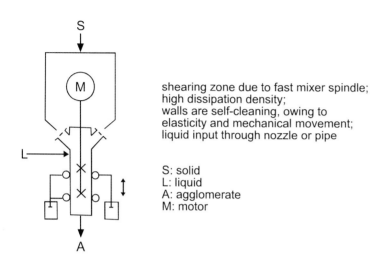

shearing zone due to fast mixer spindle;
high dissipation density;
walls are self-cleaning, owing to
elasticity and mechanical movement;
liquid input through nozzle or pipe

S: solid
L: liquid
A: agglomerate
M: motor

Figure 6.17. Schugi Flexomix [33].

The continuous mixing occurs in a vertical mixing chamber with an axial, fast-turning mixing spindle. Several independent shearing blades are attached to the spindle. The material flows through the mixing chamber from the top to the bottom; in the range of the shearing blades it enters an extremely turbulent zone and is simultaneously sprayed with liquid. The mixture of powder and liquid travels downwards on a spiral path. The product spends about a second in the mixing chamber. The wall does not become encrusted because of the particular design of the apparatus. This spray agglomeration technique yields agglomerates with grain sizes of between 0.2 and 2 mm. This grain size can be influenced by, in particular, the rotation rate of the mixing spindle, the angle of the shearing blades, and the amount of liquid added. The procedure is particularly recommended for the manufacture of detergent powders for laundry.

We should also mention here the spray mixers, in which the material is thrown upwards to form a haze inside a drum equipped with agitators and baffles, and sprayed with liquid through an axially positioned atomizer. Less than 5% of the bridge-forming liquid is sprayed onto the powder. The process produces agglomerates with grain sizes of 2 mm and under, with improved properties with respect to flow, wetting, and solubility.

This apparatus too is used most frequently in the detergent industry. The manufacture of small, loose agglomerates of the type known as clusters by spraying a cloud of powder is also important in the food industry (instant products). Some examples of apparatus used in spray-mixing agglomeration can be found in references [34, 35].

6.2.5.2 Agglomeration by Drying

In agglomerates produced by drying of moist loose materials or suspensions, solid bridges play a particularly important role as binding mechanisms, alongside van der Waals forces and capillary forces. Even a small quantity of solid impurities, at the level usually present in water, can crystallize out as solid bridges between the particles. The material to be agglomerated can either be in undivided form as powder or briquette, or in a solution or suspension, in which case it must be scattered into drops to be used. Roll dryers and freeze dryers can operate with an undivided liquid or solid phase; in contrast, spray drying and prilling require that the material (solution, suspension, melt) is divided into drops. In the case of fluidized-bed agglomeration both possibilities exist, depending on whether the process is continuous or batch.

Roll Dryer

In a roll dryer, the (usually highly concentrated) suspension is spread in a thin layer over a horizontal roll, dried, and scraped from the roll. This process yields an agglomerate in the form of scales or flocs. Reference [8] contains more information on roll drying.

6.2.6 Freeze Drying

Freeze drying is a good method of producing grainy end-products with particularly good solubility from solutions and suspensions. It is also suited to dehydration of foods. The process is also known as lyophilization because of the good solubility properties of the freeze-dried products.

Since no, or only very small, solid bridges can form from salts crystallizing out during freeze drying, and since no mobile water is present in capillaries during the process, the agglomerates formed are irregular and weak. However, as a mild drying procedure, freeze drying is ideal for retaining the flavors of foods or with temperature-sensitive substances. This includes the conversion of aqueous colloidal dispersions into finely dispersed powders while retaining the degree of dispersion. However, this is only possible if sufficient dispersion medium is present. Freeze drying is relatively expensive in comparison to, say, spray drying. Its chief industrial use is in the manufacture of instant products in the food industry and in pharmaceutical preparations. It is also occasionally used in the laboratory to dry liquid dispersions of dyes and other products gently, and to convert them into finely dispersed agglomerates that are easy to reconstitute.

The technique of freeze drying is based on the principle of drying by sublimation, that is, the direct conversion of ice to vapor without passing through the liquid phase. The process is carried out in a vacuum.

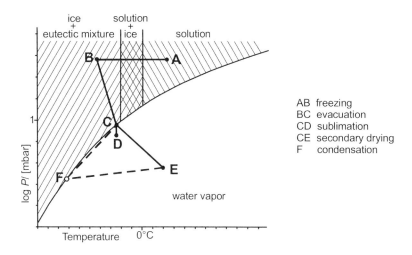

Figure 6.18. Phases of freeze drying on the phase diagram for water [36].

As can be seen from the phase diagram for water (Figure 6.18), there is a vapor pressure over ice as well as over liquid, and this pressure depends on the temperature. If the pressure of the water vapor is less than the saturated vapor pressure at temperatures under 0°C, then ice sublimes into water vapor. This vapor pressure difference is the driving force behind freeze drying, and is maintained by the removal of water vapor with the aid of vacuum pumps or by condensation on cold surfaces. One gram of water yields about 1000 L of vapor at 10^{-3} mbar and 20°C.

The freeze-drying process has three steps, which are marked on Figure 6.18:

1. Freezing at normal pressure and low temperature (A→B). Since the product is not pure water but a solution or suspension containing solutes such as dispersants, structure formers, salts, etc., it freezes at the eutectic point. This can only be determined experimentally for the phase diagram of a multicomponent system; the simplest method makes use of the electrical conductivity. The eutectic point must not be exceeded during freeze drying, so that the frozen material does not melt during the primary drying process. This way the pores originally filled with ice remain, and the dried product has a porous structure with a large surface area. This means that when water is added, the original solution or dispersion is rapidly reconstituted.

2. Primary drying by sublimation of the ice in a vacuum (B→C→D). This can only occur below the triple point. Since sublimation requires energy for the latent heat of vaporization (~2900 kJ/kg ice), heat must be added to the system to prevent further

cooling and thus retardation of the drying process. For this purpose, electrical elements or brine are used. Since the higher the vacuum, the poorer the heat transport, the vacuum is set to a value equivalent to half the saturated vapor pressure of the ice in the product at the sublimation temperature. Under these conditions the sublimation rate is maximized. The sublimed solvent condenses out as ice on the condenser, which must be kept at a lower temperature than the product throughout the sublimation.

3. Secondary drying (C→E). As soon as all the ice has been driven off, the temperature of the product rises sharply. The product still contains adsorbed water or water of hydration, and in order to remove this the vacuum is increased as much as possible and the temperature of the heating elements raised to the maximum acceptable for the particular product.

The temperature, and thus the speed at which the starting material was frozen, are of crucial importance to the properties of the dried product. Fast freezing at low temperatures creates a large number of crystal nuclei and thus an ice phase of fine crystals. Since the individual crystals are relatively small, the structure of the starting material is retained with little damage. Such a freeze-dried product has good solubility properties. In contrast, slow freezing at temperatures only slightly under the freezing point creates few nuclei, which then grow into large crystals; this can destroy the structure of the material, but, on the other hand, a coarse porous structure favors quick drying. A fine crystalline ice phase can also be formed by slow cooling in the vicinity of the freezing point. A laboratory freeze dryer is shown in Figure 6.19.

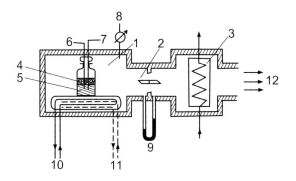

1 freeze-drying chamber
2 valve
3 condenser
4 dried material
5 frozen material
6 thermocouple
7 conductivity probe
8 manometer
9 differential pressure manometer
10 brine coolant
11 heating element
12 vacuum connection

Figure 6.19. Diagram of a laboratory freeze dryer [36].

Production plants for freeze drying, such as those constructed for the manufacture of lyophilized coffee, consist of large vacuum chambers in which the material to be dried is dehydrated on sheets in steps or semicontinuously. Inventive designs for the containers are the subject of many patents. General Foods uses containers that can actually be filled in the vacuum chamber itself and then emptied after the drying process, which makes the charging and emptying considerably easier. These product containers are fixed in position between the heating elements.

Because of the high capital and energy costs entailed, freeze drying is an expensive process. The cost of drying is 2–3 times higher than that for the same performance in spray drying. It is therefore well worth concentrating the solutions or suspensions before drying. The method chosen must be mild, in order not to lose the advantage of freeze drying, namely, the retention of flavor compounds and other temperature-sensitive substances. A suitable industrial process is freeze concentration, developed by the Dutch company Grenco N.V. [37]. Further details can be found in references [38, 39].

6.2.7 Spray Drying

The process of spray drying consists of spraying a solution, suspension, emulsion, or paste outwards and drying the droplets thus formed with a stream of hot gas, so that an agglomerate or powder is obtained. Depending on the construction of the spray dryer and the adjustment of the working parameters, round particles of up to 0.5–1 mm in diameter may be produced. The process is a gentle drying and agglomeration method particularly suitable for heat-sensitive substances such as foods and pharmaceuticals. It has the additional advantage of requiring only a single stage. The expression "Zerstäubungstrocknung", comminution drying, often used in the German literature, is incorrect; comminution, or reduction to dust, is the fine division of solids, as compared with the atomization by spraying of a solution or suspension. Only when these liquid droplets have been converted to dry particles may one refer to comminution.

6.2.7.1 The Principle of Spray Drying

The material to be dried, whether solution, suspension, emulsion, or paste, is turned into a fine mist by means of jets or rotating disks and then subjected to a stream of hot air or gases. The principle is illustrated in Figure 6.20 (spray dryer, jet dryer).

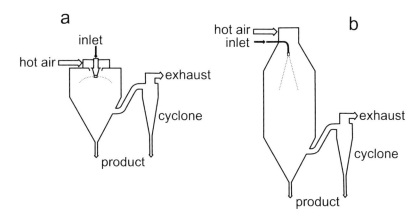

Figure 6.20. Principle of the spray dryer (a) and the jet dryer (b).

The determining factor in spray drying is the extreme fineness of the atomized liquid. The amount of heat transferred increases in proportion to the surface area of the droplets, so the heat transfer can be sharply increased and the drying times drastically shortened. In addition, the large curvature of fine droplets increases their internal pressure, creating states that would never be found in still, planar liquid surfaces. When a drop of 1 mm radius is broken into droplets of 1 μm radius, the surface area increases by a factor of 10^6, while the amount of solvent to be evaporated remains the same. Thus, the evaporation from these tiny droplets occurs in fractions of a second. The hot air is piped into the tower through special pipes.

For optimal spray drying, the spray mist and the hot air must be mixed immediately in the vicinity of the spray nozzle, and the particles must be homogeneously distributed in the drying air until drying is complete. In order to obtain a long drying path, the tower must be very large or else the hot air must be given a rotational movement through the tower. The gas and the product may move in the same direction, as countercurrents, or as a mixed flow. In the case of mixed flow or flow in the same direction, the gas and the dried product exit the tower together at the bottom, whereas in the countercurrent procedure the product is collected at the bottom of the tower while the drying gas escapes at the top. The product and the gas are separated in a cyclone or trap.

Some spray dryers are designed so that the fine particles collected in the cyclone are returned to the drying tower, where they come back into contact with the mist and collide with the droplets of spray, forming larger agglomerates. This produces a less dusty, better flowing, and more wettable or soluble product.

6.2.7.2 Apparatus

Spray apparatus: The solution or suspension can be sprayed either through nozzles or through rotating disks.

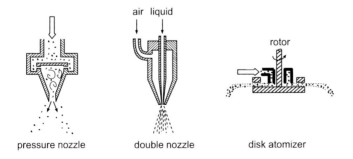

Figure 6.21. Spray equipment.

The nozzles may be single nozzles (pressure nozzles, for one substance) and double, two-fluid nozzles (pneumatic nozzles); the disks may be flat plates, nozzle wheels, bucket wheels, or some other variety. The pressure nozzle relies on the cyclonic motion which the material to be sprayed is forced to adopt. A conical mist results, usually with a hollow center. In the double nozzle, the liquid is atomized with air or steam. Generally the nozzle for the liquid is surrounded by an annular nozzle for air; this arrangement results in severe turbulence in the gas phase and thus prevents coalescence of the droplets formed. The three types of atomizer are illustrated diagrammatically in Figure 6.21.

The double nozzle may also be used to spray pastes. The material to be sprayed is extruded as a hollow thread and, on exiting the nozzle, is atomized by the compressed air.

The rotating disks use centrifugal force to atomize the liquid. When it lands on the plate, the edge of which is traveling at very high speed relative to the surrounding air, it is flung off in the form of drops, which are then torn apart by the high air resistance they meet. The rate of rotation of the disks is very high (10 000–20 000 rpm); their peripheral velocity is about 150 m/s. The size of the drops depends principally on the rate of rotation of the disk, as described in Equation 6.15 [40].

$$d_{\mathrm{m}} = 98.5 \cdot \sqrt{\frac{\sigma}{r \cdot \rho_{\mathrm{Fl}}}} \cdot \frac{1}{n} \tag{6.15}$$

(n: rate of rotation [rpm]; d_{m}: mean drop size [m]; σ: surface tension [kp/m]; r: radius of the disk [m]; ρ_{Fl}: density of the fluid [kg/m^3])

The following criteria are important in selecting the spray equipment: the size of the droplets produced, the range of droplet size, the likelihood of the nozzle becoming blocked, material wear and tear, throughput, energy consumption, and tower construction.

The pressure nozzle produces a narrow spectrum of grain size but relatively large drops, whereas the double nozzle gives very fine droplets with a broad size distribution. Therefore, if the desired product should be low in dust content, a pressure nozzle should be selected. Wear and tear is greater in pressure nozzles than in double nozzles, and lowest of all for rotating disks; most liquids that can be sprayed through nozzles can also be sprayed off rotating disks. In practice, the use of the double nozzle is restricted by its high energy requirements and low capacity to the laboratory and other special purposes. The trajectories of the droplets in the spray produced are wider for spray dryers with rotating disks than for those with nozzles, so the former require towers of greater diameter, otherwise the droplets will collide with the walls and encrust them. The disk apparatus is more mobile and can handle different products, and is also suitable for spraying thick and viscous slurries. At low speeds, disk spray dryers furnish a more consistent product. As their speed increases, the droplet size variation also grows. Reference [41] compares the various types of spraying equipment.

Hot air intake: This consists of the heater, the fan, and the air distributor. The heater is an oil- or gas-fired combustion chamber. The hot air produced is led into the drying tower through pipes; for many products that are not heat-sensitive, direct heating by gas or oil is also possible. Once the liquid droplets have dried into particles, we may refer to comminution drying.

The relative motion of the gas and the product spray may be as co-currents, countercurrents, or mixed flow. The flow patterns in disk and nozzle spray dryers are depicted schematically in Figure 6.22.

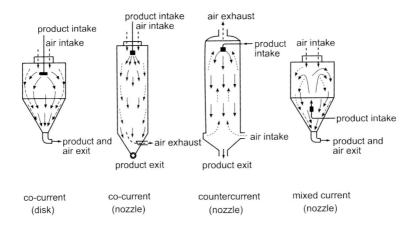

Figure 6.22. Flow patterns of product and gas in disk and nozzle spray dryers, according to Masters [42].

In a co-current dryer, the liquid is sprayed into gas at its highest temperature. The particles are swept along by the gas until they are completely dried. This means that the temperature profiles of disk and nozzle spray dryers are very different; see Figure 6.23.

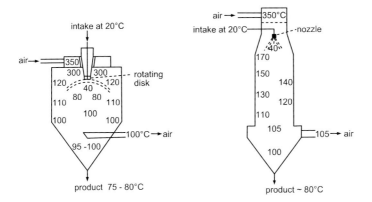

Figure 6.23. Temperature distribution in disk and nozzle spray dryers in which gas and product move together (according to Masters [42]).

In the disk spray dryer, in which the air current is a turbulent vortex, temperatures are fairly similar and low, in fact, close to the exit temperature, throughout the entire tower.

The material to be dried never gets warmer than the exhaust air. On exit, the temperature of the product is about 20–25°C lower than that of the exhaust air [42].

In contrast, in the nozzle spray dryer, which has a smaller diameter and a greater height than the disk spray dryer, the stream of gas adopts a linear flow with a greater temperature gradient down the tower. Particles that are still somewhat wet are thus exposed to higher temperatures for longer than in a disk spray dryer. The mildest drying and agglomerating procedure for heat-sensitive products is therefore in a disk spray-drying tower in which the air and product travel together.

In countercurrent drying, the gas and the product move in opposite directions (Figure 6.22). Immediately after spraying, the wet particles are exposed to a temperature lower than the exit temperature, while the dry particles experience the highest temperature, where the hot gas enters. In a mixed-flow dryer, both types of current are present.

6.2.7.3 Particle Shape and Size

Since agglomeration is generally aimed at producing and drying fairly large particles, high downward velocities must be reckoned with and therefore tall towers must be built. To summarize, the agglomerate grain size can be controlled as follows:

Most importantly, it depends on the spraying mechanism, as has already been described. Rotating disks usually produce a fine-grained product of 20–100 μm diameter, the pressure nozzle and the double nozzle a coarser agglomerate of about 50–250 μm diameter. As the rotation rate of the disk slows, the grain size becomes larger. For many products, an increase in the drying temperature also increases the size of the grains, as does a higher concentration of starting material. On the other hand, lowering the surface tension in the product to be sprayed decreases the grain size. The viscosity of the liquid product apparently has little or no influence on the particle size. For details on the effect of variation of the more important parameters on the size of the agglomerate grains, see reference [43].

According to Masters [42], the design of the drying tower and the spray equipment can be used to control the particle size. For agglomerates in the 50–120 μm range a standard dryer with a rotating disk is recommended; the ratio of the diameter of the tower to its height should be 1:0.6–1. The larger the chamber diameter, the lower the rotation rate of the disk can be and the larger the agglomerate granules formed without deposition of incompletely dried particles on the walls. For agglomerate sizes of more than about 120 μm nozzle atomizers are used, for which the ratio of tower diameter to height is about 1:1–1.3. Only nozzles can be used in narrow spray towers; in these, the diameter/height ratio is about 1:3–4. Here the largest agglomerates can be produced, with diameters of up to 250 μm. Spray-dried particles are usually spherical and hollow. The final shape of the agglomerate depends on the temperature of the air used for drying and on the material properties of the product. The combination of mechanical solidification and the forces that occur in the process produces many different shapes. A minor alteration of the formulation can be enough to change the shape of the particles.

Depending on whether the temperature of the drying air is above or below the boiling point of the solvent, hollow spheres with or without crystalline efflorescences form, or merely fragments of spheres, expanded hollow spheres, or spongy spheres. Charlesworthy and Marshall have investigated particle formation in spray agglomeration [44].

Factors which increase the rate of evaporation if, for example, the temperature of the hot air is above the boiling point of the solvent, or if a porous crust is formed, lead to larger, hollow particles. The formation of these hollow spheres can be pictured as follows: as a result of the high temperature of the gas surrounding the droplet, the solvent at the droplet surface evaporates and fresh solvent from the interior replaces it, creating a current from the center to the exterior of the droplet. The liquid current carries the dissolved or dispersed solids with it to the surface, where they are deposited. If the dissolved or suspended substance resists the flow, then porous, spongy agglomerates form. The hollow shape of the spray-dried agglomerates is responsible for their low bulk weight and also for their ready solubility, that is, for the suitability of the hollow granules for use as an instant product. The amount of moisture remaining in spray-dried agglomerates depends principally on the liquid throughput, that is, the amount of liquid added per unit time, and also on the temperature of the hot air and the air distribution in the drying tower. Herbener gives more details about low-dust powders [45].

6.2.7.4 Safety

Many powders can explode when mixed with air. There are special spray dryers for such powders, in which the drying is carried out in an inert atmosphere and the evaporated liquids, water and solvent, are separated in a condenser. These plants can have a closed ("self-inertizing") or semiclosed cycle. The closed cycle employs nitrogen as inert gas and is used for formulations containing organic solvents. In the spray drying of aqueous formulations which are only slightly pyrophoric or at low risk of explosion, it is sufficient to design explosion dampers and fire extinguishers into the plant. Another aspect of safety that is considered nowadays is the avoidance of particulate emissions into the atmosphere. Dust traps must be very effective to achieve this aim. Bag filters or cyclones are employed; cyclones can usually remove up to 98% of the dust, filter arrangements up to 99.9%. In order to remove the last remainder of the dust, a wet scrubber is fitted after the cyclone or filter.

6.2.7.5 Economic Efficiency

The amount of water evaporated per hour increases more than linearly with respect to the rise in temperature. For example, in a spray dryer of known size in which the drying gas entered at 200°C and left at 100°C, 300 kg water was evaporated per hour, while an entrance temperature of 400°C raised this to 800 kg/h.

The thermal efficiency of a spray dryer is given by Equation 6.16:

$$\eta = \frac{T_1 - T_2}{T_1 - T_a} \qquad\qquad (6.16)$$

(T_1: entrance temperature; T_2: exit temperature; T_a: ambient temperature)

A large difference between entrance and exit temperatures is therefore necessary for a high degree of thermal efficiency. An increase in thermal efficiency means that a higher proportion of the hot air is used to evaporate the liquid. This results in fuel savings and lowers the amount of air necessary to dry the product, so the size of the spray-drying tower can be reduced.

The concentration of the product to be dried also effects the economic efficiency significantly. The cost of drying per kg product can, depending on the capacity of the dryer, be reduced by more than half by an increase in concentration of the starting material from 20% to 40% solids. For this reason, attempts are often made to increase the concentration in the initial dissolved or dispersed form of the product, for example by reverse osmosis. The cost of evaporating 100 kg water in a small plant (evaporated mass of water less than 200 kg/h) with a mean temperature difference of 260°C has been estimated by Kaspar and Rosch [46] as about €30, in a large plant (more than 500 kg/h) as about €10.

Finally, we will mention a few interesting possibilities offered by spray granulation. If a suitable coating material is dissolved in a suspension of an insoluble product, the spray-dried agglomerate will receive a surface layer of this coating material. The method is suitable for the coating of pharmaceuticals with a binder, which aids the subsequent formation of tablets, or for the coating of pigments with tensides to improve their dispersion properties. It is also possible to inject a fine powder such as SiO_2, talc, or starch into the tower, where it will coat the droplets. The flow properties of agglomerates can thus be improved with suitable additives without the need for a subsequent mixing operation. Extremely adhesive materials can often be shaped into spherical particles only by means of spray agglomeration. Last but not least, spray agglomeration is an alternative means of mixing several components into a homogeneous powder, rather than dry mixing. A more detailed examination of the topic of spray drying is to be found in Masters' monograph [47].

6.2.8 Fluidized-Bed Agglomeration

Fluidized-bed agglomeration, like spray agglomeration, permits the conversion of a solution, suspension, paste, or melt into an agglomerated powder by drying in a single step. The particle accumulation occurs in a fluidized bed. Fluidized-bed agglomeration is a combination of the two techniques of fluidized-bed drying and spray drying into a new method, in which the drying is accompanied by enlargement of the particles. Larger agglomerates can be produced by this process than by spray drying, since in a fluidized bed multiple layers of droplets of the dissolved or dispersed solids can gather around the particles in the fluidized bed. The resulting product contains less dust and has a greater bulk weight than spray-dried agglomerates. As is the case for spray drying, there is a large interface between the solid or disperse phase and the gas phase, and intensive mixing, so heat transfer is optimized.

Fluidized-bed agglomeration was originally developed for batch processes and used principally for the agglomeration and coating of pharmaceutical active substances. Modern, continuous fluidized-bed agglomeration processes permit economically viable agglomeration even of relatively cheap products such as leather-tanning substances and dyes because of improved plant performance.

6.2.8.1 The Principle of Fluidized-Bed Agglomeration

A fine-grained, dry powder is suspended in a stream of hot air, forming a vortex or fluidized bed, and brought into contact with a spray of liquid droplets generated through a nozzle. The evaporation of the solvent results in the linkage by coalescence and accumulation of the particles dissolved or suspended in the liquid to those in the vortex to furnish larger, mixed particles. The hot air is supplied until the solvent has evaporated to leave the desired level of residual moisture. The particles can only be built up if the liquid rains onto the particles in the vortex in the form of a very fine spray of droplets, so the droplets must not be allowed to dry out beforehand; the atomizer must deliver them as close to the fluidized bed as possible, or even directly into it. So that solid agglomerates can form, the solid particles of the fluidized bed must be soluble in the liquid of the spray and able to form crust agglomerates, or else the liquid must contain a binder that will enable the formation of adhesive agglomerates together with the solid particles. The following possibilities exist for the production of agglomerates by the fluidized-bed agglomeration process:

– charging the fluidized-bed dryer with solid active substance and, if necessary, solid auxiliaries, and spraying with an inert agglomeration liquid (binder solution containing no active substance)
– charging with solid active substances and, if necessary, solid auxiliaries, and spraying with a liquid (solution or suspension) also containing active substances
– charging with inert solid and spraying with a solution of active substance.

The dwell time in the fluidized bed is on the order of 60 min. The heat sensitivity of the active substance is therefore very important. The temperature of the air used to create the vortex is seldom greater than 90°C, depending on the type of product to be agglomerated.

6.2.8.2 Properties of the Fluidized Bed

A hollow cylinder with a perforated plate or mesh screen in its floor is a basic fluidized-bed apparatus. When a stream of gas is piped upwards through a loose layer of fine-grained solid, the speed of the gas determines the behavior of the solid particles (Figure 6.24).

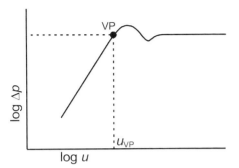

Figure 6.24. Dependence of the pressure loss Δp in the solid layer on the speed of the gas u.

The solid particles remain at rest when the gas velocity is low; the gas simply streams through the voids in the layer of loose material. As the speed u rises, the pressure loss Δp in the fluidized bed increases quadratically until it attains the vortex point VP. A logarithmic plot like that in Figure 6.24 shows a linear rise. As soon as the pressure loss of the streaming gas in the loose layer reaches the bulk weight per unit area of the solid, the solid bed begins to expand. At this point, the transition from solid bed to vortex or fluidized bed begins. After the transition has been completed, the pressure decline remains nearly constant if the gas velocity is increased further. When the minimum vortex velocity u_{VP} is reached, the particles in the layer begin to move, vibrating and colliding. This state, in which the particles are in lively motion and are mixing thoroughly, is known as the fluidized or the vortex state. The fluidized bed may now be homogeneous, bubbling, or slugging. If the stream velocity is increased still further, the entrainment point is reached, at which the particles are carried upwards and out of the bed by the gas, producing a dust cloud. Fluidized-bed technology is therefore confined to gas velocities between the minimum and the entrainment speed. The theoretical plot of the pressure drop (Figure 6.24) is only valid for a single grain size; for products with a range of grain sizes, the graph is somewhat different. Increasing particle sizes increase the viscosity of the gas/solid suspension. A low proportion of fine material in a fluidized bed of coarse particles brings the viscosity down sharply. Thus there must always be sufficient fine material in the bed to ensure good fluidization. When coarse and fine particles are mixed it is possible that the fine particles are entrained from the bed before the coarse ones have become fluidized. This can lead to filter problems, and also to agglomerates with inadequate strength.

6.2.8.3 Particle Build-Up

The first stage of fluidized-bed agglomeration is always the wetting of the particles in the bed by the droplets sprayed from the nozzle. The liquid spreads over their surface, provided that the particles are wettable. Liquid bridges form between the particles when they collide. As long as the wetted agglomerates are not too large, they remain in the fluidized state rather than sinking to the floor, and the liquid bridges solidify as they move through the bed. Whether the particle retains its form depends on the relative magnitudes of the binding and disintegrating forces. The latter arise from collisions. If the binding forces are stronger, the agglomerate granules grow uncontrollably and sink to the floor. Only if the forces are balanced can a controlled agglomeration occur and a stable grain size be attained (for details on particle build-up: reference [48]). The fluidized-bed process is also used for coating and encapsulation of granules. Particles varying in size from 1 µm (microencapsulation) to 10 mm (film coating) can be covered in the fluidized bed. One of the earliest fluidized-bed processes, the Wurster process (Section 6.2.8.4, below), was developed for this purpose.

6.2.8.4 Apparatus

Fluidized-bed agglomeration can be carried out in batches or continuously, depending on the design of the apparatus. As an example of the batch process, we will discuss the plant depicted in Figure 6.25.

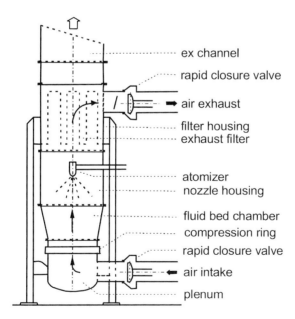

Figure 6.25. Diagram of a fluidized-bed dryer (Glatt system).

Air is passed through an intake filter and heat exchanger from a fan. The air temperature is regulated by adjustable thermostats. This purified and preheated air is sucked through a conical vessel containing the starting material by means of a grille in the base of the vessel; this grille serves as a distributor for the air. The loose material in the vessel is brought to a fluidized state by regulation of the velocity of the air passing through it. In an adjacent expansion zone, the air velocity is reduced so that the particles are no longer held suspended and sink down the walls of the expansion chamber, where they once more enter the gas vortex. This cycle continues for the duration of the process. The atomizer is set above the expansion chamber, and sprays the liquid (solution, suspension) into the fluidized powder at a controlled rate. Any fine powder entrained by the exiting gas is caught by the filters fixed above the nozzle. The dried, agglomerated product remains in the product vessel and can be emptied.

In another form [49], the vessel is cylindrical and the grille at the base is replaced with a variable-speed rotating disk which can be raised to create a gap of adjustable width into the vessel. In the lowest position of the disk, the gap is closed. When the equipment is running, air streams through the chamber in a spiral path. The nozzles are set tangentially to the powder bed. This design mixes liquid and solid more intensively (Figure 6.26).

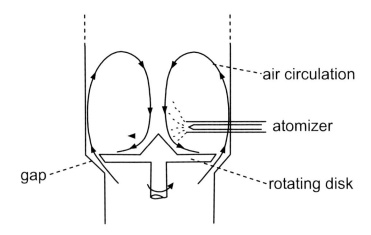

Figure 6.26. Air circulation in the rotor fluidized-bed dryer, from reference [50].

Wurster's apparatus for the manufacture of tablets from granules should be mentioned as the antecedent of fluidized-bed agglomerators. The Wurster process for coating particles was described as long ago as 1949 [51]. In it, the particles to be coated are fluidized by a current of air flowing upwards. In this fluidized bed, air is introduced at high speed from the base, accelerating and separating the particles. This zone of lively movement is limited at its outskirts by a cylindrical insert. Above this, there is a wider expansion zone in which the air velocity is reduced; this means that the particles in the space between the cylindrical liner and the exterior wall of the vessel drop downwards and return to the center in a cyclic motion. At the center of the perforated base is a nozzle through which the liquid containing the coating substance is sprayed into the chamber. The particles pass the nozzle every 6–10 seconds, and are sprayed with new coating each time. In this fashion particles of a few microns in size can be coated into large tablets, regardless of their shape. Figure 6.27 is a diagram of Wurster's apparatus.

Figure 6.27. The Wurster–Glatt apparatus.

6.2.8.5 Continuous Fluidized-Bed Agglomeration

The desire for a more efficient, productive design drove the development of continuous fluidized-bed agglomeration. There are two important variants: agglomeration of powders by addition of an agglomerating liquid through a spray, or agglomeration of slurries by spraying them onto a fluidized bed of powder already present. The apparatus consists in principle of a batch fluidized-bed dryer equipped with continuous extraction of the end product. This extractor also removes dust from the product and acts as a sorter, continuously returning the fine material to the dryer. Dry filler can be added continuously by means of another device. The extraction is adjusted so that the quantity taken is equivalent to the amount of starting material added over time, and thus the quantity of material in the bed remains constant. Unlike in the batch process, conditions in the continuous fluidized-bed dryer can always be optimized and maintained. The stationary state is achieved by adjusting the extraction, the addition of liquid, and the temperature of the air stream. The product solution, suspension, or paste is sprayed evenly over the bed through a nozzle, usually a double nozzle.

Kaspar and Rosch [46] described such a continuous design, which is depicted in Figure 6.28.

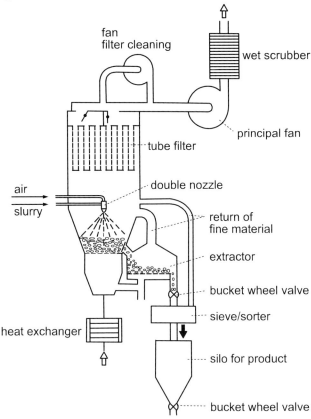

Figure 6.28. Continuous fluidized-bed granulator [46].

We should also mention the new agglomeration process developed by Niro Atomizer, which combines fluidized-bed agglomeration with spray drying. This process furnishes dustless, freeflowing agglomerates with a particle size between 0.5 and 5 mm directly [52].

For this plant to operate smoothly, the conditions in the fluidized bed (quantity of product, grain size distribution, bed temperature) must be kept as constant as possible. The amounts added are regulated as for spray drying, depending on the exit temperature of the air. The quantity of solid in the fluidized bed is maintained using the fact that the pressure loss in the air streaming through the bed is proportional to the weight of the fluidized solid and is unaffected by the quantity of the air. The measured addition of dry filler to the bed depends on the liquid added via the atomizer per unit time. Continuously operating plants for fluidized-bed agglomeration were described back in 1964, for example [53].

6.2.8.6 Practical Points

The first prerequisite for fluidized-bed agglomeration is a turbulent vortex layer. The two most important parameters for fluidized-bed agglomeration are the throughput of the spray and its total volume. The combination of these two factors to give the desired agglomerate size can be discovered empirically for any formulation. In general, agglomerates are larger if more liquid is added and faster. The spray rate is affected by the temperature of the air sent through the fluidized bed during the spray phase; the higher the entrance temperature of the air, the more liquid evaporates and the faster further liquid can be sprayed in. High air temperatures (up to 80°C) mean short drying times and a fine-grained product. The sensitivity of the product to heat must be considered in deciding on an air temperature. To achieve a particular residual moisture level in the agglomerate, the humidity of the air used must be known and controlled.

The pressure of the air in the spray nozzle is also an important influence on the end product. The agglomerate size is proportional to the drop size of the liquid spray, which becomes smaller as the pressure in the nozzle rises. There is a danger that the nozzles will become encrusted and blocked; pneumatic nozzles, in which the liquid exits below the air nozzle, are supposed to be less susceptible to such problems. The nozzles may be mounted above or below the surface of the fluidized bed.

Since the beds can easily become electrically charged, safety must be a prime consideration. The danger of a dust explosion can be reduced by appropriate pressure-resistant construction (rounded forms), the absence of sources of ignition, grounding of all metallic parts, avoidance of flammable solvents, relief apertures in the bed area, and dedicated systems for the suppression of explosions.

The grain size of the agglomerates produced by fluidized-bed drying lies, on average, between 80 μm and 2 mm [46]. In comparison, the figures for spray drying with a pressure nozzle were 50–500 μm. The same investigation found that the costs of fluidized-bed drying were lower than those for spray drying under the same conditions when the performance was less than 400–500 kg/h water evaporated. For higher rates of evaporation, spray drying was more cost-effective.

6.3 Preparation and Properties of Instant Products

Many powder formulations must be dispersed or dissolved in a liquid before use. However, very fine powders can only be dispersed in a liquid with high surface tension, such as water, with considerable difficulty. Treatment to improve the dispersibility or solubility of a powder, "instantization", favors its use as an instant product, that is, improves its "instant properties".

Instant products were first introduced in the food industry; instant coffee is the obvious example. But there is a demand for products with instant properties in other fields too, for example, pharmaceuticals and agrochemicals.

Nobody has yet come up with a precise definition of what we mean by "instant" properties of a product. However, it is normal practice to describe powdered substances and agglomerates as "instant" if they have been altered by a special technical procedure – instantized – such that they dissolve or disperse faster in liquids than did the original product.

Agglomeration is a common process in instantization. In particular, agglomeration encourages the capillary penetration of liquid into the pores of the powder. The powder must be evenly wetted without clumping. Reference [54] provides details of the basis of the processes used in instantization, and of their technical applications.

The most important of these agglomeration procedures are spray drying and freeze drying. These alter only the size and shape of the particles, so the process is a purely physical one. However, the wetting or solution behavior of the particles can also be increased by the addition of suitable auxiliaries or by heat treatment, that is, without agglomeration. This method of producing instant products can be improved further by combination with agglomeration.

6.3.1 Reconstitution of Instant Products

The dissolution or dispersion of instant products is known as reconstitution. It can be divided into stages [55]. If a powder is sprinkled over the surface of a liquid, the following stages can be distinguished:

a) moistening of the porous system
b) immersion of the agglomerate granules in the liquid
c) disintegration of the agglomerates or dispersion of the particles
d) dissolution of the particles in the liquid, in the case of a soluble substance.

These stages happen in the order given, but may overlap, in that processes b), c), and d) may occur before stage a) is complete.

For an instant product, all these stages together require only seconds. The moistening is a prerequisite for the following steps, as can be seen when a fine powder that has not been instantized is sprinkled over water: it takes a long time partially to absorb water, and, since the capillary wetting is uneven, there are some areas which water cannot penetrate. Clumps form, and can only be destroyed mechanically.

Penetration into a porous medium, as represented by a powder, is approximately described by Darcy's Equation. It is assumed that the penetrating liquid has a clearly defined front and there are no air inclusions. The speed at which this front advances can be defined by the following parameters: capillary pressure p_c, which sucks the liquid into the pores of the powder, and the resistance of the pore flux, which counteracts the capillary pressure [56].

For a one-dimensional arrangement with the y coordinate as the direction of the flow (Figure 6.29), the velocity v_A of the fluid with respect to the free cross-sectional area A is given by Equation 6.17:

$$v_A = \varepsilon \cdot \frac{dy}{dt} = \frac{B_0 \cdot p_c}{\eta \cdot y} \qquad \text{Darcy} \qquad (6.17)$$

B_0: permeability of the pore system
p_c: capillary pressure
η: dynamic viscosity
ε: porosity = hollow volume/bulk volume

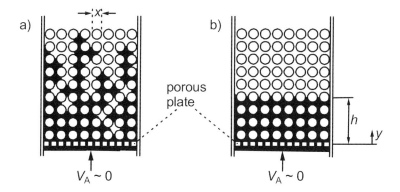

Figure 6.29. Diagram of the capillary penetration of liquid into a porous system: a) slow liquid transport; b) rapid liquid transport (from reference [56]).

Assuming that the capillary pressure remains constant, integration of the above equation over the range $y = 0$ to $y = h$ and $t = 0$ to $t = t_p$ gives, in principle, the Washburn Equation (see Section 1.7.3).

$$h^2 = \frac{2 \cdot B_0 \cdot p_c \cdot t_p}{\varepsilon \cdot \eta} \qquad (6.18)$$

For particles of approximately the same size, x, Equation 6.19 below is valid to a first approximation (for $B_0 \sim x^2$ and $p_k \sim x^{-1}$):

$$h^2 = \frac{2 \cdot x \cdot t_p}{\varepsilon \cdot \eta} \qquad (6.19)$$

This tells us that the liquid penetrates larger particles faster. Therefore, *agglomeration improves instant properties.*

The porosity ε is a significant parameter for the structure and the solubility or dispersibility of instant agglomerates. Schubert [57] was able to show that the total time required for penetration decreases as porosity increases. Nevertheless, when the porosity is too great, capillary liquid transport is interrupted, so the penetration time increases sharply once more. Therefore, there is an *optimal porosity* which should never be exceeded, and so must be determined experimentally.

Figure 6.30 shows the example of capillary penetration of water into a 4 mm high heap of powder consisting of a mixture of cocoa and sugar. The penetration time t_p is plotted as a function of porosity. For each point on the graph, a constant volume of liquid was allowed to soak into a heap of loose powder with varying porosity and the time required for penetration was measured as a function of ε. As predicted by theory, the penetration time drops as the porosity increases. After a certain porosity, the critical porosity ε_{crit}, the penetration time rises sharply again, that is, the soaking process is greatly retarded. This effect is explicable if capillary liquid transport can no longer occur once the packing has reached a certain degree of looseness. Reference [58] explains the conditions well, especially for the packing of irregularly shaped particles. It is therefore important that instant products not be allowed to reach the critical porosity.

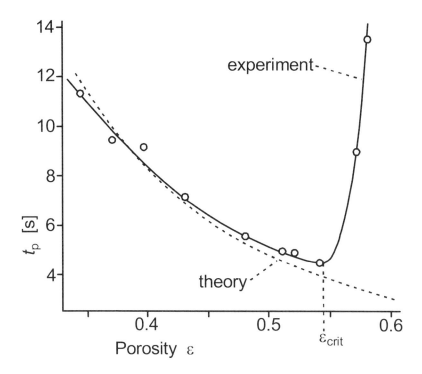

Figure 6.30. Experimental determination of the critical porosity of a cocoa/sugar mixture moistened with distilled water (V_{fl} = const., h = 4 mm, T = 293 K) [57].

It has already been determined that penetration time for a heap of loose powder decreases as the particle or agglomerate size increases. Agglomerates thus have better instant properties than does the starting material. Schubert [57] demonstrated that the penetration time can be greatly reduced if primary agglomerates of size x_2 are further agglomerated to secondary agglomerates of size x_3 in a second agglomeration stage. According to Schubert, the optimal agglomerate sizes with respect to rapid saturation are independent of the properties of the material; they are affected only by the particle size, porosity, and the height of the pile of material.

Schubert's calculations [57] showed that the production of optimal agglomerate sizes by two-stage agglomeration results in saturation three times faster than a single-stage agglomeration. If the material is not agglomerated, the saturation of an equivalent pile of powder of the same porosity takes thirty-seven times longer than that of a product that has been twice agglomerated.

6.3.2 Methods for the Manufacture of Instant Products

Processes for the manufacture of products with instant properties can be classed as those that involve agglomeration and those that do not. The latter are mostly methods that improve wettability. This can be achieved by the use of wetting agents such as lecithin in the case of edible products, or by the removal of poorly wettable substances from the surface of the product, or by heat treatment of the material to improve its solubility. For products that do not easily adopt instant properties, a combination of agglomeration and addition of wetting agents can achieve the desired end.

6.3.2.1 Agglomeration Procedures

The most important processes for manufacturing instant products are wet agglomeration processes, in which dry powders are moistened by addition of liquid, whether by condensation of vapor or with atomized liquid, or both. Capillary forces and forces resulting from the formation of links between particles are sufficient to hold the particles together. The liquid can be added to a fluidized bed, as a spray, in a drum or pan mixer, and so on. After the wet agglomeration the product is dried, resulting in the formation of solid bridges between the particles [59].

6.3.2.2 Making Instant Products by Drying

The most important methods of manufacture of instant products by drying are spray drying, combined with the fluidized bed method, and freeze drying (expensive). These have already been described in Section 6.2.6. Further references are [60–62].

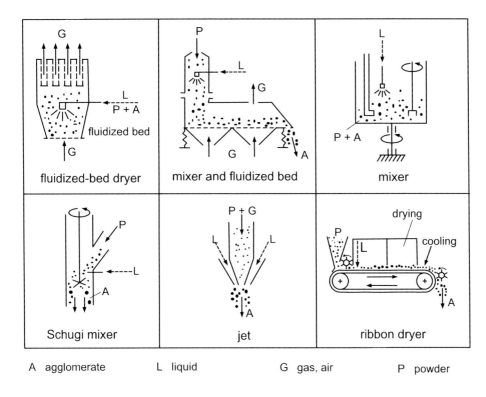

Figure 6.31. Manufacture of instant products by agglomeration with wetting and drying (build-up agglomeration); Schubert [59].

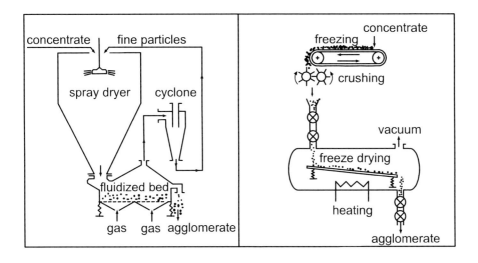

Figure 6.32. Manufacture of instant products by agglomeration with drying; Schubert [59].

Figures 6.31 to 6.33 show the most important processes diagrammatically: manufacture of instant products by agglomeration with wetting and drying, by agglomeration and drying, and by secondary agglomeration of spray-dried powders.

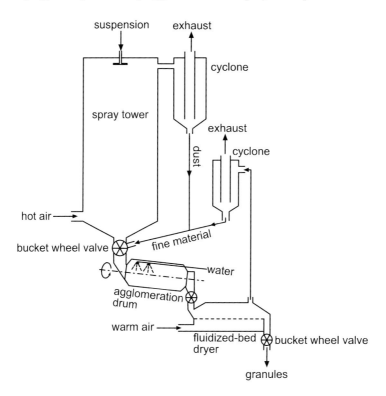

Figure 6.33. Manufacture of instant products by secondary agglomeration of spray-dried powders.

6.3.3 Measurement of Instant Properties

Many methods for the measurement of instant properties of powders (mostly in the food industry) are described in the literature. Often the insufficiencies of these methods are so considerable that they have not been adopted in practice. Nevertheless, some, most particularly the standardized measurement protocols of the International Dairy Federation (IDF) for the characterization of the properties of powdered milk, have found some application in real life. Since instant properties are a combination of the following four characteristics, the development of a generally applicable test poses a very difficult, if not almost insoluble, problem. These four characteristics, which affect the kinetics of the whole process (when a powder is scattered onto the surface of a liquid), are as follows:

a) penetration of liquid into the voids due to capillarity (known as wettability)
b) immersion of the particles in the liquid (immersibility)
c) dispersion of the powder in the liquid (dispersibility)
d) solution of the particles in the liquid in the case of a soluble substance (solubility).

For an answer to the question of which of these four steps is the rate-determining one, they must be measured separately. Specialized measuring techniques have been developed for this purpose, for which we must refer the reader to the literature, for example to Schubert's excellent survey of the treatment of powdered instant foods [63].

The methods for the measurement of these four partial processes occurring when an instant product is sprinkled onto a liquid are principally of interest for research and development. In practical applications, on the other hand, a method is required that will summarize the individual instant properties in *one* value. However, such an integral method must take into account the purpose for which the instant product is intended after dispersion and the particular requirements that it must fulfil. For instant pigment dispersions, the color and thus the particle size of the suspended solid is the most important property. This requirement is not relevant for instant foods, which must satisfy other demands. These two cases demonstrate the problems involved in developing a test independent of the products to be tested.

For powdered foods, a degree of dispersion has been defined (Schubert [63], p. 905) that takes into account not only the dispersibility but also the other partial properties mentioned. The degree of dispersion thus measures the totality of instant properties. The ratio of the concentrations of the dispersed and nondispersed portions is determined photometrically after both fractions have been ultrasonicated identically to encourage dispersion. The entire measuring process is automated and computerized. The details have been published [56]. The apparatus was built in a series by Vögtlin Messgeräte GmbH (Germany) according to reference [59]. Apparently it has now been accepted as an industrial tool [63].

While the majority of methods for the measurement of the properties of instant products have been developed by the food industry, other branches of industry have also performed studies of the instant properties of soluble and insoluble products such as dyes and pigments. However, little has been published on the subject.

There is a procedure for the measurement of the rate of solution of instant dyes. In it, the solvent is placed in a vessel and a weighed quantity of instant product added while the liquid is agitated. Simultaneously, a micropump adjusted to a specified throughput is started; this pumps the solution through a glass frit into a fractional collector, and the color concentration of the fractions is measured in a photosedimentometer. This time-consuming batch method can be improved and rendered continuous in that the solution or suspension sucked through the frit can be diluted by a constant flow of water in a mixing zone, so that the color intensity can be monitored continuously in a flow cell with a photometer (unpublished). Instead of a glass frit, which easily becomes blocked, a fine-mesh metal sieve with pores of 140 μm (depending on the product) can be used.

The process of penetration of a solvent into a bed of powder can be measured with an Enslin cell (Section 1.7.4). However, only loose powders, that is, highly porous ones, can be measured by this means. Carino and Mollet [64] altered the method such that it can also cope with compacted powders. The powder is compressed to the desired porosity in a Plexiglass cylinder resting on the glass frit of the Enslin apparatus (Figure 6.34). The volumes of liquid absorbed as a function of time, read off the scale on the horizontal pipette, can be converted to linear rates of penetration given the volume of the powder bed and the porosity. In a method for the measurement of the wetting of powdered substances by liquid published by Crowl and Wooldridge [65], the height of the liquid front rising in the powder bed is measured as a function of time. However, this is difficult, since usually there is no defined front showing the progress of the wetting clearly and enabling it to be measured. In contrast, the Enslin cell measures the volume of liquid absorbed, which is possible to a satisfactory degree of accuracy. The method is not suitable for the measurement of the instant properties of water-soluble products, but is applicable to formulations of substances that are used in organic media.

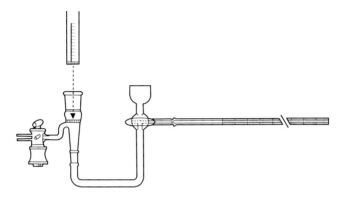

Figure 6.34. Measurement of the rate of rise of a liquid in a powder bed by means of a modified Enslin cell, from reference [64].

6.4 Microencapsulation

The technology of microencapsulation is aimed at the coating or enclosure of gases, liquids, or solids with natural or synthetic solids. The average particle size of these capsules lies in a range of 1–5000 μm. Microcapsules can be smooth spheres, aggregates like a bunch of grapes, or irregular structures with smooth, rough, or wrinkled surfaces. The capsule coat usually gathers directly onto its contents, so that irregular solids yield irregular microcapsules.

The manner in which the capsule contents are released depends on the polymer type and the thickness of the capsule coating. Depending on the application, the coating needs different properties as regards its permeability. The most important of these properties are:

- The capsule wall must be impermeable to both the capsule contents and to its external surroundings. To release the contents, the capsule must be opened, whether mechanically from outside by bursting, from inside by heating, or by dissolution, melting, or burning of the coating.
- The capsule wall must be permeable to the contents. The liberation of the contents is controlled by the wall thickness and pore size. The contents are released more or less slowly depending on this obstacle. This property is important for slow-release formulations.
- The capsule wall must be semipermeable, for example, impermeable to the contents but permeable to low molecular mass liquids from the surrounding phase. If these are miscible with the core material and diffuse into the capsule, an interior osmotic pressure builds up that can burst the capsule if it is not able to withstand the pressure. This liberates the contents.
- The permeability of the capsule wall can also be dependent on its surroundings. In an impermeable microcapsule surrounded by a liquid in which the coating material swells, the polymeric network expands and the capsule wall can become permeable to its core material. This is how encapsulated pharmaceutical and agrochemical active substances, for example, are dissolved out of their capsules. pH-sensitive materials also belong to this class (for example, swelling in the high pH experienced in the intestine).

Microencapsulation can be desirable for a number of reasons, which fall into one of the following five categories:
- to protect reactive materials from their environment until the time of their use
- to enable safe and practical handling of toxic or perishable substances
- to achieve steady, controlled release of material
- to enable liquids to be handled like solids
- to prevent mixing of substances.

The topic of microencapsulation is reviewed in references [66–71].

6.4.1 Coating Materials

The current state of the art in microencapsulation makes use of a huge selection of coating materials with which to enclose liquids and solids and which permit the mechanism of release of the contents to be varied at will. The permeability of the capsule wall is determined as much by its material as by the dimensions of the capsule. For details of the influence of the latter on the release of the core material, see Figure 6.37.

The first material to be used for capsule walls was gelatin, which is still in use today. Other proteins are also employed, as well as agar-agar and gum arabic. There are also many suitable synthetic polymers like polyamides, polysulfonamides, polyesters, polycarbonates, and urea compounds. Cellulose ethers such as ethylcellulose and methylcellulose should also be mentioned. Other polymers that can be used for encapsulation are listed in Section 6.4.2.1.

The first patents for the coating of a solid suspension are in the name of B. K. Green of the National Cash Register Co., 1953 [72].

In the selection of a coating material for the required combination of core content and continuous phase, the solubility parameter δ (Chapter 8) is useful [67]. In order to manufacture impermeable capsules, a coating material should be chosen that has a solubility parameter as different from that of the contents as possible. As a rule, solvents are miscible if their solubility parameters δ differ by no more than 2–3 units. Table 6.3, below, lists the solubility parameters for some polymers.

Table 6.3. Solubility parameters of polymers.

Polymer	$\delta\,[\text{MPa}^{1/2}]$
polytetrafluoroethylene	12.7
polychlorotrifluoroethylene	14.7
polydimethylsiloxane	14.9–15.5
ethylene–propylene rubber	16.2
polyethylene	16.2–16.6
polystyrene	17.6–18.6
poly(methyl methacrylate)	19.0
poly(vinyl chloride)	19.4–19.8
amine resins	19.6–20.7
epoxy resins	19.8–22.3
polyurethane	20.5
ethylcellulose	21.1
poly(vinyl chloride–acetate)	21.3
poly(ethylene terephthalate)	21.9
cellulose acetate (sec.)	21.3–23.5
cellulose nitrate	19.8–23.5
phenol–formaldehyde resin	23.5
poly(vinylidene chloride)	25.0
Nylon 6-6	27.8

6.4.2 Microencapsulation Processes

Chemical, physical, and mechanical processes may be used in microencapsulation. The most important are covered below, but the list is by no means complete.

6.4.2.1 Coacervation

The term coacervation (Latin coacervare – to pile up, heap together) was coined in 1926 by H. G. Bungenberg de Jong and H. R. Kruyt [69], and is used to refer to the process of visible phase separation in polymer solutions. Under the influence of various factors such as pH, temperature, or addition of another substance, a homogeneous, single-phase polymer solution (of macromolecules or association colloids) can be observed to divide into two or more phases. One of these solutions is rich in colloid and is described as the coacervate; the other is low in colloid (equilibrium liquid). However, this is not a case of precipitation of pure polymer; rather, small particles of a phase rich in polymer develop, surrounded by a phase poor in polymer. Both phases are macro- and microscopically homogeneous, and are in a state of thermodynamic equilibrium. If, in addition, suspended solid particles are present and have a surface that can be wetted by the coacervate drops, then these particles become coated with a film of material rich in polymers. When the solvent has been removed and, if necessary, the film has been hardened, microencapsulated solid particles remain. Figure 6.35 shows the course of microencapsulation schematically.

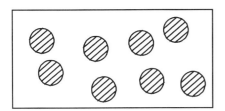

Phase I:
dispersed particles of core material
in gelatin solution

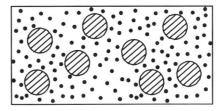

Phase II:
coacervation begins,
microcoacervate separates out

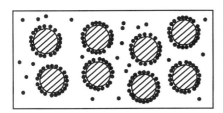

Phase III:
microcoacervate settles on surface
of droplets of core material

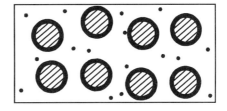

Phase IV:
microcoacervate droplets coalesce
to form a coating

Figure 6.35. Microencapsulation by coacervation.

A list of some of the materials suitable for microencapsulation by coacervation follows.

General:
– natural and synthetic substances, especially polymers which have ionized groups when in solution

Positively charged substances:
- gelatin $\leq$ pH 5.0
- other proteins, e.g. casein
- amino acids
- imines
- polymers with amino groups
- polyvinylpyrrolidone

Negatively charged substances:
- water-soluble or dispersible polymers with carboxyl groups, for example gum arabic, pectins, starches
- carboxymethylcellulose
- styrene–maleic anhydride copolymer
- ethylene–maleic anhydride copolymer
- ethylene–alkyl acrylate copolymer

It should be borne in mind that positively charged substances can undergo simple or complex coacervation, whereas negatively charged substances only undergo complex coacervation.

Procedure for microencapsulation (from reference [73])

Two solutions are prepared:
Solution A: 22 g sodium sulfate in 90 mL distilled water, at 37°C
Solution B: 8 g gelatin in 80 mL distilled water, at 37°C.

The material to be coated is suspended or dispersed in the warm gelatin solution. Coacervation is initiated by addition of the sodium sulfate solution and agitation. When the system is well mixed, it is cooled to 30°C in an ice bath at 10°C. The microcapsules formed are separated after settling and washed, depending on their type, with cold water, alcohol, concentrated formaldehyde solution, or a combination of these solvents. The best procedure comprises the addition of 1 mL formaldehyde solution for every mL of capsules, followed by 5 min agitation. Ethanol is then added (2 mL for every mL formaldehyde solution) and the mixture is stirred for another 5 min. The white flocs formed on addition of the alcohol are filtered off through paper and dried to give a fine powder which is washed, filtered, and dried again.

The microencapsulation of Aspirin will serve as an actual example: 5 g Aspirin is suspended in a solution of 2 g gelatin and 100 mL distilled water at 37°C. This system is coacervated by addition of 26 g sodium sulfate dissolved in 57 mL distilled water at 37°C. The drops of coacervate containing aspirin are gelled as described and obtained as a powder.

6.4.2.2 Complex Coacervation

In complex coacervation, there are two polymers or sols in solution with opposite charges. For example, a gelatin solution is mixed with a solution of gum arabic, and the pH is adjusted to 4.5 by dilution. The isoelectric point of gelatin, it being an amphoteric polymer, is at pH 8. Acidification of the solution to pH 4.5 causes the polymer to carry a positive charge. It reacts with the (always) negatively charged gum arabic to form a complex; this neutralization of the two polymers in the pH range 3.8–4.6 results in coacervation. The active substance (the capsule core) must be present in emulsified form in the gelatin solution and in the desired size, say 1–5 µm. Since the gelatin solution sets at room temperature, complex coacervation must be carried out above the gelling temperature of 37°C. The reaction mixture is stirred throughout. When it is cooled to 5–10°C, the complex coacervate of gelatin and gum arabic gels around the core material where it separated out. Addition of formaldehyde and sodium hydroxide to a pH value of 9–10 yields microcapsules that are agglomerated into clusters resembling bunches of grapes [67]. This agglomeration can be prevented by suitable measures, for instance those described in reference [74]. This is the procedure used, for example, for the manufacture of microcapsules for use in Green's reactive tracing paper [72]. Figure 6.36 shows color formers as examples of microcapsules formed by complex coacervation.

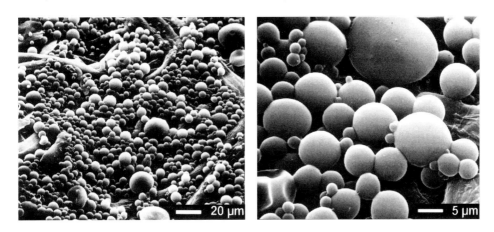

Figure 6.36. Microcapsules of color former.

6.4.2.3 Spray Drying

Spray drying is the atomization from a nozzle under pressure of an emulsion or a dispersion in a stream of hot gas. The particles of the material to be enclosed in the core, say as a pigment dispersion or dye solution, are emulsified in excess in a solution of film-forming polymer. This emulsion is spray-dried, to yield free-flowing microcapsules of 1–10 µm in size after evaporation of the water [75]. The procedure can also be used to coat the capsules a second and a third time, thus changing the permeability of the capsule wall.

6.4.2.4 Fluidized-Bed Coating

In a fluidized-bed dryer, the fine particles of the active substance are held suspended by a stream of hot air and sprayed with a solution of the coating material. This surrounds the solid particles and, after the solvent has evaporated, leaves a solid skin around the core material. The process can be used to coat solid particles with diameters from around 40 μm right up to tablet size. It is principally applied for pharmaceuticals, as well as chemicals and foods. Coating materials include gelatin, sugar, resins, waxes, synthetic polymers, and cellulose derivatives.

The Wurster process (Section 6.2.8.4) is a special form of this procedure known since 1949 in the USA [76].

6.4.2.5 Physical Methods of Microencapsulation

Physical methods of microencapsulation are *electrostatic microencapsulation* and *microencapsulation in a centrifuge*, the latter now being combined with extrusion nozzles in a later development. More details are given in reference [77].

6.4.2.6 Chemical Methods of Microencapsulation

In chemical methods of microencapsulation, the capsule wall can be formed by interfacial condensation polymerization from monomeric or oligomeric starting materials. In this process, one monomer, for example ethylenediamine, is dissolved in water, and a solution of the second component, for example an isocyanate or polyisocyanate in toluene, is layered over the top. From these components, a polyurea film insoluble in both solvents forms at the interface between the two immiscible phases [67]. If the mixture is agitated, the method yields capsules with toluene in the interior. A corresponding procedure with terephthaloyl dichloride as the material of the wall is described in the literature [78].

Microencapsulation of agrochemical substances by interfacial polymerization is covered in Chapter 14. Monomer 1, dissolved in the organic phase, is stirred together with monomer 2 in aqueous solution. The monomer solution also contains the agrochemical in finely divided form. The interfacial polymerization reaction is as follows:

$$H_2N\text{-}R_2\text{-}NH_2 + O\text{=}C\text{=}N\text{-}R_1\text{-}N\text{=}C\text{=}O \longrightarrow \cdots HN\text{-}R_2\text{-}NH\text{-}\underset{\underset{O}{\|}}{C}\text{-}NH\text{-}R_1\text{-}NH\text{-}\underset{\underset{O}{\|}}{C}\text{-}NH\text{-}R_2\text{-}NH\cdots$$

Aqueous solutions too can be microencapsulated by chemical means. Triamines can be used to form crosslinks that decrease the permeability of the capsule walls. Injection of an aqueous solution of diethylenetriamine into a solution of terephthaloyl dichloride in benzene furnishes capsules containing the aqueous solution as the core liquid [67].

Water or glycerin droplets can be coated with a partially saponified copolymer of ethylene–vinyl acetate. The coating material is dissolved in toluene, heated, and stirred together with a solution of polydimethylsiloxane in toluene (1:1), and mixed together with the glycerin to be encapsulated. The glycerin phase, the core of the microcapsule, is surrounded by the solution of coating materials; the continuous phase is made up of polydimethylsiloxane in toluene. The microcapsules are cooled and the wall material crosslinked with toluene diisocyanate, and then they are separated [79].

An example of a frequently used interfacial polymerization is the reaction between sebacoyl chloride and 1,6-hexamethylenediamine to give Nylon 6-10. This and other systems, such as the reaction of 1,6-hexamethylenediamine and phthaloyl dichloride, are described in reference [80].

6.4.3 The Suspension Medium

Microencapsulation can be classified into two types with respect to suspension medium:
1. Microencapsulation in an aqueous medium
– The film-former consists of proteins, particularly various types of gelatin.
– The phase that is to be coated can comprise water-insoluble liquids such as perfume oils, flavorings, organic solvents, fat-soluble vitamins, silicones, and so on.
– The core material may also be a water-insoluble solid; most water-insoluble organic compounds are suitable, but inorganic basic salts that can dissolve when the pH alters, for example, are not.
– Solutions and dispersions of hydrophobic solids in water-insoluble liquids can also be coated, for instance pigment dispersions and dye solutions.
2. Microencapsulation in organic media
– Modified cellulose is the film-forming polymer.
– The phase to be coated is restricted by practical considerations to solids either soluble in water or insoluble in water as long as they are not soluble in the solvent. Many pharmacologically active compounds and other organic substances belong to this group.

6.4.4 Controlled Release of the Microcapsule Contents

Multiple coating of a core of active ingredient lowers the rate of dissolution and thus the rate of release of the contents. In pharmaceutical formulation, slow release of the medicament from the microcapsule is used with the intention of achieving a depot effect. For the walls of the capsules, the most commonly used substances are gelatin, gum arabic, sodium alginate, ethylcellulose, and carboxymethylcellulose. Figure 6.37 shows the release of penicillic acid for various active substance/coating material mass ratios as an example.

The topic of slow- and controlled-release formulations is an important one that can only be mentioned briefly here. A detailed discussion can be found in reference [81].

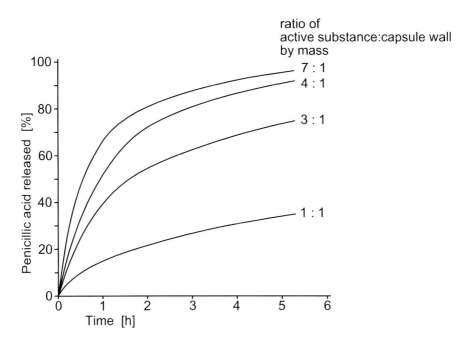

Figure 6.37. Characteristics of release of penicillic acid in vitro for different mass ratios of active substance:capsule wall.

6.4.5 Factors Influencing the Characteristics of Microcapsules

The amount of substance that diffuses through the wall of a microcapsule, dm/dt [mol/s], follows Fick's Law, Equation 6.20 (A: surface area; dc/dw: concentration coefficient; D: diffusion coefficient [cm^2/s]):

$$\frac{dm}{dt} = -D \cdot A \cdot \frac{dc}{dw} \tag{6.20}$$

According to this equation, the amount of substance diffusing through the wall of microcapsule dm/dt is proportional to the surface area of the wall and to the concentration gradient dc/dw, where dc is the difference in concentration between the inside and outside of the wall and dw is the wall thickness. It also depends on the diffusion coefficient D. This takes into account the properties of the coating material and its interactions with its surroundings, and their temperature dependence. D/dw is the permeability coefficient, or the permeability [cm/s].

Influence of grain size and phase ratio:
It is useful to know how the thickness of the capsule wall depends on the particle size of the core material and on the mass ratio of the encapsulating material to the core material. A simplified model yields some qualitative information.

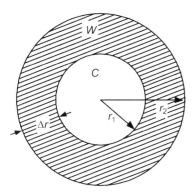

Figure 6.38. Geometry of an ideal coated particle.

Assuming that the particle is perfectly spherical, simple geometrical consideration of the cross-section of a coated particle, as depicted in Figure 6.38, yields Equation 6.21 (V: volume; C: core of coated material; W: wall of capsule; r_1: core radius; r_2: capsule radius; Δr: wall thickness):

$$\frac{V_{C+W}}{V_C} = \frac{\frac{4}{3}\pi \cdot r_2^3}{\frac{4}{3}\pi \cdot r_1^3} = \frac{r_2^3}{r_1^3} \tag{6.21}$$

or:

$$r_2 - r_1 = \Delta r = \left[\sqrt[3]{\frac{V_{C+W}}{V_C}} - 1\right] \cdot r_1 \tag{6.22}$$

The wall thickness Δr is therefore a linear function of particle size and only proportional to the cube root of the volume ratio $V_w:V_c$ or of the mass ratio. Consequently, in order to decrease the rate of diffusion it is better to enlarge the core than to increase the amount of coating material.

References for Chapter 6:

[1]   D. Geldart, Powder Technology 7, 285 (1973).
[2]   M. Glor, Chimia 5, 210 (1997).
[3]   A. W. Jenike, Flow of Bulk Solids, Bulletin 108, Univ. Utah, Salt Lake City, 1961.
[4]   A. W. Jenike, Bulletin 123, Univ. Utah, Salt Lake City, 1964.

[5]   H. Wilms, J. Schwedes, in Fließverhalten von Stoffen und Stoffgemischen (W. M. Kulicke, Ed.), Hüthig und Wepf, Basel, Heidelberg, New York, 1986.
[6]   N. Pilpel, Endeavour *28*, 73 (1969).
[7]   A. M. H. Andreasen, Kolloid Z. *86*, 70 (1939).
[8]   C. E. Capes, Particle Size Enlargement, Elsevier, Amsterdam, Oxford, New York, 1980, chapter 8.
[9]   H. Rumpf, Chem.-Ing.-Tech. *30*, 144 (1958).
[10] P. V. Dankwerts, Research (London) *6*, 365 (1953).
[11] P. M. C. Lacey, J. Appl. Chem. *4*, 257 (1954).
[12] K. Stange, Chem.-Ing.-Tech. *26*, 361 (1954).
[13] K. Stange, Chem.-Ing.-Tech. *35*, 580 (1963).
[14] H. B. Ries, Aufbereitungstechnik *1*, 1 (1969).
[15] Perry's, Chem. Eng. Handbook, 50th edition, McGraw Hill, New York, 1984.
[16] I. A. Peschl, Seminar Powder Technol., IPT, Vaduz, 1986.
[17] H. Rumpf, W. Herrmann, Aufbereitungstechnik *3*, 117 (1970).
[18] H. Sucker, S. Speiser, Pharmazeutische Technologie, Thieme, Stuttgart, 1978.
[19] H. Rumpf, Chem.-Ing.-Tech. *30*, 144 (1958); *46*, 1 (1974).
[20] H. Rumpf, in Agglomeration (W. A. Knepper, Ed.), Symp. Philadelphia, 1961.
[21] W. Pietsch, H. Rumpf, Proc. Int. Colloq. C.N.R.S. 213 (1966).
[22] P. M. Newitt, J. Conway-Jones, Trans. Inst. Chem. Eng. *39*, 422 (1958).
[23] W. Pietsch, H. Rumpf, Chem.-Ing.-Tech. *39*, 885 (1967).
[24] H. Rumpf, Chem.-Ing.-Tech. *46*, 1 (1974).
[25] H. Rumpf, Pharm. Ind. *34*, 270 (1972).
[26] H. Schubert, Powder Technol. *11*, 107 (1975).
[27] B. M. Hunter, J. Pharm. Pharmacol. *25* Suppl., 11 (1967).
[28] H. R. Komarex, Chem. Eng. *74*, 154 (1967).
[29] H. B. Ries, Aufbereitungstechnik *3*, 147 (1970); *5*, 262 (1970); *10*, 615 (1970); *12*, 745 (1970).
[30] C. E. Capes, Particle Size Enlargement, Elsevier, Amsterdam, Oxford, New York, 1980, chapter 5.
[31] DB Patent 937 495, 8.1.1956.
[32] C. E. Capes, Particle Size Enlargement, Elsevier, Amsterdam, Oxford, New York, 1980, chapter 3.
[33] C. M. Ginneken, Agglomeration, 3rd Int. Symp., Nürnberg, 1981.
[34] O. Pfrengle, Seifen - Öle - Fette - Wachse *99*, 358 (1973).
[35] H. Zilske, Seifen - Öle - Fette - Wachse *68*, 972 (1966).
[36] H. Rahm, Ausgewählte Grundlagen der pharmazeutischen Technologie, GSIA-Fortbildungskurs 1979/1980, p. 31.
[37] S. Barnett, 4th Int. Symp. Agglomeration (C. E. Capes, Ed.), Toronto, 1985.
[38] O. Krischner, Die Wissenschaftlichen Grundlagen der Trocknungstechnik, Springer-Verlag, Berlin, 1962.
[39] R. Rey, Advances in Freeze-Drying, Editions Herrmann, Paris, 1966.
[40] P. Bär, Über die physikalischen Grundlagen der Zerstäubungstrocknung, Diss. TH Karlsruhe, 1935.
[41] Y. Mori, A. Suganuma, Kagaku Kogaku (Japan) *4*, 228 (1966).

[42] K. Masters, American Ink Maker, Dec. 1979, 27.

[43] E. J. Crosby, W. Marshall, Chem. Eng. Progress *54*, 56 (1958).

[44] D. H. Charlesworthy, W. M. Marshall, J. Am. Inst. Chem. Eng. *6*, 9 (1960).

[45] R. Herbener, Chem.-Ing.-Tech. *59*, 112 (1987).

[46] J. Kaspar, M. Rosch, Chem.-Ing.-Tech. *45*, 736 (1973).

[47] K. Masters, in Spray Drying (G. Goodwin Ltd., Ed.), London, 1979.

[48] A. Maroglou, A. W. Nienow, 4th Int. Symp. Agglomeration (C. E. Capes, Ed.), Toronto, 1985, p. 465.

[49] K. H. Bauer, 3rd Int. Symp. Agglomeration (NMA, Ed.), Nürnberg, 1981, p. F91.

[50] D. M. Jones, 4th Int. Symp. Agglomeration (C. Capes, Ed.), Toronto, 1985, p. 435.

[51] D. E. Wurster, US Patent 3 089 824 (1963).

[52] S. Mortensen, A. Kristiansen, 3rd Int. Symp. Agglomeration (NMA, Ed.), Nürnberg, 1981, p. F77.

[53] M. W. Scott, H. A. Liebermann, A. S. Rankell, J. V. Battista, J. Pharm. Sci. *53*, 314, 321 (1964).

[54] H. Schubert: Kapillarität in porösen Feststoffsystemen,Springer-Verlag, Berlin, 1982.

[55] L. Pfalzer, W. Bartusch, R. Heiss, Chem.-Ing.-Tech. *45*, 510 (1973).

[56] H. Schubert, Verfahrenstechnik *12*, 296 (1978).

[57] H. Schubert, Chem.-Ing.-Tech. *47*, 86 (1975).

[58] J. van Brakel, Capillary liquid transport in porous media, Diss. Delft, 1975.

[59] H. Schubert, 4th Int. Symp. Agglomeration (C. E. Capes, Ed.), Toronto, 1985, p. 519.

[60] S. Mortensen, A. Kristiansen, 3rd Int. Symp. Agglomeration, Nürnberg, 1981, p. F77.

[61] K. Masters, A. Stoltze, Food Eng. Feb. 1973, p. 64.

[62] J. D. Jensen, Food Technol., June 1975, p. 60.

[63] H. Schubert, Chem.-Ing.-Tech. *62*, 892 (1990).

[64] L. Carino, H. Mollet, Ber. VI. Int. Kongr. grenzflächenakt. Stoffe, Zürich, 1972, p. 563.

[65] V. T. Crowl, W. D. Wooldridge, Research Memorandum No. 283, Paint Research Station Teddington, *Vol. 12*, No.16.

[66] J. R. Nixon (Ed.), Microencapsulation, Marcel Dekker, New York, Basel, 1976.

[67] W. Sliwka, "Mikroenkapsulierung"/"Microencapsulation", Angew. Chem. *87*, 556 (1975); Angew. Chem. Int. Ed. Engl. *14*, 539 (1975).

[68] M. Gutcho, Capsule Technol. and Microencaps., Noyes Data Corp., Park Ridge, NY, 1972.

[69] H. G. Bungenberg de Jong, H. R. Kruyt, "Koazervation", Kolloid-Zeitschr. *50*, 39 (1930).

[70] A. Kondo, in Microcapsule Processing and Technology (J.Wade van Valkenbourg, Ed.), Marcel Dekker, New York, 1979, chapter 8, p. 70.

[71] A. Kondo, in Microcapsule Processing and Technology (J.Wade van Valkenbourg, Ed.), Marcel Dekker, New York, 1979, chapter 9, p. 95.

[72] National Cash Register Co., US Patent 2 800 457 and 2 800 458 (1953).

[73] R. E. Phares, G. J. Sperandio, J. Pharm. Sci. *53*, 516 (1964).

[74]   National Cash Register Co., Belgian Patent 695 911 (1966).

[75]   N. Macauley, US Patents 3 016 308 (1957), 2 799 241 (1953).

[76]   D. E. Wurster, Wisconsin Alumni Research Foundation, US Patent 2 648 609 (1949).

[77]   J. T. Goodwin, G. R. Sommerville, "Microencapsulation by physical methods", Chemtech, October 1974, 623.

[78]   P. Vandegaer, US Patent 3 577 515 (1963).

[79]   R. G. Bayless et al., National Cash Register Co., US Patent 3 674 704 (1971).

[80]   R. Nixon, Endeavour, New Series, *Vol. 9*, No. 3, 123 (1985).

[81]   J. R. Robinson, Ed., Sustained and Controlled Release Drug Delivery, Marcel Dekker, New York, 1979.

7 Rheology

7.1 Basic Principles

Rheology is about the deformation and flow characteristics of materials under the influence of external forces. In particular, the flow behavior of fluid formulations like liquids or pastes is an important property from the point of view of their applications; as part of the characterization of such formulations, their viscosity is measured. The dynamic viscosity η (the tackiness) describes the tendency of a fluid to resist (by internal friction) the "mutually nonaccelerated laminar displacement of two adjacent layers".

7.1.1 Gases

In theories of fluid viscosity, the equations from the kinetic theory of gases work well for simple gases. For liquids or more complex fluids, however, there are no adequate theories.

For dilute gases, the viscosity equation is as given in Equation 7.1:

$$\eta = 0.499 \, m\bar{v} \, / \left(\sqrt{2} \, \pi \sigma^2 \right) \qquad (7.1)$$

(η: viscosity; m: molecular mass; $\bar{v}$: mean molecular velocity; σ: molecular diameter)

Measurements carried out with CO_2 at 40°C at pressures of $1.57 \cdot 10^{-5}$, $1.69 \cdot 10^{-5}$, and $4.83 \cdot 10^{-5}$ Pa·s (1, 23.8, and 100 atm) illustrate the fact that the viscosity of gases is practically independent of their density, as predicted by Equation 7.1. Deviations from the equation occur only at high pressures.

7.1.2 Viscosity

In older literature both viscosity and shear stress are quoted in cgs rather than SI units. In the cgs system the unit of viscosity is the *poise* (1 P = 1 dyn·s·cm^{-2}); in the SI system it is the *pascal·s* (1 Pa·s = 1 N·s·m^{-2}); the conversion factors are:

Viscosity η:	1 P = 0.1 Pa·s	(7.2)
Shear stress τ:	1 dyn·cm^{-2} = 0.1 Pa	(7.3)

To illustrate viscosity, consider a fluid between two parallel, plane plates. The lower plate remains still while the upper is displaced sideways at constant speed. The layers of fluid directly in contact with the plates have the same speed as the plates. The layers in between experience shearing forces as a result of friction, and therefore move with a speed proportional to the height y (Figure 7.1).

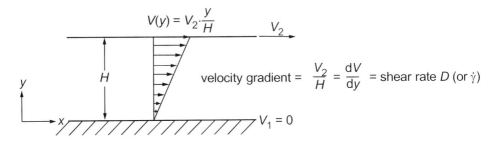

Figure 7.1. Speed in the liquid layers between two plates.

The shear stress τ (force per unit area on a plane, direction within the plane) transferred from layer to layer obeys the *Newtonian law of laminar friction*; it is proportional to the rate of change of speed, or the *shear rate D* (also called $\dot{\gamma}$, units: s^{-1}; Figure 7.1). The proportionality factor is the viscosity, η (Equation 7.4):

$$\tau = \eta \cdot \frac{dV}{dy} = \eta \cdot D \qquad (7.4)$$

7.1.3 Flow Behavior

Fluids for which Equation 7.4 remains valid regardless of time and shearing are known as *Newtonian fluids*. Fluids for which the viscosity, in particular, depends on time or shearing, such as assorted melts, solutions, latices, and pastes of polymers, are said to show non-Newtonian behavior on shearing; the phrase used is "apparent viscosity". Non-Newtonian fluids are classified as dilatant, structurally viscous, thixotropic, and rheopectic fluids. Figure 7.2 shows various types of shear diagrams.

In *structurally viscous* fluids, the apparent viscosity τ/D decreases with increasing shear stress; in *dilatant* fluids, in contrast, it increases. As D tends to zero, both pseudoplastc and dilatant fluids behave like Newtonian fluids. Structural viscosity is seen when asymmetrical, rigid particles orient themselves in a flow current, or when flexible, tangled polymers are deformed by a flow speed gradient. Dilatancy occurs occasionally in dispersions and rarely in polymer melts and solutions; most frequently it is observed in certain pastes once the critical pigment volume concentration has been reached or exceeded. In this case, the viscosity increases with increasing shear and the paste solidifies.

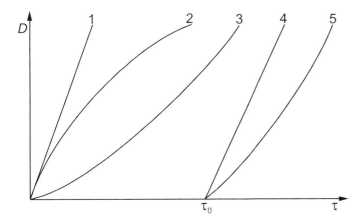

Figure 7.2. Typical shear diagrams for fluids: 1: Newtonian; 2: dilatant (shear-thickening); 3: pseudoplastic (shear-thinning or structurally viscous); 4: ideal plastic; 5: nonideal plastic; τ_0 = initial yield stress.

Plastic bodies have an initial yield stress. They flow only above this value. Ideal plastic fluids, also known as Bingham plastic fluids, behave like Newtonian fluids over this threshold, while nonideal plastic systems exhibit structurally viscous behavior over the initial yield stress. The initial yield stress is interpreted as the point at which associated structures disintegrate. Structurally viscous fluids without an initial yield stress are called *pseudoplastic* fluids.

For Newtonian, structurally viscous, and dilatant fluids, the viscosity or apparent viscosity is given by Equation 7.5:

$$\eta = \frac{\tau}{D} \tag{7.5}$$

For plastic bodies (viscosity η'), the initial yield stress must be included, as it is in Equation 7.6:

$$\eta' = \frac{\tau - \tau_0}{D} \tag{7.6}$$

For example, shear stress measurements of an ideal plastic fluid made at different shear rates: $\tau = 174$ Pa at $D = 10$ s^{-1}, and $\tau = 867$ Pa at $D = 100$ s^{-1}.

By substituting the two sets of measurements into the equation for viscosity, we can find τ_0:

$$\eta' = \frac{174 - \tau_0}{10} = \frac{867 - \tau_0}{100} \quad \rightarrow \quad \tau_0 = \frac{174 \cdot 100 - 867 \cdot 10}{100 - 10} = 97 \text{ Pa}$$

It follows that the viscosity is:

$$\eta' = \frac{174 - 97}{10} = 7.7 \text{ Pa} \cdot \text{s}$$

The following table, Table 7.1, gives an example for each of structurally viscous (emulsion) and dilatant (PVC plastisol) behavior.

Table 7.1. Parameters measured for a structurally viscous and a dilatant system (from reference [1]).

	Shear stress τ [Pa]	Shear rate D [s^{-1}]	Viscosity $\eta = \tau/D$ [Pa·s]
emulsion	28	7	4.0
	50	29	1.7
	62.5	72	0.9
PVC plastisol	71	36	2.0
	143	58	2.5
	213	77	2.8

When the shear stress changes, Newtonian, structurally viscous, dilatant, and Bingham fluids all adopt the corresponding speed gradient almost instantaneously. However, for some fluids a noticeable relaxation time is required. If the apparent viscosity at constant shear rate or constant shear stress decreases over time, the fluid is called *thixotropic*; if the apparent viscosity increases, the term used is *rheopexy*.

Thixotropic systems at rest have a gel structure which is disrupted by the effects of shearing forces, and a final structure dependent on shear rate is formed. When the system returns to rest, the gel structure rebuilds itself. Figure 7.3 shows a typical plot of the flow of a structurally viscous thixotropic fluid.

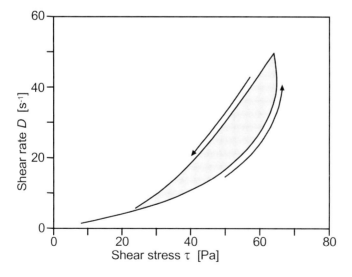

Figure 7.3. Typical shear diagram for a thixotropic liquid.

The time taken for the restoration of the equilibrium structure is represented in Figure 7.4. The fluid was sheared for 5 min at 232 s⁻¹, and the reformation occurs at a very small rate of 0.023 s⁻¹.

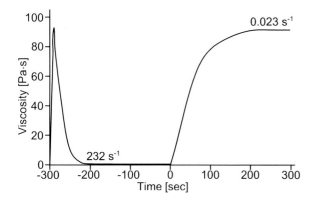

Figure 7.4. Restoration of the equilibrium structure after shearing of a thixotropic fluid (from reference [2]).

Rheopexy, the opposite of thixotropy, that is, the increase in viscosity at constant shear rate, is less common. It has occasionally been observed for very concentrated emulsions, for example w/o emulsions with concentrations higher than the closest possible spherical packing [3]. Rheopectic behavior has also been measured for latex emulsions in which the disperse phase makes up less than 30% [4]; see Figure 7.5.

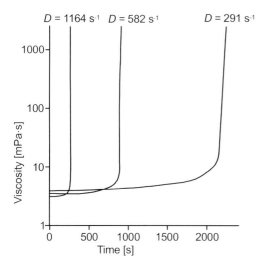

Figure 7.5. Viscosity of polystyrene latex as a function of the duration of shearing for different shear rates (polymer concentration 28%); from reference [4].

There are plenty of qualitative descriptions for consistency, such as "runny", "creamy", "sticky", without precise correspondence to a given viscosity. A map of their relationships with the initial yield stress and viscosity is therefore sketched in Figure 7.6.

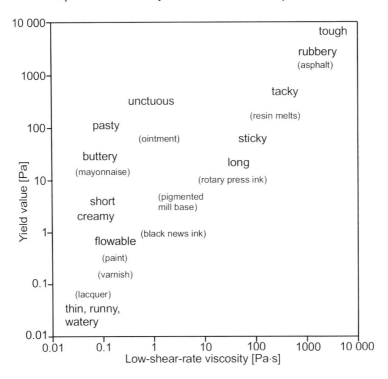

Figure 7.6. Map of terms to describe flow behavior related to yield value and low-shear-rate viscosity (from Patton [1]).

7.1.4 Temperature Dependence of Viscosity

Like other properties of materials dependent on their structure, viscosity too depends upon the temperature. Usually, viscosity sinks as temperature rises; the relationship can be represented by an Arrhenius Equation that yields a straight line in a (log η–1/T) plot (Equation 7.7):

$$\eta = A \cdot e^{B/T} \tag{7.7}$$

(A, B: constants; T: absolute temperature)

Thus, for example, the data given for linseed oil in Table 7.2 yields the following constants (least-squares fit over log η–1/T): A = 5.58·10^{-6} Pa·s; B = 2610 K.

Table 7.2. Temperature dependence of the viscosity of linseed oil.

Temperature [°C]	Viscosity [Pa·s]	
	experimental	calculated from Equation 7.7
10	0.060	0.059
30	0.033	0.032
50	0.018	0.019
90	0.0071	0.0077
150	0.0029	0.0028

7.1.5 Viscoelasticity

Ideal elastic materials are deformed under stress but return to their original shape and give up the energy absorbed after the applied force ceases. In ideal viscous materials, the energy of the deformation is completely lost. Many materials are viscoelastic, that is, they exhibit both viscous and elastic behavior, which can be the cause of relaxation and retardation in technical processes and, depending on the application, may be useful or damaging. Paints and creams are examples of viscous fluids with elastic characteristics; chocolate and gels are examples of elastic materials with viscous properties.

Viscoelastic behavior is measured by dynamic, oscillating loading. Relevant parameters [2] are the complex shear modulus G^*, the storage modulus G', the loss modulus G'', and the components η' and η'' of the complex viscosity η^*. The relationships are given by:

$$G^* = G' + iG'' \tag{7.8}$$
$$\eta^* = \eta' + i\eta'' \tag{7.9}$$

and connected thus (ω = angular velocity):

$$\eta' = \frac{G''}{\omega} \quad \text{and} \quad \eta'' = \frac{G'}{\omega} \tag{7.10}$$

Viscoelasticity is observed, for instance, in emulsions such as cosmetic lotions and creams. In these cases, dynamic measurements furnish important information on haptic properties of application like the change in elasticity when a product is rubbed into the skin.

7.2 Viscosity of Dispersions and Emulsions

7.2.1 Newtonian Systems

The viscosity of dilute dispersions can be described by Einstein's relation (Equation 7.11). This holds for dispersions of spherical particles up to a volume fraction

ϕ of 0.01. For higher concentrations up to $\phi = 0.55$, the viscosity is covered by Thomas's empirical formula [5], Equation 7.12.

Einstein $\eta = \eta_c \cdot (1 + 2.5 \cdot \phi)$ (7.11)

Thomas $\eta = \eta_c \cdot (1 + 2.5\phi + 10.05\phi^2 + 2.73 \cdot 10^{-3} \exp(16.6\phi))$ (7.12)

(η: viscosity of the dispersion; η_c: viscosity of the continuous phase, e.g. the binder solution; ϕ: volume fraction of the disperse phase (spherical particles of solid))

Emulsions differ from dispersions in that the droplets therein can be deformed by shearing. The viscosity not only of the continuous phase but also of the disperse phase must be considered. Frankel and Acrivos [6] extended Einstein's relation to give Equation 7.13 ($\eta_{dc} = \eta_d / \eta_c$; η_d = viscosity of the disperse phase; η_c = viscosity of the continuous phase):

$$\eta = \eta_c \cdot \left[1 + \left(\frac{1 + 2.5\eta_{dc}}{1 + \eta_{dc}} \right) \cdot \phi \right]$$ (7.13)

Like Einstein's relation, Equation 7.13 is also only valid for dilute systems. Choi and Schowalter [7] extended it further to cover concentrated systems.

When $\eta_d = \infty$ (solid spheres), Equation 7.13 becomes Einstein's equation for dispersions. When $\eta_d = 0$, the equation applies to spherical foams. Even in this case the viscosity is greater than for the continuous phase alone.

7.2.2 Non-Newtonian Systems

The flow behavior of real dispersions containing binders, for instance paints or inks, cannot be unambiguously classified as one single rheological type [8]. For dispersions, in the absence of particles the binder system may well have Newtonian or structurally viscous properties, with the exception of formulations for specialized purposes such as gel varnishes and thixotropic paints. However, as the proportion of disperse phase present increases, so do the deviations from expected behavior. These structurally viscous systems are often not only shear-thinning but also thixotropic; frequently there is also an initial yield stress t_0. The shearing forces often have the effect of orienting the polymer molecules adsorbed on the dispersed solid particles, so that the effective cross-section and thus the flow resistance decrease.

Thixotropy is observed more or less distinctly for dispersions as the solid content increases. Very concentrated dispersions contain particle–binder gel structures when they are at rest. These are destroyed by shear forces, but recover partially when the shear forces are small and completely when the shearing ceases. Thixotropy is therefore a time-dependent, reversible change in consistency which can be observed when shear forces are applied and when they cease. This is illustrated by the example of a pigmented alkyd resin paint in Figure 7.7.

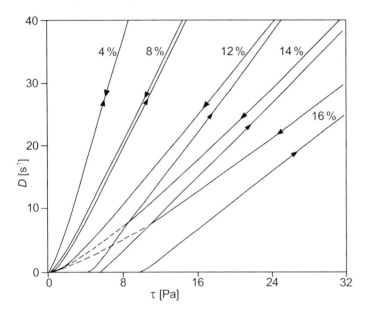

Figure 7.7. Pseudoplastic thixotropic behavior of a pigmented alkyd resin paint (from [9]).

Various properties relevant to the intended application of the system depend directly on its rheological behavior. We can set guidelines for the desirable rheological properties from empirically determined values. Figure 7.8 shows the recommended viscosity profile for the formulation of dispersion paints. Sedimentation, coverage, and dripping are just some of the aspects important for this application that are determined by the rheological properties.

For offset printing, shortness (how far a thread of ink can be pulled from the body of the fluid) or its reciprocal, tack, is a primary factor in determining the printability and therefore the commercial application of a printing ink. These properties depend on the initial yield stress of the ink and the viscosity extrapolated to very high shear rates, as in Equation 7.14.

$$\text{shortness} = \frac{\tau_0}{\eta_\infty} \qquad (7.14)$$

The influence of a very high shear rate such as that produced by high-speed printing machines in rotogravure for illustrations, particularly in the gap between the scraper and the intaglio cylinder, can be examined with ease by modern high-pressure capillary viscometers for shear rates of 10^6 s^{-1} over intervals of 10^{-4} s. (Practical requirements: $D = 10^7$ s^{-1}; $t = \sim 10^{-4}$ s.)

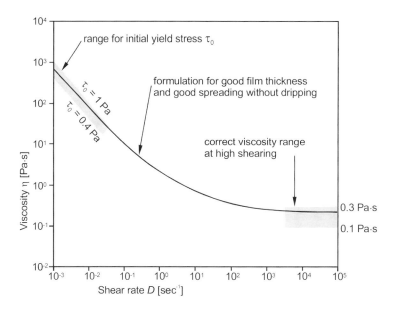

Figure 7.8. Recommended viscosity profile for dispersion paints (according to reference [10]).

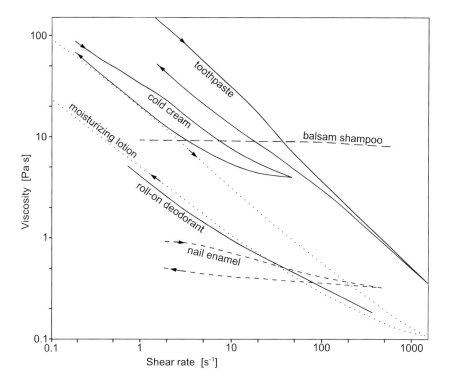

Figure 7.9. Viscosity ranges of cosmetics (data from reference [2]).

The field of cosmetics, with its many different solutions and emulsions, contains many structurally viscous and thixotropic formulations. This can be seen in Figure 7.9, in which plots of the viscosity are gathered for representative cosmetic products.

7.3 Viscosity of Polymer Melts and Solutions

Figure 7.10 shows the viscosity change of structurally viscous fluids over a fairly wide shearing range, of the type that occurs for many polymer solutions and polymer melts. For very small shearing effects, these solutions behave like Newtonian fluids (zero-shear viscosity, η_0). At very high shear rates a second Newtonian range is usually observed (high-shear viscosity η_∞). In the range between η_0 and η_∞, the fluid behaves in a non-Newtonian manner. The relationship between shear rate and shear stress in this range can often be calculated from the empirical relation developed by Ostwald and de Waele, Equation 7.15. n is the flow exponent, and has a value of 2–3 for polymer melts. The point of inflection in the shear diagram $\log D$–$\log \tau$ is used as a measure of non-Newtonian behavior.

$$D = \left(1/\eta\right)^n \tag{7.15}$$

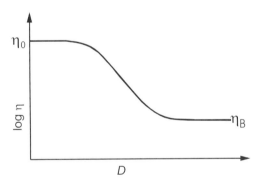

Figure 7.10. Viscosity change as a function of shear rate for structurally viscous fluids.

In the first range of Newtonian behavior, the viscosity of polymer systems depends heavily on structural parameters such as the molecular mass, the distribution of molecular mass, and the degree of branching. Figure 7.11 shows how the viscosity of polymer melts is related to their molecular mass. For all of these polymers, the dependence of η_0 on the mean mass $\overline{M}_\mathrm{w}$ is linear below a critical molecular mass M_c. Above M_c, η_0 increases in proportion to $\overline{M}_\mathrm{w}$ to the power 3.4 (Equation 7.16). The sudden transition can be ascribed to the start of chain entanglement, for which the polymer molecule must have a certain length.

$$\eta_0 = K \cdot \overline{M}_{\mathrm{w}}^{3.4}$$

(7.16)

Concentrated polymer solutions behave in a similar fashion [11].

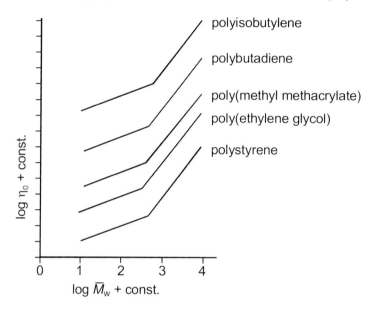

Figure 7.11. Viscosity η_0 of polymer melts as a function of $\overline{M}_{\mathrm{w}}$.

The intrinsic viscosity $[\eta]$, also called the Staudinger index, is an important characteristic of polymer solutions. It is derived from the Einstein relation (Equation 7.11) by substituting particles other than unsolvated rigid spheres, for example polymer coils. If we define the specific viscosity η_{sp} of a polymer solution according to Equation 7.17 (η_0: solvent viscosity),

$$\eta_{\mathrm{sp}} = \frac{\eta}{\eta_0} - 1$$

(7.17)

then the intrinsic viscosity is the value of the ratio of specific viscosity and polymer concentration extrapolated to concentration 0. It is determined graphically from measurements of a series of concentrations (c is the polymer concentration in g/cm^3):

$$[\eta] \equiv \lim_{c \to 0} \frac{\eta_{\mathrm{sp}}}{c}$$

(7.18)

The intrinsic viscosity depends not only on molecular mass but also on hydrodynamic volume [12]. Thus, certain proteins are present in water as solvated spheres in which the water of hydration only exchanges at a negligible rate with the surrounding water during measurements of viscosity. Some examples are ribonuclease ($[\eta]$ = 3.3), β-lactoglobulin (3.1), and hemoglobin (3.6). In contrast, for coils through which the solvent flushes, the relative speed of the solvent inside and outside the coils is the same. Such free movement of solvent can be expected when the polymers have relatively rigid chains and the solvent is a good one; in the case of very flexible molecules in a poor solvent, solvent does not flow through the coils. For more on intrinsic viscosity, see Chapter 13 (the connection with zero-shear viscosity in polysaccharide solutions).

7.4 Viscometers

A variety of different viscometers and measuring techniques is now available to measure rheological properties [13]. An assortment of norms have been agreed upon in different fields to indicate the rheology of fluid formulations in connection with parameters relevant to the application. So, for example, a norm exists for rheology measurements of paints with rotational viscometers or standard cone-and-plate instruments at shear rates of 5000–20 000 s^{-1} under conditions resembling those experienced during painting [14]. Other norms are described in references [15–17]; reference [18] covers rheological measurements for high molecular mass polymers, and references [19, 20] give more details on capillary viscometers.

Viscosity measurements are only reliable if the flow measured is laminar. Good viscometers are designed to take this requirement into account. These flow conditions are favored by high viscosity and low shear rate. In most measuring systems, the flow nevertheless becomes turbulent at high shear rates, a fact that should be borne in mind. The Reynolds number is a measure of this possibility, which in the case of tubes, for instance, depends on the diameter, the flow rate, and the kinematic viscosity.

Computerized systems have the advantage that measurements can be carried out in accordance with various programs, so that different methods may be applied rapidly, for instance for shear diagrams, low-shear measurements, thixotropy cycles, or, for appropriate systems, dynamic oscillatory measurements. Further advantages are data storage, the possibility of using different methods of evaluation, and instant graphical representation.

7.4.1 Capillary Viscometers

Capillary viscometers are based on flow through a cylindrical tube caused either by application of gas pressure or by hydrostatic pressure from the fluid column. This type of viscometer is most suitable for Newtonian fluids with low viscosity. The following formula, somewhat simplified, is applied (in the case of hydrostatic pressure):

$$\eta = K \cdot \rho \cdot t \qquad (7.19)$$

(K: apparatus constant; ρ: fluid density; t: measuring period)

For concentrated polymer solutions and for melts, on the other hand, there are better viscometers, in which the fluid is forced through the capillary at a controlled rate by means of a piston. These viscometers can work at shear rates of up to 20 000 s^{-1} and are particularly suitable for the investigation of non-Newtonian fluids.

Precision measurement with capillary viscometers demands that not only the Hagen–Poseuille Law (the first term in Equation 7.20) be considered but also the acceleration and deceleration at the ends of the capillary (the second term, which can be neglected for long measuring periods or at high viscosity).

$$\eta = \frac{\pi r^4 \cdot P \cdot t}{8V \cdot \left(l + n \cdot r\right)} - \frac{m \cdot \rho \cdot V}{8\pi \cdot \left(l + n \cdot r\right) \cdot t} \qquad (7.20)$$

(r: capillary radius; P: mean pressure drop in the capillary; V: volume flow over time t; l: length of the capillary; n, m: coefficients related to the flow at the ends of the capillary; η: viscosity; ρ: density)

Usually the constants of each term in Equation 7.20 are combined and the two resulting values determined experimentally with calibration fluids, as in the *Bingham* viscometer (which works under pressure from a gas) or *kinematic viscometers*. The latter work under the hydrostatic pressure of the fluid to be measured; they owe their name to the fact that they measure the kinematic viscosity v (η/ρ) directly (ρ: density). Figure 7.12 shows various designs of kinematic viscometers.

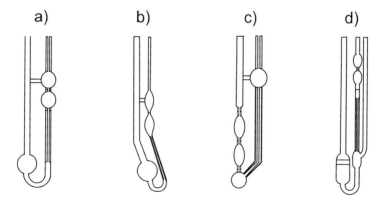

Figure 7.12. Kinematic viscometers: a) Ostwald; b) Cannon–Fenske; c) Cannon–Fenske, modified; d) Ubbelhode.

7.4.2 Rotation Viscometers

In *rotation viscometers* (Figure 7.13), the liquid is sheared between two surfaces on the stator–rotor principle, whether by means of coaxial cylinders (Couette viscometer, cylinder–container), a cone and plate, or a disk and plate. Normally the rotation rate is controlled and the torsional moment measured. However, there are also rheometers in which the applied force is controlled, which are used to determine initial yield stress. Rotation viscometers are well suited for carrying out measurements on non-Newtonian fluids.

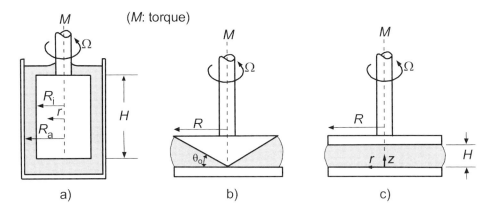

Figure 7.13. Rotation viscometers: a) coaxial cylinders (Couette); b) cone-and-plate; c) disk-and plate.

In the *coaxial cylinder system*, the inner cylinder usually rotates while the outer container remains still. The flow in the annular gap is therefore tangential. If the cylinder is too short in the cylinder–container viscometer, the shear rate calculation may be erroneous, especially if there is insufficient space between the cylinder base and the container base. The base of the cylinder is therefore often conical, so that the local conditions are similar to those in the cone-and-plate viscometer. Other artifacts occur at too great a rotation rate, not only owing to turbulence when the critical Reynolds number is exceeded, but also to the Taylor instability at a much lower rotation rate. This is a set of toroidal cells in which the flow rotates vertically to the tangential flow. For Newtonian fluids, the viscosity is given by Equation 7.21 (see also Figure 7.13).

$$\eta = \frac{M \cdot \left(1 - \dfrac{R_i^2}{R_a^2}\right)}{4\pi \cdot H \cdot \Omega \cdot R_i^2} \tag{7.21}$$

In the *cone-and-plate viscometer*, the liquid occupies the gap between a stationary, flat plate and a rotating cone, although the tip of the cone is usually truncated in order to avoid direct contact. If the cone is shallow enough, the shear stress τ and the shear rate D are constant over the radius and depend only on the cone angle, the cone radius, the rate of rotation, and the viscosity (Equations 7.22, 7.23; see also Figure 7.13). Shear diagrams for $\eta(D)$ can therefore be obtained directly without conversion calculations. Another advantage is that it is possible to measure the elasticity of viscoelastic fluids. A disadvantage of the viscometer is its limited range of shear rates, which is considerably smaller than that of capillary viscometers, Couette coaxial cylinder viscometers, and disk-and-plate viscometers. The reason for this limitation is the existence of secondary flow movements and other irregularities at the edges, even at low rotation rates. For highly fluid systems the upper limit lies at 1000 s^{-1}, for polymer melts, 100 s^{-1}.

$$D = \frac{\Omega}{\theta_o} \tag{7.22}$$

$$\tau = \frac{3M}{2\pi \cdot R^3} \tag{7.23}$$

In the *disk-and-plate viscometer*, in which the gap width is constant, the shear rate is not homogeneous over the radius of the disk. Its advantage over the cone-and-plate viscometer is that the gap width can be altered by a large amount so that measurements can be made over a wide range of shear rates in the same cycle up to 10^5 s^{-1}. This system too can be used to obtain information on elasticity, though less directly than the cone-and-plate system.

The *Brookfield viscometer* is a simple rotation viscometer in which the rotating body dips into the fluid to be measured at quite a distance from the vessel walls. Strictly, the Brookfield viscometer gives correct viscosity values only for Newtonian fluids, but Williams [21] developed a method for the determination of moderate shear rates and shear stresses in non-Newtonian fluids. For Newtonian fluids the viscosity can be calculated from Equation 7.24.

$$\eta = \frac{0.09375\,M}{R^3 \cdot \Omega} \tag{7.24}$$

7.4.3 Other Measuring Apparatus

Stokes' Law describes the descent of a ball-bearing in the *falling-ball viscometer*; the viscosity is calculated from Equation 7.25. In cylindrical pipes such as that in the Höppler viscometer an additional correction must be made for the effect of the wall.

$$\eta = \frac{2\,r^2}{9\,v} \cdot \left(\rho_K - \rho\right) \cdot g \;\; = K \cdot \left(\rho_K - \rho\right) \cdot t \tag{7.25}$$

(*r*: radius of the ball; ρ_K, ρ: densities of the ball and the fluid; *v*: rate of fall; *t*: time taken for the fall; *g*: gravitational constant; *K*: apparatus constant)

In the *bubble viscometer* (for example, the Gardner–Holdt viscometer), the liquid occupies a standardized, calibrated, sealed tube. When the tube is inverted, an air bubble rises. The length of the bubble is greater than the diameter of the tube; the rate of the rise is then independent of the bubble size. Tubes of this sort are used, for example, in the paints industry.

Other methods are based on the *efflux viscometer*, in which the time taken for fluid to flow through a hole is measured, or the *spindle viscometer*, used in the paints industry to monitor viscosity.

A special class of viscometers relies on the principle of *forced oscillatory shearing* caused by vibrators. These dynamic techniques permit simultaneous investigation of both the dynamic viscous behavior and the elastic properties of fluid systems such as polymer solutions or dispersions [13]. They are of particular interest in the case of o/w emulsions to which hydrocolloid thickeners have been added. These systems are viscoelastic. The measurement of creep and relaxation behavior of viscous pastes and other thick materials also belongs in this group. It is relatively simple; the samples are sandwiched between two plates and sheared.

References for Chapter 7:

[1] T. C. Patton, Paint flow and pigment dispersion, Wiley-Interscience Publishers, 2nd ed., New York, 1979.
[2] D. Laba, in Rheological Properties of Cosmetics and Toiletries (D. Laba, Ed.), Marcel Dekker, Inc., New York, Basel, Hong Kong, 1993, p. 403.
[3] K. J. Lissant, Colloid and Interface Science, Vol. IV (M. Kerker, Ed.), Academic Press, New York, 1976, p. 473.
[4] A. J. De Vries, in Rheology of Emulsions (P. Sherman, Ed.), Pergamon, Elmsford, New York, 1963, p. 43.
[5] D. G. Thomas, J. Colloid Sci. *20*, 267 (1965).
[6] N. A. Frankel, A. Acrivos, J. Fluid Mech. *56*, 401 (1970).
[7] S. J. Choi, W. R. Schowalter, Phys. Fluids *18*, 420 (1975).
[8] W. M. Kulicke, Fließverhalten von Stoffen und Stoffgemischen, Hüthig & Wepf Verlag, Basel, Heidelberg, New York, 1986.
[9] W. Herbst, K. Hunger, Industrial Organic Pigments, VHC Verlagsgesellschaft mbH, Weinheim, 1983.
[10] J. Schröder, Farbe + Lack *104* (3), 26, (1998).
[11] G. C. Berry, T. G. Fox, Adv. Polym. Sci. *5*, 261 (1968).

[12]   H.-G. Elias, Makromoleküle, 3rd ed., Hüthig & Wepf Verlag, Basel, Heidelberg, 1975.

[13]   J. Greener, Viscosity and its measurement, in Physical Methods of Chemistry, 2nd ed., Vol. 6: Determination of Thermodynamic Properties (B. W. Rossiter, R. C. Baetzold, Eds.), John Wiley & Sons, Inc., New York, 1992.

[14]   DIN 53 229: Bestimmung der Viskosität bei hoher Schergeschwindigkeit.

[15]   DIN 53 018/1: Messung der dynamischen Viskosität Newtonscher Flüssigkeiten mit Rotationsviskosimetern: Grundlagen.

[16]   DIN 53 018/2: Fehlerquellen und Korrekturen bei Zylinder-Rotations-viskosimetern.

[17]   DIN 53 222: Bestimmung der Viskosität mit dem Fallstabviskosimeter.

[18]   J. Schurz, Viskositätsmessungen an Hochpolymeren, Verlag Berliner Union Kohlhammer, Stuttgart, 1972.

[19]   J. E. McKie, J. F. Brandts, in High Precision Capillary Viscosimetry, Methods in Enzymology, Vol. XXVI, Enzyme Structure, Part C (C. H. W. Hirs, S. N. Timasheff, Eds.), Academic Press, New York, 1972, p. 257.

[20]   D. K. Carpenter, L. Westerman, Viscosimetric Methods for Studying Molecular Weight and Molecular Weight Distribution, in Polymer Molecular Weights, Part II (E. Slade, Ed.), Marcel Dekker, New York, 1975, chapter 7 P.

[21]   R. W. Williams, Rheol. Acta, *18*, 345 (1979).

8 Solubility Parameters, Log *P*, LSER, M Numbers

Solubility parameters and log *P*, as well as other parameters, are used to predict the behavior of substances, for example by means of quantitative structure–activity relationships (QSARs). Solubility parameters are more commonly used in the technical sector, log *P* and linear solvation energy relationships (LSERs) more in biological fields. M numbers provide information on the miscibility of solvents.

Amongst the uses of solubility parameters is the prediction of the behavior of substances both in the manufacture of formulations and in use. Hildebrand originally developed the concept to describe the thermodynamics of mixtures of nonelectrolytes, and then applied it to the solubility of gases and solids in solvents, the miscibility of liquids, metallic solutions, polymer solutions, interfacial phenomena, and critical properties [1]. Since then, the concept has been developed further to include polar effects and "chemical" interactions such as association, solvation, electron donor–acceptor complexation, and hydrogen bonding. The best known and most frequently used model is Hansen's three-dimensional solubility parameter model, in which Hildebrand's solubility parameter δ is split into terms for dispersion, δ_D, polarity, δ_P, and hydrogen bonding, δ_H.

This degree of precision is, however, not good enough for relevant quantitative results. It is adequate for qualitative use, though, particularly for formulations that contain polymers. For example, the good agreement of some of the measured and calculated solubilities of naphthalene quoted in Table 8.1 is partly due to the small differences in the Hildebrand solubility parameters, and partly to the canceling effect of enthalpic and entropic errors. The deviations are considerably larger for Hansen's three-dimensional solubility parameters, but the trend is more reliable, which is important for qualitative comparisons. The calculation according to Hildebrand's method employed Equation 8.1 (variables are defined in the next section). As can be seen, solubility parameters take deviations from ideal behavior into account ($\delta_2 = \delta_1$).

$$x_2 = x_2^{\text{ideal}} \cdot \exp\left(\frac{V_2 \phi_1^2 \cdot (\delta_2 - \delta_1)^2}{RT} \right) \tag{8.1}$$

Table 8.1. Solubility of naphthalene at 20°C (observed values taken from reference [1]).

Solvent	$\delta\,[\text{MPa}^{1/2}]$	$V\,[\text{cm}^3]$	Solubility x_2 (mole fraction)	
			observed	calculated
naphthalene (ideal)	20.3	111.5	–	0.261
chlorobenzene	19.4	102.1	0.256	0.257
acetone	20.3	74.0	0.183	0.261
n-butanol	23.3	91.5	0.0495	0.203
methanol	29.7	40.7	0.0180	0.0051

8.1 Hildebrand Solubility Parameters

Obviously, solubility parameters have something to do with solubility. As will be shown in Chapter 9, the solubility of a solid can be derived thermodynamically from the mixing of two liquids, namely, the solid, hypothetically melted and then cooled to the mixing temperature, and the solvent. The solubility parameters refer to this mixing process. Other properties quoted in connection with solubility parameters too, though, can in the end be traced back to mixing processes (amongst other things). For example, the relationship between the energy of mixing ΔE^M and the solubility parameters for the mixing of two liquids is approximated by Equation 8.2 [1] (ϕ_i: volume fraction, x_i: mole fraction, V_i: molar volume, δ_i: Hildebrand solubility parameter of the i-th liquid):

$$\Delta E^M = \phi_1\phi_2\left(x_1V_1 + x_2V_2\right)\left(\delta_1 - \delta_2\right)^2 \tag{8.2}$$

In this equation, the change in the intermolecular interactions in the mixture compared with those in the separate substances is taken into account.

In condensed material these interactions can be expressed by the molar *cohesion energy density* c; this is the molar energy of the molecules in the condensed material compared with that in the gas state, that is, the molar *energy of vaporization* ΔE_V divided by the molar volume V (Equation 8.3):

$$c = \frac{\Delta E_V}{V} \tag{8.3}$$

On the basis of cohesion energy densities, Scatchard [2] obtained the following relation for the energy of mixing of two substances (Equation 8.4); it contains the cohesion energy density of the pure substances and of the mixture:

$$\Delta E^M = \phi_1\phi_2\left(x_1V_1 + x_2V_2\right)\left(c_1 + c_2 - 2c_{12}\right) \tag{8.4}$$

If the cohesion energy density for the mixture c_{12} is *empirically* treated as the geometric mean of the pure substances' cohesion energy densities (Equation 8.5),

$$c_{12} = \sqrt{c_1 \cdot c_2} \tag{8.5}$$

then the last term of Equation 8.4 becomes:

$$\left(c_1 + c_2 - 2c_{12}\right) = \left(\sqrt{c_1} - \sqrt{c_2}\right)^2 \tag{8.6}$$

leading to the definition of the solubility parameter given in Equation 8.7. Combining Equations 8.6 and 8.7 with Equation 8.4, we obtain Equation 8.2.

$$\delta = \sqrt{c} = \sqrt{\frac{\Delta E_V}{V}} \tag{8.7}$$

Solubility parameters can be obtained from the experimentally easily determined values of energy of vaporization and molar volume, by means of Equation 8.7. It must be borne in mind that in the older literature, solubility parameters are quoted in units of $cal^{1/2}\, cm^{-3/2}$. The conversion factor for the SI unit of megapascal$^{-1/2}$ is:

$$1\, cal^{1/2}\, cm^{-3/2} = 2.0455\, MPa^{1/2} \tag{8.8}$$

8.2 Multicomponent Solubility Parameters

Error compensation based on the geometric mean of cohesion energies on analogy to Equation 8.5 is probably good enough for apolar materials, but not for polar substances, as can be seen from Table 8.1. There has been no shortage of attempts to deal with this aspect by modifying the Hildebrand model. See Section 1.4 concerning intermolecular binding forces.

Bagley et al. [3] broke δ down on the basis of thermodynamics into a "physical" parameter δ_V, which takes dispersion forces and nonpolar interactions into account, and a "chemical" parameter δ_r. This model is apparently superior to Hansen's three-component system; however, it is less frequently used.

Prausnitz [4] divided the cohesion energy density into polar and apolar terms. However, the geometric mean only works reliably for the apolar part. Keller et al. [5] pointed out that polar interactions must be further subdivided into those between permanent dipoles (δ_{or}), which are oriented and for which the geometric mean of Equation 8.5 is valid as it is for the dispersion term δ_D, and induced interactions (δ_{in}), which are asymmetrical and for which geometric averaging is unreliable.

Other approaches also take into account specific interactions that can be grouped together under the general heading of hydrogen bonds [6–8].

The most frequently applied model is Hansen's three-component system [9, 10], with a dispersion term δ_D, a polar term δ_P, and a hydrogen-bonding term δ_H, which together make up the total solubility parameter δ_t, roughly equivalent to Hildebrand's solubility parameter δ.

$$\delta_t^2 = \delta_D^2 + \delta_P^2 + \delta_H^2 \tag{8.9}$$

As in the Hildebrand model, the change in the cohesion energy density Δc on mixing two substances can be expressed as in Equation 8.10:

$$\Delta c = \left(\delta_{D1} - \delta_{D2}\right)^2 + \left(\delta_{P1} - \delta_{P2}\right)^2 + \left(\delta_{H1} - \delta_{H2}\right)^2 \tag{8.10}$$

As has already been said, the geometric mean for the polar component is not reliable. In addition, the fact that hydrogen bonds form only if both donor and acceptor groups are present is also ignored in Equation 8.10. The Hansen model is therefore rather unsuitable for quantitative calculations. Its practical significance lies in its usefulness for the qualitative comparison of the influence of different components such as polymers or solvents on the properties of a formulation. A plot of a measured property as a function of the solubility parameters of the varying components can yield pointers to desirable changes to the formulation.

Some solubility parameters taken from reference [11] are gathered in Table 8.2.

Table 8.2. Molar volumes and Hansen solubility parameters, from reference [11].

Solvent	Molar volume [cm^3/mol]	δ_D	δ_P [MPa$^{1/2}$]	δ_H	δ_t
	Alkanes				
n-butane	101.4	14.1	0.0	0.0	14.1
n-pentane	116.2	14.5	0.0	0.0	14.5
isopentane, 2-methylbutane	117.4	13.7	0.0	0.0	13.7
n-hexane	131.6	14.9	0.0	0.0	14.9
n-heptane	147.4	15.3	0.0	0.0	15.3
n-octane	163.5	15.5	0.0	0.0	15.5
2,2,4-trimethylmethane, isooctane	166.1	14.3	0.0	0.0	14.3
n-nonane	179.7	15.8	0.0	0.0	15.8
n-decane	195.9	15.8	0.0	0.0	15.9
n-dodecane	228.6	16.0	0.0	0.0	16.0
n-hexadecane	294.1	16.4	0.0	0.0	16.4
n-eicosane	359.8	16.6	0.0	0.0	16.6
cyclohexane	108.7	16.8	0.0	0.2	16.8
methylcyclohexane	128.3	16.0	0.0	1.0	16.0
cis-decahydronaphthalene (Decalin)	156.9	18.8	0.0	0.0	18.8
trans-decahydronaphthalene	159.9	18.0	0.0	0.0	18.0
	Aromatic hydrocarbons				
benzene	89.4	18.4	0.0	2.0	18.6
toluene	106.8	18.0	1.4	2.0	18.2
naphthalene	111.5	19.2	2.0	5.9	20.3
styrene	115.6	18.6	1.0	4.1	19.0
o-xylene	121.2	17.8	1.0	3.1	18.0
ethylbenzene	123.1	17.8	0.6	1.4	17.8
1-methylnaphthalene	138.8	20.7	0.8	4.7	21.2
mesitylene	139.8	18.0	0.0	0.6	18.0
tetrahydronaphthalene, tetralene	136.0	19.6	2.0	2.9	20.0
biphenyl	154.1	21.5	1.0	2.1	21.6
	Halogenated hydrocarbons				
methyl chloride	55.4	15.3	6.1	3.9	17.0
dichloromethane	63.9	18.2	6.3	6.1	20.3
bromochloromethane	65.0	17.4	5.7	3.5	18.6
chlorodifluoromethane, Freon 22	72.9	12.3	6.3	5.7	14.9
dichlorofluoromethane, Freon 21	75.4	15.8	3.1	5.7	17.1
ethyl bromide	76.9	15.8	3.1	5.7	17.0
1,1-dichloroethylidene	79.0	17.0	6.8	4.5	18.8
1,2-dichloroethane	79.4	19.0	7.4	4.1	20.9
chloroform	80.7	17.8	3.1	5.7	19.0

(Table 8.2, continued)

Solvent	Molar volume [cm^3/mol]	δ_D	δ_P [MPa$^{1/2}$]	δ_H	δ_t
1,1-dichloroethane	84.8	16.6	8.2	0.4	18.5
1,2-dibromoethane	87.0	19.6	6.8	2.1	23.9
bromoform	87.5	21.5	4.1	6.1	22.7
n-propyl chloride	88.1	16.1	7.8	2.0	17.8
trichloroethylene	90.2	18.0	3.1	5.3	19.0
dichlorodifluoromethane, Freon 12	92.3	12.3	2.0	0.0	12.5
trichlorofluoromethane, Freon 11	92.8	15.3	2.1	0.0	15.4
bromotrifluoromethane	97.0	9.6	2.5	0.0	9.9
carbon tetrachloride	97.1	17.8	0.0	0.6	17.8
1,1,1-trichloroethane	100.4	17.0	4.3	2.1	17.7
perchloroethylene	101.1	19.0	6.5	2.9	20.3
chlorobenzene	102.1	19.0	4.3	2.0	19.6
n-butyl chloride	104.9	16.4	5.5	2.1	17.4
1,1,2,2-tetrachloroethane	105.2	18.8	5.1	9.4	21.6
bromobenzene	105.3	20.5	5.5	4.1	21.7
o-dichlorobenzene	112.8	19.2	6.3	3.3	20.5
benzyl chloride	115.0	18.8	7.2	2.7	20.3
1,1,2,2-tetrabromoethane	116.8	22.7	5.1	8.2	24.7
1,2-dichlorotetrafluoroethane, Freon 114	117.0	12.7	1.8	0.0	12.8
1,1,2-trichlorotrifluoroethane, Freon 113	119.2	14.7	1.6	0.0	14.7
cyclohexyl chloride	121.3	17.4	5.5	2.1	18.4
1-bromonaphthalene	140.0	20.3	3.1	4.1	20.9
trichlorobiphenyl	187.0	19.2	5.3	4.1	20.3
perfluoromethylcyclohexane	196.0	12.5	0.0	0.0	12.5
Ethers					
furan	72.5	17.8	1.8	5.3	18.6
epichlorohydrin	79.9	19.0	10.2	3.7	21.9
tetrahydrofuran	81.7	16.8	5.7	8.0	19.4
1,4-dioxane	85.7	19.0	1.8	7.4	20.5
diethyl ether	104.8	14.5	2.9	5.1	15.8
di-(2-chloroethyl) ether	117.6	18.8	9.0	5.7	21.6
anisol, methoxybenzene	119.1	17.8	4.1	6.8	19.5
di-(2-methoxyethyl) ether	142.0	15.8	6.1	9.2	19.3
dibenzyl ether	192.7	17.4	3.7	7.4	19.3
di-(2-chloroisopropyl) ether	146.0	19.0	8.2	5.1	21.3
di-(*m*-phenoxyphenyl) ether	373.0	19.6	3.1	5.1	20.5
Ketones					
acetone	74.0	115.5	10.4	7.0	20.0
methyl ethyl ketone	90.1	16.0	9.0	5.1	19.0
cyclohexanone	104.0	17.8	6.3	5.1	19.6
diethyl ketone	106.4	15.8	7.6	4.7	18.1
mesityl oxide	115.6	16.4	6.1	6.1	18.9
acetophenone	117.4	19.6	8.6	3.7	21.8
methyl isobutyl ketone	125.8	15.3	6.1	4.1	17.0
methyl isoamyl ketone	142.8	16.0	5.7	4.1	17.4
isophorone	150.5	16.6	8.2	7.4	19.9
diisobutyl ketone	177.1	16.0	3.7	4.1	16.9
Aldehydes					
acetaldehyde	57.1	14.7	8.0	11.3	20.2
furfural	83.2	18.6	14.9	5.1	24.4
butyraldehyde	88.5	14.7	5.3	7.0	17.1
benzaldehyde	101.5	19.4	7.4	5.3	21.5

(Table 8.2, continued)

Solvent	Molar volume [cm³/mol]	δ_D	δ_P [MPa$^{1/2}$]	δ_H	δ_t
Esters					
ethylene carbonate	66.0	19.4	21.7	5.1	29.6
γ-butyrolactone	76.8	19.0	16.6	7.4	26.3
methyl acetate	79.7	15.5	7.2	7.6	18.7
ethyl formate	80.2	15.5	7.2	7.6	18.7
propylene 1,2-carbonate	85.0	20.0	18.0	4.1	27.3
ethyl acetate	98.5	15.8	5.3	7.2	18.1
trimethyl phosphate	99.9	16.8	16.0	10.2	25.3
diethyl carbonate	121.0	16.6	3.1	6.1	17.9
diethyl sulfate	131.5	15.8	14.7	7.2	22.8
n-butyl acetate	132.5	15.8	3.7	6.3	17.4
isobutyl acetate	133.5	15.1	3.7	6.3	16.8
2-ethoxyethyl acetate	136.2	16.0	4.7	10.6	20.0
isoamyl acetate	148.8	15.3	3.1	7.0	17.1
isobutyl isobutyrate	163.0	15.1	2.9	5.9	16.5
dimethyl phthalate	163.0	18.6	10.8	4.9	22.1
ethyl cinnamate	166.8	18.4	8.2	4.1	20.6
triethyl phosphate	171.0	16.8	11.5	9.2	22.3
diethyl phthalate	198.0	17.6	9.6	4.5	20.6
di-*n*-butyl phthalate	266.0	17.8	8.6	4.1	20.2
n-butylbenzyl phthalate	306.0	19.0	11.3	3.1	22.3
isopropyl palmitate	330.0	14.3	3.9	3.7	15.3
di-*n*-butyl sebacoate	339.0	13.9	4.5	4.1	15.2
methyl oleate	340.0	14.5	3.9	3.7	15.5
dioctyl phthalate	377.0	16.6	7.0	3.1	18.2
dibutyl stearate	382.0	14.5	3.7	3.5	15.4
Nitrogen compounds					
acetonitrile	52.6	15.3	18.0	6.1	24.4
acrylnitrile	67.1	16.4	17.4	6.8	24.8
propionitrile	70.9	15.3	14.3	5.5	21.7
butyronitrile	87.0	15.3	12.5	5.1	20.4
benzonitrile	102.6	17.4	9.0	3.3	19.9
nitromethane	54.3	15.8	18.8	5.1	25.1
nitroethane	71.5	16.0	15.5	4.5	22.7
2-nitropropane	86.9	16.2	12.1	4.1	20.6
nitrobenzene	102.7	20.0	8.6	4.1	22.2
ethanolamine	60.2	17.2	15.6	31.3	31.5
ethylenediamine	67.3	16.6	8.8	17.0	25.3
1,1-dimethylhydrazine	76.0	15.3	5.9	11.0	19.8
2-pyrrolidone	76.4	19.4	17.4	11.3	28.4
pyridine	80.9	19.0	8.8	5.9	21.8
n-propylamine	83.0	17.0	4.9	8.6	19.7
morpholine	87.1	18.8	4.9	9.2	21.5
aniline	91.5	19.4	5.1	10.0	22.6
N-methylpyrrolidone	96.5	18.0	12.3	7.2	22.9
n-butylamine	99.0	16.2	4.5	8.0	18.6
diethylamine	103.2	14.9	2.3	6.1	16.3
diethylenetriamine	108.0	16.8	13.3	14.3	25.8
cyclohexylamine	115.2	17.4	3.1	6.6	18.9
quinoline	118.0	19.4	7.0	7.6	22.0
di-*n*-propylamine	136.9	15.3	1.4	4.1	15.9
formamide	39.8	17.2	26.2	19.0	36.6

(Table 8.2, continued)

Solvent	Molar volume [cm^3/mol]	δ_D	δ_P [MPa$^{1/2}$]	δ_H	δ_t
N,N-dimethylformamide	77.0	17.4	13.7	11.3	24.8
N,N-dimethylacetamide	92.5	16.8	11.5	10.2	22.7
tetramethylurea	120.4	16.8	8.2	11.1	21.7
Sulfur compounds					
carbon disulfide	60.0	20.5	0.0	0.6	20.5
dimethyl sulfoxide	71.3	18.4	16.4	10.2	26.7
dimethyl sulfone	75.0	19.0	19.4	12.3	29.8
diethyl sulfide	108.2	17.0	3.1	2.1	17.4
Alcohols					
methanol	40.7	15.1	12.3	22.3	29.6
ethanol	58.5	15.8	8.8	19.4	26.5
1-propanol	75.2	16.0	6.8	17.4	24.5
2-propanol	76.8	15.8	6.1	16.4	23.5
3-chloropropanol	84.2	17.6	5.7	14.7	23.6
furfuryl alcohol	86.5	17.4	7.6	15.1	24.3
1-butanol	91.5	16.0	5.7	15.8	23.1
2-butanol	92.0	15.8	5.7	14.5	22.2
2-methyl-1-propanol, isobutanol	92.8	15.1	5.7	16.0	22.7
benzyl alcohol	103.6	18.4	6.3	13.7	23.8
cyclohexanol	106.0	17.4	4.1	13.5	22.4
1-pentanol	109.0	16.0	4.5	13.9	21.7
2-ethyl-1-butanol	123.2	15.8	4.3	13.5	21.2
diacetone alcohol	124.2	15.8	8.2	10.8	20.8
1,3-dimethyl-1-butanol	127.2	15.3	3.3	12.3	19.9
ethyl lactate	115.0	16.0	7.6	12.5	21.6
n-butyl lactate	149.0	15.8	6.5	10.2	19.9
ethylene glycol monomethyl ether	79.1	16.2	9.2	16.4	24.8
ethylene glycol monoethyl ether	97.8	16.2	9.2	14.3	23.5
diethylene glycol monomethyl ether	118.0	16.2	7.8	12.7	22.0
diethylene glycol monoethyl ether	130.9	16.2	9.2	12.3	22.3
ethylene glycol mono-*n*-butyl ether	131.6	16.0	5.1	12.3	20.8
2-ethyl-1-hexanol	157.0	16.0	3.3	11.9	20.2
1-octanol	157.7	17.0	3.3	11.9	21.0
2-octanol	159.1	16.2	4.9	11.1	20.2
diethylene glycol mono-*n*-butyl ether	170.6	16.0	7.0	10.6	20.4
1-decanol	191.8	17.6	2.7	10.0	20.4
1-tridecanol	242.0	14.3	3.1	9.0	17.2
"nonyl"phenoxyethanol	275.0	16.8	10.2	8.4	21.4
oleyl alcohol	316.0	14.3	2.7	8.0	16.6
triethylene glycol monooleyl ether	418.5	13.3	3.1	8.4	16.0
Acids					
formic acid	37.8	14.3	11.9	16.6	24.9
acetic acid	57.1	14.5	8.0	13.5	21.4
benzoic acid	100.0	18.2	7.0	9.8	21.8
n-butanoic acid	110.0	14.9	4.1	10.6	18.8
caprylic acid	159.0	15.1	3.3	8.2	17.5
oleic acid	320.0	14.3	3.1	5.5	15.6
stearic acid	326.0	16.4	3.3	5.5	17.6
Phenols					
phenol	87.5	18.0	5.9	14.9	24.1
resorcinol	87.5	18.0	8.4	21.1	29.0
m-cresol	104.7	18.0	5.1	12.9	22.7

(Table 8.2, continued)

Solvent	Molar volume [cm³/mol]	δ_D	δ_P	δ_H	δ_t
			[MPa$^{1/2}$]		
o-methoxyphenol	109.5	18.0	8.2	13.3	23.8
methyl salicylate	129.0	16.0	8.0	12.3	21.7
"nonyl"phenol	231.0	16.6	4.1	9.2	19.4
water	18.0	15.6	16.0	42.3	47.8

8.3 Incremental Methods

Often, solubility parameters are not available for the substances involved in a formulation problem. In such cases, values can be estimated by incremental methods.

The contributions of different groups to the cohesion energy and the molar volume have been published by Fedors [12], amongst others. Although these are not very precise, they can be used in the calculation of approximate Hildebrand solubility parameters. Some data from reference [11] is collected in Table 8.3. These data can be used to furnish Hildebrand parameters by means of Equation 8.11, following the definition of solubility parameters given in Equation 8.7.

$$\delta = \left(\sum_i \Delta E_{Vi} \Big/ \sum_i V_i \right)^{1/2} \tag{8.11}$$

One application of the group increments is the estimation of the changes in properties of a substance that will result from the introduction of new substituents.

Example:	phenylpiperidine	
5 –CH=	5× 4.31	5× 13.5
1 >C=	4.31	−5.5
5 –CH₂–	5× 4.94	5× 16.1
1 –N<	4.2	−9
3 conjugated double bonds	3× 1.67	3× −2.2
2 six-membered rings	2× 1.05	2× 16.0
	61.87 kJ/mol	158.9 cm³/mol

$\delta = (61.87 \cdot 10^{-3} \text{ MJ·mol}^{-1} / 158.9 \cdot 10^{-6} \text{ m}^3 \cdot \text{mol}^{-1})^{1/2} = 19.73 \text{ MPa}^{1/2}$

Incremental methods have also been developed for the calculation of Hansen solubility parameters. These are based on group attraction constants F_{Di} and F_{Pi}, for dispersion and polar components, and group cohesion energies E_{Hi}; Equations 8.12–8.14. These group constants are given in Table 8.4.

$$\delta_D = \frac{\sum_i F_{Di}}{\sum_i V_i} \qquad \delta_P = \frac{\sqrt{\sum_i F_{Pi}^2}}{\sum_i V_i} \qquad \delta_P = \sqrt{\frac{\sum_i E_{Hi}}{\sum_i V_i}} \tag{8.12–8.14}$$

Table 8.3. Group increments for cohesion energies and molar volumes (source: [11]).

Group	ΔE_{Vi} [kJ/mol]	V_i [cm^3/mol]	Group	ΔE_{Vi} [kJ/mol]	V_i [cm^3/mol]
$-CH_3$	4.71	33.5	$-N=$	11.7	5.0
$-CH_2-$	4.94	16.1	$-NHNH_2$	22.0	–
$>CH-$	3.43	-1.0	$-NHNH-$	16.7	16.0
$>C<$	1.47	-19.2	$-N_2$ (diazo)	8.4	23.0
$=CH_2$	4.31	28.5	$-N=N-$	4.2	–
$-CH=$	4.31	13.5	$>C=N-N=C<$	20.1	0.0
$>C=$	4.31	-5.5	$-N=C=N-$	11.47	–
$HC\equiv$	3.85	27.4	$-NC$	18.8	23.1
$-C\equiv$	7.07	6.5	$-NF_2$	7.66	33.1
ring closure, 5 or more atoms	1.05	16.0	$-NF-$	5.07	24.5
ring closure, 3 or 4 atoms	3.14	18.0	$-CONH_2$	41.9	17.5
conjugation in ring per =	1.67	-2.2	$-CONH-$	33.5	9.5
$-F$	4.19	18.0	$-CON<$	29.5	-7.7
$-F$ (disubst.)	3.56	20.0	$HCON<$	27.6	11.3
$-F$ (trisubst.)	2.30	22.0	$HCONH-$	44.0	27.0
$-CF_2-$ (perfluoro compounds)	4.27	23.0	$-NHCOO-$	26.4	18.5
$-CF_3$ (perfluoro compounds)	4.27	57.5	$-NHCONH-$	50.2	–
$-Cl$	11.55	24.0	$-NHCON<$	41.9	–
$-Cl$ (disubst.)	9.63	26.0	$>NCON<$	20.9	-14.5
$-Cl$ (trisubst.)	7.53	27.3	NH_2COO-	37.0	–
$-Br$	15.49	30.0	$-NCO$	28.5	35.0
$-Br$ (disubst.)	12.4	31.0	$-ONH_2$	19.1	20.0
$-Br$ (trisubst)	10.7	32.4	$>C=NOH$	25.1	11.3
$-I$	19.05	31.5	$-CH=NOH$	25.1	24.0
$-I$ (disubst.)	16.7	33.5	$-NO_2$ (aliphatic)	29.3	24.0
$-I$ (trisubst.)	16.3	37.0	$-NO_2$ (aromatic)	15.36	32.0
halogen on C with =	$0.8 \times \Delta E_{VHal.}$		$-NO_3$	20.9	33.5
$-CN$	25.5	24.0	$-NO_2$ (nitrite)	11.7	33.5
$-OH$	29.8	10.0	$-NHNO_2$	39.8	28.7
$-OH$ (disubst. or on adjacent C)	21.9	13.0	$-NNO$	27.2	10.0
$-O-$	3.35	3.8	$-SH$	14.44	28.0
$-CHO$	21.4	22.3	$-S-$	14.15	12.0
$-CO-$	17.4	10.8	$-S_2-$	23.9	23.0
$-COO-$	18.0	18.0	$-S_3-$	13.40	47.2
$-CO_3-$ (carbonate)	17.6	22.0	$>SO$	39.1	–
$-C_2O_3-$ (anhydride)	30.6	30.0	SO_3	18.8	27.6
$HCOO-$	18.0	32.5	SO_4	28.5	31.6
$-O_2CCO_2-$ (oxalate)	26.8	37.3	$-SO_2Cl$	37.1	43.5
$-COF$	13.4	29.0	$-SCN$	20.1	37.0
$-COCl$	17.6	38.1	$-NCS$	25.1	40.0
$-COBr$	24.2	41.6	PO_3	14.2	22.7
$-COI$	29.3	48.7	PO_4	20.9	28.0
$-NH_2$	12.6	19.2	$PO_3(OH)$	31.8	32.2
$-NH-$	8.4	4.5	SiO_4	21.8	20.0
$-N<$	4.2	-9.0	BO_3	0.0	20.4

The molar volumes needed in Equations 8.12–8.14 can be calculated by means of the group increments in Table 8.3 or according to Exner's method [13]. The values of group attraction constants F_{Di} and F_{Pi} given by Krevelen and Hoftyzer [14] and those of group cohesion energies E_{Hi} given by Hansen and Beerbower [15] are listed in Table 8.4. When several polar groups are present, the calculated polar term δ_p must be multiplied by an appropriate symmetry factor (shown in Table 8.4); however, caution is necessary: in phthalate esters, for example, the polar component is multiplied by four as a result of the presence of the second –COO– group [15]. For hydrogen-bonding terms, too, it must be remembered that interactions only occur in the presence of both donor and acceptor groups, as has already been mentioned.

Table 8.4 Group increments for calculation of Hansen solubility parameters [11].

Group	F_{Di} [$J^{1/2}cm^{3/2}mol^{-1}$]	F_{Pi} [$J^{1/2}cm^{3/2}mol^{-1}$]	E_{Hi} (aliph.) [$Jmol^{-1}$]	E_{Hi} (arom.) [$Jmol^{-1}$]
–CH$_3$	420	0	–	–
–CH$_2$–	270	0	–	–
>CH–	80	0	–	–
>C<	–70	0	–	–
=CH$_2$	400	0	–	–
=CH–	200	0	–	–
=C<	70	0	–	–
cyclohexyl	1620	0	–	–
phenyl	1430	110	–	–
phenylene (o, m, p)	1270	110	–	–
–F	(220)	–	~0	~0
–Cl	450	500	400 ± 80	400 ± 80
–Br	(550)	–	2100 ± 400	2100 ± 400
–I	–	–	4000 ± 800	–
–CN	430	1100	2100 ± 800	2300 ± 800
–OH	210	500	19 500 ± 1700	19 500 ± 2100
(–OH)$_n$	–	–	n(19 500 ± 1700)	n(19 500 ± 1700)
–O–	100	400	4800 ± 1200	5200 ± 1200
–CHO	470	800	–	–
>C=O	290	770	3300 ± 600	3300 ± 600
–COOH	530	420	11 500 ± 1000	9400 ± 1000
–COO–	390	490	5200 ± 600	3300 ± 600
HCOO–	530	–	–	–
–NH$_2$	280	–	5600 ± 800	9400 ± 800
–NH–	160	210	3100 ± 800	–
–N<	20	800	–	–
–NO$_2$	500	1070	1700 ± 200	1700 ± 200
–S–	440	–	–	–
ring	190	–	–	–
one plane of symmetry	–	× 0.5	–	–
two planes of symmetry	–	× 0.25	–	–
several planes of symmetry	–	0	–	–

8.4 Solvent Mixtures

The closer the solubility parameters of two substances are, the greater the compatibility of those substances. In many formulations containing solvent, one solvent is not sufficient to meet all the demands made on it, whether with regard to the solubility of the different components or to its volatility, viscosity, environmental effects, and so on. For these mixtures, the Hansen solubility parameters are calculated from the solubility parameters in proportion to the volume fractions ϕ_i of the individual solvents.

$$\delta_{D\,(\text{Mischung})} = \sum_i \phi_i \cdot \delta_{Di} \tag{8.15}$$

$$\delta_{P\,(\text{Mischung})} = \sum_i \phi_i \cdot \delta_{Pi} \tag{8.16}$$

$$\delta_{H\,(\text{Mischung})} = \sum_i \phi_i \cdot \delta_{Hi} \tag{8.17}$$

The huge change in properties possible can be illustrated by the example of nitrocellulose, which is insoluble in diethyl ether ($\delta_{D/P/H}$ 14.5/2.9/5.1) and in ethanol ($\delta_{D/P/H}$ 15.8/8.8/19.4), but soluble in a mixture of the two.

8.5 Polymer Solutions

8.5.1 Flory–Huggins Parameters

The behavior of polymer solutions is very far from ideal. For instance, a mixture of polystyrene of mean molar mass $\overline{M}_n$ 290 000 and 3.4 g toluene has a vapor pressure of 0.32 kPa at 25°C instead of the 3.7 kPa predicted for an ideal mixture [16].

According to the Flory–Huggins theory, the interaction between polymer and solvent can be expressed in terms of the Flory–Huggins interaction parameter χ. For the free energy of mixing ΔG_M, therefore, we obtain Equation 8.18 (definitions are in Section 8.1; the index 2 corresponds to the polymer):

$$\Delta G_M = RT \left(x_1 \ln \phi_1 + x_2 \ln \phi_2 + \chi x_1 \phi_2 \right) \tag{8.18}$$

χ is not constant; rather, it depends on the composition of the solution. The first two terms in Equation 8.18 result from the combined entropy of the polymer solution taking into account the different sizes of the solvent and polymer molecules. On the other hand, the third term results from the enthalpy of mixing according to the original Flory–Huggins theory.

In order for polymer and solvent to be miscible, ΔG_M must be negative. Since the contributions of the first two terms are both negative, the third must be smaller than these combined entropy contributions when χ is positive. Flory showed that a polymer and a solvent are miscible in all proportions under the following conditions [17, 18]:

$$\chi \le \tfrac{1}{2}\left[1+\left(V_1/V_2\right)^{\tfrac{1}{2}}\right]^2 \tag{8.19}$$

If this condition is not fulfilled, the mixture separates into two phases, a "solution of polymer in solvent" and a "solution of solvent in polymer". This is illustrated in Figure 8.1 for various ratios of the molecular masses of polymer and solvent.

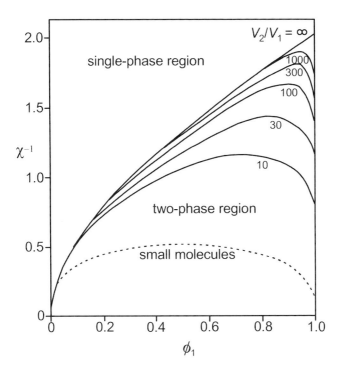

Figure 8.1. Phase diagram for polymer/solvent systems as a function of the reciprocal Flory–Huggins interaction parameters χ^{-1} and the volume fraction of the solvent ϕ_1 for various ratios of polymer and solvent molecular masses, according to reference [19].

If χ is known as a function of the concentration and the temperature for a particular polymer/solvent pair, quantitative predictions about the thermodynamics of that system can be made. Since, however, this is usually not the case, we can estimate polymer solution behavior using the concept of solubility parameters just as we can for smaller molecules. Another aspect is the compatibility of different polymers in liquid formulations, which Flory–Huggins interaction parameters do not cover at all.

8.5.2 Polymer Solubility Parameters

The concept of solubility parameters is widely applied in the paints and printing ink industries and also finds uses in adhesive manufacture. However, Hildebrand solubility parameters for polymers cannot be calculated from vaporization energies and molar volumes as they can for smaller molecules, since the polymers cannot be evaporated. Instead, they must be empirically determined by solution experiments. The same is true for Hansen solubility parameters. For resins and other binders, the solubility parameters δ_D, δ_P, and δ_H can be determined by solution experiments in selected solvents; good solvents cluster together on the solubility parameter scale. The center of this range characterizes the three solubility parameters sought for the binder under investigation.

In Hansen's method, the solution experiments are carried out with polymer concentrations of 10%. Mixtures which yield very viscous solutions are heated and the interactions assessed after cooling (soluble, incompletely soluble, very swollen but poorly soluble, swollen, slightly swollen, no effect).

According to Hansen, the area of good solubility can be defined as a sphere in the three-dimensional space of the solubility parameters if the δ_D axis is expanded by a factor of 2 (Figure 8.2).

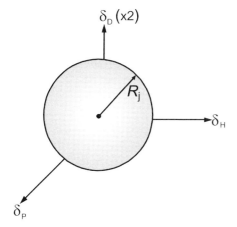

δ_D (x2)

R_j

δ_H

δ_P

Figure 8.2. Spherical solubility volume of a polymer in the three-dimensional space of the solubility parameters. The δ_D axis has been expanded by a factor of 2.

The corresponding interaction radius for solvent i and polymer j is therefore given by Equation 8.20:

$$R_{ij} = \left[4\left(\delta_{Di} - \delta_{Dj}\right)^2 + \left(\delta_{Pi} - \delta_{Pj}\right)^2 + \left(\delta_{Hi} - \delta_{Hj}\right)^2 \right]^{1/2} \tag{8.20}$$

Beerbower et al. [20] prefer block-shaped volumes for solvents that swell the polymer by more than 25%; in addition they define the area of greatest solubility (centroid). For both procedures, the solubility parameters may be found in reference [11]. Some examples are shown in Table 8.5.

Table 8.5. Solubility parameter ranges for polymers [11].

Polymer	25 % swelling [MPa$^{1/2}$]			"Centroid" [MPa$^{1/2}$]		
	δ_D	δ_P	δ_H	δ_D	δ_P	δ_H
natural rubber	13.5/20.1	0.0/12.3	0.0/11.3	18.4	2.1	7.2
styrene–butadiene rubber	13.5/22.5	0.0/12.3	0.0/8.2	17.4	2.9	6.8
nitrile–butadiene rubber	14.3/21.9	0.0/19.2	0.4/9.2	18.0	4.1	4.1
chloroprene	14.1/21.9	0.0/12.5	0.4/7.8	19.4	3.1	2.7
vinylsilicone rubber	14.3/20.7	0.0/12.3	0.0/4.1	16.4	3.7	4.5
fluorosilicone rubber	14.9/19.4	1.4/14.9	3.9/8.0	15.3	7.2	6.1
ethylene–propylene copolymer	13.5/19.4	0.0/8.8	0.0/6.1	18.0	0.8	2.1

Another method works on the assumption that the δ_D values of most solvents vary considerably less than do the δ_P and δ_H values. Therefore, the solution behavior may justifiably be represented in a two-dimensional space that considers only the polar and hydrogen-bonding components. However, the effect of δ_D can still be considered in the form of height contours, as shown in Figure 8.3.

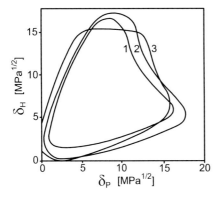

Figure 8.3. Solubility diagram for the alkyd resin Alphthalate AF342 from Hoechst (20% concentration; 23°C). Plot 1: $\delta_D = 14.7–15.3$ MPa$^{1/2}$; 2: $\delta_D = 16.2$ MPa$^{1/2}$; 3: $\delta_D = 16.8$ MPa$^{1/2}$.

The center of gravity of the shape inside the contours can be regarded as the set of solubility parameters (δ_P, δ_H). Such two-dimensional diagrams can, of course, be used for polymer mixtures as well as single polymers.

Another method for representing polymer–solvent systems was developed by Teas [22]. He plots fractional parameters from Equations 8.21–8.23 in a triangular frame (Figure 8.4).

$$f_D = \frac{\delta_D}{\delta_D + \delta_P + \delta_H} \qquad f_P = \frac{\delta_P}{\delta_D + \delta_P + \delta_H} \qquad f_H = \frac{\delta_H}{\delta_D + \delta_P + \delta_H} \qquad (8.21\text{–}8.23)$$

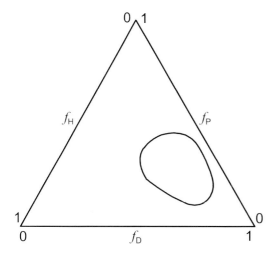

Figure 8.4. Triangular representation of the solubility limits of poly(methyl methacrylate) with Teas' fractional solubility parameters [23].

If no solubility parameters for the polymer are available, they can be estimated roughly according to the incremental methods described in Section 8.3 just as for smaller molecules. The solubility parameters are calculated for the basic units as described by Ahmad and Yaseen in reference [24]. It must be remembered that the polar components turn out too high for some polymers (for example polyacrylonitrile 1.7×, poly(vinyl chloride) 1.5×).

8.6 Application of Solubility Parameters

Many problems in formulation require the selection of a substance from a range of available possibilities to optimize the formulation properties. Some substances are tested and a choice is made on the basis of the results. In this procedure, the test results are classified according to the discrete variable <substance>, for which no direct correlation will be visible. However, if solubility parameters are assigned to the individual substances, it is easier to see a correlation between test results and solubility parameter and thus to make a selection.

The important point is therefore that the discrete variable <substance> is transformed into the continuous variable <solubility parameter> and the latter is correlated with the dependent variable <test result>. An example is shown in Figure 8.5.

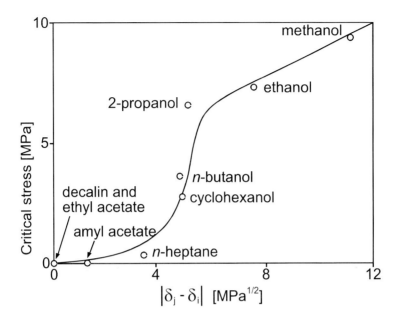

Figure 8.5. Conversion of the discrete variable <solvent> into the continuous variable <solubility parameter> and correlation with the measured variable: relationship between critical stress for crazing of polystyrene under the influence of solvents and the difference between the Hildebrand solubility parameters of the polymer and the solvent (according to reference [25]).

Many such correlations with solubility parameters have been established for polymers. The mechanical properties of solids depend, amongst other things, on their intermolecular interactions, which are expressed in the cohesion energy density. Correlations between mechanical properties and solubility parameters can therefore be applied successfully to the solution of optimization problems, for instance in the estimation of the effect of molecular structures on the properties of the solid. So, for example, there is a linear relationship between the maximum tensile strength σ_{max} of polymers and the cohesion energy density δ^2, expressed in Equation 8.24. Flexibility, glass transition temperature, melting point, creep, hardness, and elasticity are some other properties that have been investigated in relation to solubility parameters for various polymers [11].

$$\sigma_{max} = 0.25\,\delta^2 \tag{8.24}$$

The mechanical properties of polymers are particularly heavily influenced by the surrounding solvent. Examples include changes in the glass transition temperature due to interfacial plasticization, and also interfacial hardening, by solvents. Figure 8.5 shows the fracture behavior of polystyrene affected by solvent correlated with solubility parameters.

There are many other applications, including, but not limited to, microencapsulation [26], problems in the manufacture of asymmetric membranes [27], resistance to fluids of elastomers in hydraulic systems [28], volatility of solvents in polymer melts [29], polymer compatibility in blends [30], solvents for film-formers [31], the oxygen permeability of poly(methyl methacrylate) [32], interactions between pigment and binder [33], interactions between filler and polymer [34], controlled release of pesticides [35], solubility of pesticides [36], distribution in bilayer membranes [37], lipid extraction from biological tissues [38], steroid transport by polymers [39], absorption of active substances by transcutaneous administration [40], surface energetics of artificial blood substitutes [41], and anesthetics [42].

Solubility is of great importance in biochemical processes. We shall mention here aspects such as the transport of pharmaceuticals and agrochemicals in biological tissues, structure–activity relationships, and the formulation of active substances, which are reviewed in references [43–45]. It is known that the effects of many active substances are correlated with their distribution coefficient between oil and water; this can be expressed in terms of Hansch's lipophilicity parameter, Equation 8.25 [44].

$$\pi_{\mathrm{X}} = \log D_{\mathrm{RX}} - \log D_{\mathrm{RH}} \tag{8.25}$$

π_{X} is the Hansch parameter for the substituent X; D_{RX} and D_{RH} are the molar distribution coefficients for the substituted and the unsubstituted compound, respectively, between 1-octanol and water. Tables of Hansch parameters can thus be used to estimate the effect of substituents on the octanol/water distribution coefficient of a substance. Instead of Hansch parameters, solubility parameters can of course be used equally successfully for the calculation of distribution coefficients. The relationship between Hansch parameters and Hildebrand solubility parameters is defined in Equation 8.26; the group increments for the cohesion energy F_{X} and volume V_{X} can be found in Table 8.3.

$$2.303\, RT \cdot \pi_{\mathrm{X}} = V_{\mathrm{X}} \cdot \left(\delta_{\mathrm{Wasser}} - \delta_{\mathrm{Octanol}}\right)^2 + 2F_X\left(\delta_{\mathrm{Octanol}} - \delta_{\mathrm{Wasser}}\right) \tag{8.26}$$

Relationships that can be plotted graphically are not the only ones suitable for investigation by means of solubility parameters. Qualitative descriptions of visual, haptic, olfactory, and other properties can be represented in a two-dimensional solubility parameter diagram of the substance varied, such as a solvent. Such ordering – solvent mapping – is illustrated by the electron micrographs of a pigment recrystallized from various solvents reproduced in Figure 8.6. These representations make it plain whether there is a continuous relationship or whether additional, special effects occur in certain areas. Figure 8.7 shows the solvents used.

Problems of interfacial chemistry too can be attacked with solubility parameters, whether the adsorption of gases at solid surfaces or their wetting with liquids. Thus the wetting behavior and flocculation tendencies of pigments can be characterized by solubility parameter diagrams just like polymer solubility. An interaction radius R_{j}, as in Equation 8.20, can be determined. Pigment samples are shaken and the tendency to settle is assessed [9]. Some data are listed in Table 8.6.

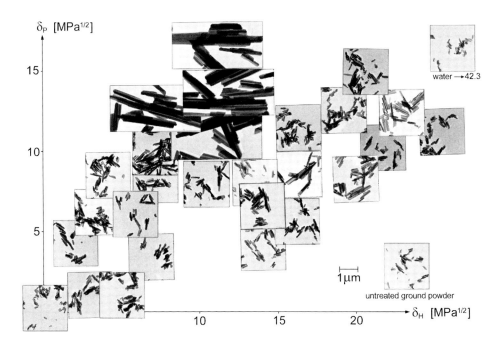

Figure 8.6. Solvent mapping: recrystallization of the pigment shown in Figure 9.6. Ground pigment was recrystallized from different solvents over 2 h at 60°C.

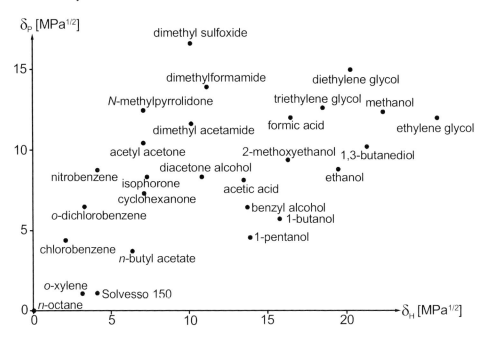

Figure 8.7. Solvents used in the recrystallizations in Figure 8.6.

Table 8.6. Solubility parameters and interaction radii R_j for pigments (source: reference [11]).

Pigment	δ_D	δ_P	δ_H	δ_t	R_j
		[MPa$^{1/2}$]			[MPa$^{1/2}$]
titanium dioxide (Kronos RN 57)	24.1	14.9	19.4	34.4	17.2
graphite Printex V (Degussa)	21.1	12.3	11.3	26.8	11.3
phthalocyanine blue BG (Du Pont)	19.0	6.3	7.6	21.5	4.7
Isolbona red 7522 (Koge)	17.8	7.2	7.2	20.5	5.1
Reflex blue TBK Ext. (Hoechst)	22.1	7.8	13.5	27.0	14.3
Hansa yellow 10G (Hoechst)	18.6	8.2	6.8	21.5	6.8
Heliogen blue B powder (BASF)	22.1	7.2	8.2	24.6	10.6
permanent yellow H 10G (Hoechst)	17.2	3.1	4.7	18.0	4.5

8.7 QSAR, Octanol/Water Distribution Coefficient

Quantitative structure–activity relationships (QSARs) were developed to provide an insight into the influence of substances in complex relationships; they reflect the quantitative relationship between measurable variables and structural parameters. The US EPA Office of Pollution Prevention and Toxins has developed 49 different QSARs to predict the toxicity of substances to water-dwelling organisms [46]. The governing structural parameter in the biological context is the octanol/water distribution coefficient, or its logarithm, log P. With the aid of the measured log P value and the biological effect of known substances, the relationships can be revealed empirically by regression analysis. These relationships or expressions, the QSARs, can then be applied to the prediction of substances that have not been investigated, if their log P values can be estimated. In the fields of toxicology, pharmacology, and environmental science, in fact in the biological sciences generally, QSARs are often applied as a function of log P.

Several approaches to the calculation of log P from the molecular structure have been developed. The method of summation of the Hansch substituent lipophilicity parameters (defined in Equation 8.25) is well known (Equation 8.27).

$$\log P = \sum_i \pi_i \tag{8.27}$$

Work on this topic includes that of Meylan and Howard [47], Hansch and Leo [48], Rekker and de Kort [49], Niemi et al. [50], Klopman et al. [51], Suzuki and Kudo [52], Ghose et al. [53], Bodor and Huang [54], and Broto et al. [55].

Meylan and Howard's method has a better statistical foundation than the others, being validated with data for 7167 substances. Although the method of Hansch and Leo is also good, its validation excluded 10% of the available data on the grounds that the behavior of these substances could not be predicted well. The Rekker method correlated well with a training set of data for 1054 substances, but was not practically validated.

Log P calculations are usually performed by commercial computer software, since without its assistance, a great deal of experience is required to obtain reliable estimates owing to the many corrections necessary. More details are given in reference [46].

8.8 LSER

Interactions in solution are important in all fields of chemistry and biology. Conformation, association, reactivity, and solubility are aspects to consider. Parameters describing the substance and its environment are used in attempts to represent these interactions quantitatively and to estimate their effect on specified properties. As was mentioned in the previous section, this is done by empirical quantification (QSAR) of the complex relationship between such parameters and a measured variable such as, in the biological field, toxicity to fish, or in the field of technology, tribofragmentation. When log P is the substance parameter, it should be possible to represent these QSARs as nonlinear regression lines; in the case of LSERs, we deal instead with multiple linear regression lines.

The LSER (linear solvation energy relationship), a model for solvent effects developed by Kamlet et al. [56], is based on the idea that a cavity must be formed in the solvent which a molecule separated from the solute mass can occupy and where it is subject to forces of attraction. The interactions are represented by linear multiparameter equations (as shown for solubility in Chapter 9).

An interaction or property X can be represented as an LSER as shown in Equation 8.28:

$$X = X_0 + \frac{m \cdot V_1}{100} + s \cdot \pi^* + b \cdot \beta_{\mathrm{m}} + a \cdot \alpha_{\mathrm{m}} \qquad (8.28)$$

where:

$m \cdot V_1/100$:   Cavity term, measure of the endergonic process of solvent removal to make room for the solute molecule. V_1 is the intrinsic molar volume of the substance to be fitted into the cavity. The factor of 1/100 brings this variable to a size similar to that of the others.

$s \cdot \pi^*$:   Dipolarity term: a measure of the exergonic effects of interactions between molecule and solvent, dipole and dipole, dipole and induced dipole.

$b \cdot \beta_{\mathrm{m}}$:   Exergonic hydrogen-bonding term for the molecule as an H acceptor.

$a \cdot \alpha_{\mathrm{m}}$:   Exergonic hydrogen-bonding term for the molecule as an H donor (the index m means "monomer"; special values of α and β must be used for self-associating molecules).

LSER increments $V_1/100$, π^*, β_{m}, and α_{m} for basic structures and substituents are listed in Tables 8.7 and 8.8.

Table 8.7. LSER increments for basic structures (from [57]).

Fragment	$V_1/100$	π^*	β_m	α_m	Fragment	$V_1/100$	π^*	β_m	α_m
Aliphatic basic structures									
n-butane	0.455	0.00	0.00	0.00	cyclohexane	0.598	0.00	0.00	0.00
n-pentane	0.553	0.00	0.00	0.00	cycloheptane	0.690	0.00	0.00	0.00
2-methylbutane	0.543	0.00	0.00	0.00	cyclooctane	0.815	0.00	0.00	0.00
n-hexane	0.648	0.00	0.00	0.00	octahydro-1H-indene	0.884	0.02	0.00	0.00
2-methylpentane	0.638 0.00	0.00	0.00	0.00	decahydronaphthalene	0.982		0.00	0.00
n-heptane	0.745	0.00	0.00	0.00	ethylene oxide	0.249	0.56	0.50	0.00
2-methylhexane	0.735	0.00	0.00	0.00	tetrahydrofuran	0.455	0.58	0.51	0.00
n-octane	0.842	0.00	0.00	0.00	tetrahydropropane	0.553	0.51	0.50	0.00
2-methylheptane	0.832	0.00	0.00	0.00	1,4-dioxane	0.508	0.55	0.41	0.00
2,2,4-trimethylpentane	0.812	0.00	0.00	0.00	tetrahydrothiophene	0.509	0.44	0.27	0.00
cyclopropane	0.310	−0.02	0.00	0.00	pyrrolidine	0.460	0.14	0.70	0.00
cyclobutane	0.450	−0.01	0.00	0.00	imidazolidine	0.431	0.27	0.54	0.00
cyclopentane	0.500	0.00	0.00	0.00	piperidine	0.556	0.17	0.70	0.00
Aromatic basic structures									
benzene	0.491	0.59	0.14	0.00	dibenzofuran	1.581	0.60	0.30	0.00
indan	0.784	0.52	0.14	0.00	dibenzo-*p*-dioxin	1.616	0.45	0.60	0.00
tetrahydronaphthalene	0.883	0.50	0.14	0.00	thiophene	0.445	0.70	0.25	0.00
biphenyl	0.920	1.20	0.28	0.00	pyrrole	0.428	0.74	0.69	0.41
9H-fluorene	0.960	1.18	0.25	0.00	imidazole	0.401	0.87	0.64	0.41
naphthalene	0.753	0.70	0.20	0.00	pyridine	0.472	0.87	0.43	0.00
azulene	0.753	0.90	0.35	0.00	pyridazine	0.451	0.35	0.54	0.00
anthracene	1.015	0.81	0.20	0.00	pyrimidine	0.451	0.87	0.64	0.00
phenanthrene	1.015	0.81	0.20	0.00	pyrazine	0.541	0.92	0.69	0.00
furan	0.370	0.40	0.35	0.00	triazine	0.430	1.15	0.72	0.00

Table 8.8. LSER increments for substituents (from [57]).

Fragment		$V_1/100$	π^*	β_m	α_m	Fragment		$V_1/100$	π^*	β_m	α_m
HC						*Unsaturated and salts*					
1–3 –CH$_3$	ar	0.098	−0.04	0.01	0.00	olefin		−0.026	0.10	0.10	0.05
4–6 –CH$_3$	ar	0.098	−0.04	0.02	0.00	alkyne		−0.036	0.20	0.20	0.13
1–3 –CH$_2$–	ar	0.098	−0.02	0.01	0.00	–C≡CCH$_3$		0.315	0.20	0.17	0.13
4–6 –CH$_2$–	ar	0.098	−0.02	0.02	0.00	ion pair (+,−)		0.000	0.50	0.50	0.00
–CH$_3$, –CH$_2$–, >CH–,											
>C< (prim., sec.)	al	0.098	0.00	0.00	0.00						
(tert., quart.)	al	0.088	0.00	0.00	0.00						
–C$_6$H$_5$		0.491	0.59	0.14	0.00						
–C(C$_6$H$_5$)$_3$		1.485	1.45	0.40	0.00						
Halogens											
–F	al	0.030	0.08	0.19	0.06	–CCl$_3$	ar	0.368	0.35	−0.10	0.15
	ar	0.030	0.03	−0.05	0.08		py	0.368	0.35	−0.10	0.10
	py	0.030	0.04	0.09	0.00	–Br	al	0.131	0.43	0.17	0.05
–CF$_3$		0.188	0.25	−0.25	0.15		ar	0.131	0.20	−0.08	0.10
–Cl	al	0.090	0.35	0.15	0.06		ar + os	0.131	0.04	−0.04	0.00
	ar	0.090	0.12	−0.04	0.00		py	0.131	0.04	0.07	0.07
	ar + os	0.090	0.05	−0.05	0.00	–CH$_2$Br	al	0.257	0.38	0.05	0.00
	py	0.090	0.04	0.09	0.00		ar	0.257	0.05	−0.05	0.00

(Table 8.8, continued)

Fragment		$V_I/100$	π^*	β_m	α_m
–CCl$_3$	al	0.368	0.35	–0.15	0.15
–CBr$_3$	al	0.491	0.40	–0.10	0.12
	ar	0.491	0.40	–0.10	0.15
	py	0.491	0.40	–0.10	0.10

Oxy, hydroxy, carboxy derivatives

Fragment		$V_I/100$	π^*	β_m	α_m
–OH	al	0.045	0.40	0.47	0.33
	al + on	0.045	0.45	0.51	0.31
	ar	0.045	0.13	0.23	0.60
–OOH		0.080	0.41	0.36	0.40
–O–	al	0.045	0.27	0.45	0.00
	al + in	0.045	0.54	0.51	0.00
	ar	0.045	0.10	0.22	0.06
–OO–		0.080	0.28	0.34	0.00
:C=O		0.098	0.81	0.65	0.00
–C(=O)–	al	0.098	0.67	0.48	0.00
	al + os	0.098	0.30	0.35	0.00
	al + in	0.098	0.76	0.52	0.00
	ar	0.098	0.39	0.39	0.06
HC(=O)H		0.140	0.69	0.43	0.00
–C(=O)H	al	0.115	0.65	0.41	0.00
	al + os	0.115	0.33	0.33	0.00
	ar	0.115	0.33	0.42	0.00
HC(=O)OH		0.224	0.65	0.38	0.65
HC(=O)SH		0.294	0.55	0.25	0.05
–OC(=O)H		0.225	0.62	0.37	0.00
–SC(=O)H		0.294	0.55	0.30	0.00
–C(=O)OH	al	0.139	0.60	0.45	0.55
	ar	0.149	0.15	0.30	0.59
–C(=O)SH		0.294	0.55	0.25	0.05
–C(=O)OOH		0.174	0.61	0.34	0.62
–C(=O)O–	al	0.139	0.55	0.45	0.12
	al + on	0.139	0.55	0.49	0.12
	la + in	0.139	0.68	0.51	0.12
	ar	0.139	0.17	0.29	0.12
–C(=O)S–		0.294	0.50	0.30	0.00
–C(=O)OO–		0.174	0.56	0.34	0.12
–C(=O)OC(=O)–		0.395	0.65	0.55	0.00

Sulfur compounds

Fragment		$V_I/100$	π^*	β_m	α_m
–SH	al	0.117	0.35	0.16	0.03
	ar	0.117	0.35	0.02	0.23
–S–		0.117	0.36	0.28	0.00
–SS–		0.234	0.58	0.10	0.00
–S(=O)–	al	0.150	1.00	0.78	0.00
	ar	0.154	1.00	0.62	0.00

Amines, imides, nitriles

Fragment		$V_I/100$	π^*	β_m	α_m
–C(=O)NH$_2$	al	0.185	0.95	0.74	0.56
	ar	0.185	0.35	0.65	0.49
–C(=O)NH–	al	0.183	0.85	0.74	0.25
	la + in	0.183	0.72	0.70	0.28

Fragment		$V_I/100$	π^*	β_m	α_m
–I	al	0.181	0.45	0.18	0.04
	ar	0.181	0.22	0.02	0.10
	ar + os	0.181	0.04	–0.04	0.00
	py	0.181	0.04	0.05	0.05
–OC(=O)OOC(=O)O–		0.522	0.46	0.30	0.12
–C(=O)OOC(=O)–		0.430	0.66	0.44	0.00
–OC(=O)O–		0.185	0.45	0.38	0.12
–OC(=O)OH		0.185	0.55	0.48	0.55
HOC(=O)OH		0.185	0.45	0.60	0.65
–OC(=O)SH		0.339	0.45	0.35	0.15
–OC(=O)S–		0.339	0.42	0.30	0.14
–SC(=O)OH		0.339	0.35	0.38	0.50
HSC(=O)SH		0.374	0.35	0.45	0.10
–SC(=O)SH		0.374	0.45	0.38	0.05
–SC(=O)S–		0.374	0.35	0.33	0.00
–OC(=O)NH$_2$		0.202	0.48	0.78	0.55
–SC(=O)NH$_2$		0.312	0.52	0.65	0.38
–HNC(=O)OH		0.202	0.78	0.82	0.70
–HNC(=O)SH		0.312	0.50	0.63	0.23
–OC(=O)NH–	al	0.202	0.76	0.62	0.36
	ar	0.202	0.76	0.57	0.36
–OC(=S)NH–		0.270	0.56	0.42	0.36
–HNC(=O)S–		0.312	0.50	0.63	0.19
–SC(=O)NH–		0.312	0.53	0.40	0.39
>NC(=O)OH		0.202	0.76	0.84	0.60
>NC(=S)OH		0.312	0.48	0.62	0.05
>NC(=O)O–		0.228	0.75	0.65	0.00
>NC(=O)S–		0.312	0.48	0.60	0.00
H$_2$NC(=O)NH$_2$		0.265	0.90	0.74	0.76
–HNC(=O)NH$_2$		0.265	0.89	0.75	0.65
>NC(=O)NH$_2$		0.265	0.88	0.77	0.38
–HNC(=O)NH–		0.265	0.87	0.77	0.38
–HNC(=S)NH–		0.307	0.67	0.55	0.38
>NC(=O)NH–		0.265	0.85	0.78	0.19
>NC(=O)N<		0.265	0.83	0.74	0.00
–S(=O)$_2$O–		0.221	0.85	0.55	0.00
–OS(=O)O–		0.250	0.70	1.02	0.00
–S(=O)$_2$–	al	0.170	1.00	0.48	0.00
	ar	0.174	1.00	0.42	0.00
–OS(=O)$_2$O–		0.270	0.70	0.72	0.00
–SO$_3$H		0.266	1.00	0.76	0.75
–C(=S)NH–	al	0.225	0.65	0.52	0.31
	ar	0.225	0.65	0.48	0.44
–C(=O)N<	al	0.183	0.76	0.66	0.00
	la + in	0.183	0.74	0.80	0.00

(Table 8.8, continued)

Fragment	$V_1/100$	π^*	β_m	α_m	Fragment	$V_1/100$	π^*	β_m	α_m
	ar 0.183	0.30	0.65	0.30		ar 0.185	0.35	0.65	0.00
HC(=O)NH$_2$	0.185	0.95	0.65	0.49	–N=C<	0.152	0.30	0.75	0.00
–HNC(=O)H	0.185	0.91	0.67	0.25	>C=NOH	0.197	0.55	0.45	0.32
>NC(=O)H	0.185	0.80	0.66	0.00	–N=C=O	0.206	0.75	0.35	0.00
–C(=O)NHC(=O)–	0.430	0.70	0.60	0.33	–N=C=S	0.278	0.63	0.22	0.00
–C(=O)N(R)C(=O)–	0.430	0.65	0.70	0.00	–C≡N	al 0.100	0.65	0.44	0.22
–N=CH$_2$	0.152	0.45	0.80	0.15		ar 0.099	0.20	0.37	0.22
–N=CH–	0.152	0.35	0.78	0.10		py 0.099	0.20	0.37	0.20
Amines, hydrazines									
–NH$_2$	al 0.080	0.32	0.69	0.00	–NHO–	al 0.130	0.52	0.90	0.05
	ar 0.080	0.13	0.38	0.26		ar 0.120	0.35	0.80	0.22
–NH–	al 0.080	0.25	0.70	0.00	>NO–	al 0.128	0.35	0.90	0.05
	ar 0.080	0.13	0.30	0.17		ar 0.116	0.35	0.70	0.05
–N<	al 0.080	0.15	0.65	0.00	>NOH	al 0.130	0.35	0.90	0.14
	ar 0.080	0.13	0.73	0.00		ar 0.116	0.35	0.70	0.14
–NH–NH$_2$	al 0.150	0.75	0.90	0.15	–NHSH	al 0.200	0.65	0.80	0.10
	ar 0.138	0.55	0.90	0.45		ar 0.188	0.55	0.60	0.25
–NH–NH–	al 0.150	0.60	0.85	0.05	–NHS–	al 0.200	0.60	0.75	0.00
	ar 0.138	0.45	0.85	0.35		ar 0.188	0.45	0.55	0.17
>N–NH$_2$	al 0.150	0.65	0.90	0.15	>NS–	al 0.198	0.45	0.85	0.00
	ar 0.138	0.65	0.90	0.20		ar 0.178	0.40	0.45	0.00
>N–NH–	al 0.150	0.50	0.85	0.00	>NSH	al 0.198	0.45	0.70	0.05
	ar 0.138	0.33	0.85	0.17		ar 0.178	0.45	0.50	0.05
>N–N<	al 0.150	0.30	0.80	0.00	–N=O	0.100	0.50	0.15	0.00
	ar 0.138	0.30	0.75	0.00	–NO$_2$	al 0.140	0.79	0.25	0.12
–N=N–	0.125	0.15	0.15	0.00		al + os 0.140	0.35	0.20	0.00
–NHOH	al 0.130	0.56	0.93	0.26		ar 0.140	0.42	0.20	0.16
	ar 0.130	0.26	0.83	0.46		ar + os 0.140	0.10	0.25	0.16
Inorganic groups									
R$_3$P	al 0.160	0.30	0.65	0.00	R$_3$P=S	0.237	0.75	0.47	0.00
	ar 0.160	0.30	0.75	0.00	(–O)$_3$P=S	al 0.387	0.60	0.38	0.00
R$_3$P=O	al 0.195	0.90	1.05	0.00		ar 0.387	0.60	0.92	0.00
	ar 0.195	0.90	0.92	0.00	(–O)(–S)$_2$P=O	0.459	0.55	0.90	0.00
(–O)$_3$P	al 0.295	0.45	0.72	0.00	(–O)$_2$(–S)P=S	0.459	0.55	1.02	0.00
	ar 0.295	0.45	0.50	0.00	(–O)(–S)$_2$P=S	0.531	0.50	1.07	0.00
(–O)$_2$(R)P=O	al 0.270	0.75	0.75	0.00	(–O)(R)(R')P=O	0.235	0.85	0.70	0.00
	ar 0.270	0.75	0.55	0.00	(–O)$_x$(HO)$_y$P=O	0.270	0.85	0.60	0.75
(–O)$_3$P=O	al 0.315	0.65	0.77	0.00	(>N)$_3$P=O	0.420	0.95	1.87	0.00
	ar 0.315	0.65	0.62	0.00	(>N)$_3$P=S	0.462	1.40	2.55	0.00
R$_4$Si	al 0.208	0.00	0.00	0.00	R$_4$Sn	al 0.240	0.10	0.05	0.00
	ar 0.188	0.00	0.00	0.00		ar 0.220	0.10	0.05	0.00

al: aliphatic; ar: aromatic; py: pyridine; os: other substituents; on: on a ring; in: in a ring; la: lactone.

To calculate LSER parameters for a molecule, one imagines it broken down into fragments and sums the values for the individual fragments. Thus, for example, the β_m value for 3-trifluoromethyl-5-cyanophenol is 0.14 + 0.23 + 0.37 – 0.25 = 0.49 (sum of the β_m values for benzene, -OH, -CN, -CF$_3$).

For heavily substituted molecules, the calculated values for π^*, β_m, and α_m may be unrealistic. Reference [57] describes possibilities for correction.

Kamlet et al. used LSER in areas as diverse as substance distribution in blood and tissue, solubility in polymers, toxicity to various species such as bacteria, tadpoles, and goldfish, the effect of irritant substances on the respiratory tracts of mice, adsorption on active charcoal, stationary and mobile phase effects in HPLC [56].

As an example of an LSER equation, we reproduce Equation 8.29 from reference [57] for *Pimephales Promelas* (a minnow). This equation was obtained from measurement data and LSER parameters as given in Tables 8.7 and 8.8 by multiple linear regression.

$$\log LC_{50} = -0.38 - 5.24 \cdot V_1 / 100 - 0.76 \cdot \pi^* + 3.93 \cdot \beta_m - 0.83 \cdot \alpha_m \qquad (8.29)$$

In biology, interactions with water are of particular interest. Table 8.9 shows the degree to which individual terms contribute to these interactions for various molecules. The cavity term (molecular size) is of especial importance, as is hydrogen bonding. The calculated solution energies and solubilities agree with the measured values within the range of experimental error expected for solubility measurements.

Table 8.9. Solubility in water: contributions to the free energy of solution for various substances at 25°C according to the LSER model; from reference [56].

Substance	Cavity	Dipol-arity	H-Bond acceptor	H-Bond donor	Melting term	Total [kcal/mol] calc.	Total [kcal/mol] exp.
pyridine	+3.38	−0.33	−3.51	0	0	−0.46	−0.64
p-nitroaniline	+5.06	−0.33	−2.63	−0.98	+1.65	+2.77	+3.11
benzoic acid	+4.68	−0.33	−2.19	−1.66	+1.32	+1.82	+2.15
benzyl alcohol	+4.57	−0.33	−2.85	−0.73	0	+0.66	+0.54
1,4-dinitrobenzene	+5.55	−0.33	−2.74	0	+2.12	+4.60	+4.54
dibenz[ah]acridine	+10.05	−0.33	−4.06	0	+2.74	+9.34	+8.51
2,2'-biquinoline	+11.82	−0.33	−7.01	0	+2.30	+6.78	+7.35

$V_1/100$, π^*, β_m, and α_m for each substance are obtained by summation of the increments for segments of the molecule. These data, taken from reference [57], are listed in Tables 8.7 and 8.8.

8.9 M Numbers

When it is only necessary to know roughly whether organic liquids are miscible, M numbers are quicker than the lengthy methods described above. Godfrey [58] ordered a selection of 31 standard solvents according to their lipophilicity or affinity for "oil-like substances". Their position in the order (1–31) is their M (miscibility) number, and the following points regarding miscibility have been established:

- Any pair of standard solvents with a difference in M number of less than 16 units is miscible.
- Any pair whose M numbers differ by 16 has a critical demixing temperature between 25 and 75°C.
- A difference of 17 or more implies immiscibility or a demixing temperature greater than 75°C.

When the solvents are arranged in a diagram of the type shown in Figure 8.8, two distinct regions for miscibility and immiscibility are clear. The squares along the hypotenuse edge of the triangle represent the pure solvent and those at the intersections of each row and column correspond to mixtures of those solvents.

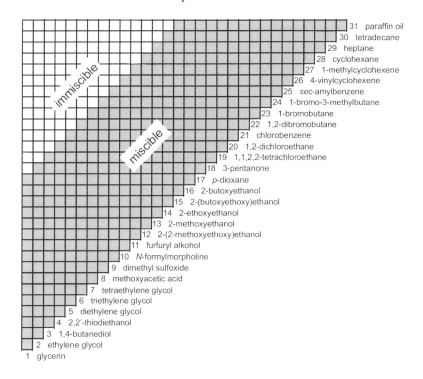

immiscible

miscible

31   paraffin oil
30   tetradecane
29   heptane
28   cyclohexane
27   1-methylcyclohexene
26   4-vinylcyclohexene
25   *sec*-amylbenzene
24   1-bromo-3-methylbutane
23   1-bromobutane
22   1,2-dibromobutane
21   chlorobenzene
20   1,2-dichloroethane
19   1,1,2,2-tetrachloroethane
18   3-pentanone
17   *p*-dioxane
16   2-butoxyethanol
15   2-(butoxyethoxy)ethanol
14   2-ethoxyethanol
13   2-methoxyethanol
12   2-(2-methoxyethoxy)ethanol
11   furfuryl alkohol
10   *N*-formylmorpholine
9   dimethyl sulfoxide
8   methoxyacetic acid
7   tetraethylene glycol
6   triethylene glycol
5   diethylene glycol
4   2,2'-thiodiethanol
3   1,4-butanediol
2   ethylene glycol
1   glycerin

Figure 8.8. M numbers: miscibility of the standard solvents [58].

The standard solvents listed here were used to test many other solvents for miscibility; M numbers were then assigned to these too. The data are given in Table 8.10. For these liquids too a difference in M number of less than 16 implies miscibility. For some of the liquids in the table two numbers are quoted. Such liquids are immiscible with solvents at both ends of the lipophilicity scale; the larger number defines the miscibility limit with solvents of a lower M number, and vice versa. Diethylene glycol acetate (M numbers 12, 19) is thus miscible with solvents of M number between (19 − 15) and (12 + 15), that is, 4–27. An extreme case would be that of the pair (0, 32); this liquid would be immiscible with all the standard solvents.

Table 8.10. M numbers of organic liquids (from [58]).

14	acetic acid	23	isobutyl isobutyrate	19	diethyl adipate
12, 19	acetic anhydride	26	*n*-butyl ether	21	diethyl carbonate
15, 17	acetone	19	butyl formate	5	diethylene glycol
11, 17	acetonitrile	23	butyl methacrylate	12, 19	diethylene glycol diacetate
15, 18	acetophenone	28	butyl oleate	9	diethylenetriamine
11	*N*-acetylmorpholine	26	butyl sulfide	23	diethyl ether
14, 18	acrylonitrile	10	butyrolactone	18	diethyl ketone
8, 19	adiponitrile	14, 19	butyronitrile	14, 20	diethyl oxalate
14	allyl alcohol	26	carbon disulfide	13, 20	diethyl phthalate
22	allyl ether	24	carbon tetrachloride	12, 21	diethyl sulfate
13	2-allyloxyethanol	25	castor oil	17	2,5-dihydrofuran
2	2-aminoethanol	21	chlorobenzene	17	1,2-dimethoxyethane
5	aminoethylethanolamine	19	chloroform	26	dimethoxydimethylsilane
2	2-(2-aminoethoxy)ethanol	20	α-chlorotoluene	13	*N,N*-dimethylacetamide
6	1-amino-2-propanol	29	coconut oil	10	*N,N*-dimethylacetoacetamide
25	*sec*-amylbenzene	14	*p*-cresol	14	2-dimethylaminoethanol
12	aniline	28	cyclohexane	14, 19	dimethyl carbonate
20	anisole	16	cyclohexanecarboxylic acid	12	dimethylformamide
15, 19	benzaldehyde	16	cyclohexanol	12, 19	dimethyl maleate
21	benzene	17	cyclohexanone	11, 19	dimethyl malonate
15, 19	benzonitrile	26	cyclohexene	12, 19	dimethyl phthalate
13	benzyl alcohol	29	cyclooctane	16	1,4-dimethylpiperazine
15, 21	benzyl benzoate	27	cyclooctene	16	2,5-dimethylpyrazine
29	bicyclohexyl	29	decalin	22	dimethyl sebacoate
6	bis(2-hydroxypropyl) maleate	29	decane	12, 17	2,4-dimethylsulfolane
15, 17	bis(2-methoxyethyl) ether	18	1-decanol	9	dimethyl sulfoxide
11, 19	bis(2-methoxyethyl) phthalate	29	1-decene	24	dioctyl phthalate
21	bromobenzene	14	diacetone alcohol	17	*p*-dioxane
6	1,2-butanediol	21	diallyl adipate	15,19	*p*-dioxene
4	1,3-butanediol	25	1,2-dibutoxy ethane	26	dipentene
3	1,4-butanediol	17	*N,N*-dibutylacetamide	26	dipentyl ether
12, 17	2,3-butanedione	23	diisobutyl ketone	22	diphenyl ether
16	butanoic acid	22	dibutyl maleate	23	diphenylmethane
15	1-butanol	22	dibutyl phthalate	25	diisopropylbenzene
16	2-butanol	13	dichloroacetic acid	11	dipropylene glycol
16	butoxyethanol	21	*o*-dichlorobenzene	26	diisopropyl ether
15, 17	isobutoxyethanol	20	1,2-dichloroethane	23	diisopropyl ketone
15	2-(2-butoxyethoxy)ethanol	20	dichloromethane	12, 17	dipropylsulfone
22	*n*-butyl acetate	12	1,3-dichloro-2-propanol	29	dodecane
21	isobutyl acetate	26	dicyclopentadiene	18	1-dodecanol
22	*sec*-butyl acetate	26	didecyl phthalate	29	1-dodecene
15	isobutyl alcohol	1	diethanolamine	14, 19	epichlorohydrin
16	*tert*-butyl alcohol	14	*N,N*-diethylacetamide	15, 19	epoxyethylbenzene

(Table 8.10, continued)

5	ethanesulfonic acid	28	1-heptene	19	methyl isoamyl ketone
14	ethanol	26	hexachlorobutadiene	27	2-methyl-1-butene
14	2-ethoxyethanol	30	hexadecane	26	2-methyl-2-butene
13	2-(2-ethoxyethoxy)ethanol	29	1-hexadecene	19	methyl isobutyl ketone
15, 19	2-ethoxyethyl acetate	15	hexamethyl phosphoramide	13, 19	methyl chloroacetate
19	ethyl acetate	29	hexane	8, 17	methyl cyanoacetate
13, 19	ethyl acetoacetate	5	2,5-hexanediol	29	methylcyclohexane
24	ethylbenzene	12, 17	2,5-hexanedione	27	1-methylcyclohexene
21	ethyl benzoate	2	1,2,6-hexanetriol	28	methylcyclopentane
17	2-ethylbutanol	17	1-hexanol	17	methyl ethyl ketone
22	ethyl butyrate	27	1-hexene	14, 19	methyl formate
6, 17	ethylene carbonate	6	3,3'-hydroxydipropionitrile	8	2,2'-methyliminodiethanol
9	ethylene diamine	2	2-hydroxyethyl carbamate	20	methyl methacrylate
2	ethylene glycol	1	2-hydroxyethylformamide	13	methyl methoxyacetate
12, 19	ethylene glycol diacetate	12	2-hydroxyethyl methacrylate	16	N-methylmorpholine
8, 17	ethylene glycol diformate	3	2-hydroxypropyl carbamate	22	1-methylnaphthalene
10, 19	ethylene monothiocarbonate	14, 17	hydroxypropyl methacrylate	26	methyl oleate
9	ethylformamide	22	iodobenzene	7	5-methyloxazolidinone
15, 19	ethyl formate	22	iodoethane	29	2-methylpentane
14, 17	2-ethyl-1,3-hexanediol	21	iodomethane	29	3-methylpentane
17	2-ethylhexanol	18	isophorone	14	2-methyl-2,4-pentanediol
23	ethyl hexanoate	25	isoprene	17	4-methyl-2-pentanol
14	ethyl lactate	30	kerosene	28	4-methyl-1-pentene
16	N-ethylmorpholine	9	2-mercaptoethanol	27	cis-4-methyl-2-pentene
23	ethyl orthoformate	24	mesitylene	13	1-methyl-2-pyrrolidinone
21	ethyl propionate	18	mesityl oxide	26	methyl stearate
13	ethylthioethanol	15, 19	methacrylonitrile	23	α-methylstyrene
21	2-ethyl trichloroacetate	4	methanesulfonic acid	10, 17	3-methylsulfolane
20	fluorobenzene	12	methanol	14	morpholine
21	1-fluoronaphthalene	8	methoxyacetic acid	14, 20	nitrobenzene
3	formamide	11, 19	methoxyacetonitrile	13, 20	nitroethane
5	formic acid	14	3-methoxybutanol	10, 19	nitromethane
10	N-formylmorpholine	13	2-methoxyethanol	15, 20	2-nitropropane
20	furan	12	2-(methoxyethoxy)ethanol	17	1-nonanol
11, 17	furfural	14, 17	2-methoxyethyl acetate		nonylphenol
11	furfuryl alcohol	5	3-methoxy-1,2-propanediol		1-octadecene
1	glycerin	15	1-methoxy-2-propanol		1,7-octadiene
3	glycerin carbonate	11, 17	3-methoxypropionitrile		octane
13, 19	glycidyl phenyl ether	15	3-methoxypropylamine		1-octanethiol
29	heptane	10	3-methoxypropylformamide		1-octanol
17	1-heptanol	15, 17	methyl acetate		2-octanol
22	3-heptanone	19	methylal		2-octanone
23	4-heptanone	11	2-methylaminoethanol		1-octene

(Table 8.10, continued)

28	*trans*-2-octene	19	propyl acetate	28	tributylamine
7	PEG-200	19	isopropyl acetate	29	triisobutylene
7	PEG-300	24	isopropylbenzene	18	tributyl phosphate
8	PEG-600	9, 17	propylene carbonate	24	1,2,4-trichlorobenzene
25	1,3-pentadiene	17	propylene oxide	22	1,1,1-trichloroethane
7	pentaethylene glycol	16	pyridine	19	1,1,2-trichloroethane
9	pentaethylenehexamine	10	pyrrolidinone	20	trichloroethylene
9	pentafluoroethanol	22	styrene	20	1,2,3-trichloropropane
3	1,5-pentanediol	9, 17	sulfolane	21	tricresyl phosphate
12, 18	2,4-pentanedione	13, 19	1,1,2,2-tetrabromoethane	2	triethanolamine
17	1-pentanol	19	1,1,2,2-tetrachloroethane	26	triethylamine
23	pentyl acetate	25	tetrachloroethylene	25	triethylbenzene
16	*tert*-pentyl alcohol	30	tetradecane	6	triethylene glycol
20	phenetole	29	1-tetradecene	9	triethylenetetramine
12	2-phenoxyethanol	7	tetraethylene glycol	14	triethyl phosphate
13, 17	1-phenoxy-2-propanol	9	tetraethylene pentamine	16	trimethylborate
12, 19	phenylacetonitrile	23	tetraethylorthosilicate	12	trimethylnitrilotripropionate
10	phenylethanolamine	17	tetrahydrofuran	29	2,2,4-trimethylpentane
16	2-picoline	13	tetrahydrofurfuryl alcohol	27	2,4,4-trimethyl-1-pentene
14	PPG-400	21	tetrahydrothiophene	27	2,2,4-trimethyl-2-pentene
14, 23	PPG-1000	24	tetralin	10	trimethyl phosphate
11	propanediamine	29	tetramethylsilane	26	tripropylamine
4	1,2-propanediol	15	tetramethylurea	12	tripropylene glycol
3	1,3-propanediol	29	tetrapropylene	20	vinyl acetate
15	1-propanol	4	2,2'-thiodiethanol	22	vinyl butyrate
15	2-propanol	8	1,1'-thiodi-2-propanol	26	4-vinylcyclohexene
7, 19	propanosultone	6, 19	3,3'-thiodipropionitrile	26	vinylidenenorbornene
19	isopropenyl acetate	20	thiophene	23	*m*-xylene
15	propionic acid	23	toluene	23	*o*-xylene
13, 17	propionitrile	11, 19	triacetin	24	*p*-xylene

References for Chapter 8:

[1]   J. H. Hildebrand, R. L. Scott, The Solubility of Nonelectrolytes, 3rd ed., Dover Publications, Inc., New York, 1964.

[2]   G. Scatchard, J. Am. Chem. Soc. *56*, 995 (1934).

[3]   E. B. Bagley, T. B. Nelson, J. M. Scigliano, J. Paint Technol. *43*, 35 (1971); J. Phys. Chem. *77*, 2794 (1973).

[4]   R. F. Blanks, J. M. Prausnitz, Ind. Eng. Chem. Fundam. *3*, 1 (1964).

[5]   R. A. Keller, B. L. Karger, L. R. Snyder, Gas Chromatogr. Proc. Int. Symp. (Eur.) *8*, 125 (1971).

[6]   H. Burrell, Interchem. Rev. *14*, 3, 31 (1955).

[7]   J. D. Crowley, G. S. Teague, J. W. Lowe, J. Paint Technol. *38*, 269 (1966).

[8]   R. C. Nelson, R. W. Hemwall, G. D. Edwards, J. Paint Technol. *42*, 636 (1970).

[9]   C. M. Hansen, J. Paint Technol. *39*, 104, 505 (1967).

[10]   C. M. Hansen, K. Skaarup, J. Paint Technol. *39*, 511 (1967).

[11]   A. F. M. Barton, CRC Handbook of Solubility Parameters and Other Cohesion
       Parameters, CRC Press, Boca Raton, FL, 1983.

[12]   R. F. Fedors, Polym. Eng. Sci. *14*, 147, 472 (1974).

[13]   O. Exner, Collect. Czech. Chem. Commun. *32*, 1 (1967).

[14]   D. W. van Krevelen, P. J. Hoftyzer, Properties of Polymers: Their Estimation and
       Correlation with Chemical Structure, 2nd ed., Elsevier, Amsterdam, 1976.

[15]   C. M. Hansen, A. Beerbower, Solubility Parameters, in Kirk–Othmer, Encyclopedia
       of Chemical Technology Supp. Vol., 2nd ed., Interscience, New York, 1971.

[16]   R. A. Orwoll, Rubber Chem. Technol. *50*, 451 (1977).

[17]   P. J. Flory, J. Chem. Phys. *10*, 51 (1942).

[18]   P.J. Flory, Principles of Polymer Chemistry, Cornell University Press, Ithaca, NY, 1953.

[19]   P. A. Small, J. Appl. Chem. *3*, 71 (1953).

[20]   A. Beerbower, J. R. Dickey, Am. Soc. Lubric. Eng. Trans. *12*, 1 (1969).

[21]   G. Walz, G. Emrich, Kunstharz Nachr. (Höchst) *34*(8), 19 (1975).

[22]   J. P. Teas, J. Paint Technol. *40*, 19 (1968).

[23]   J. L. Gardon, J. P. Teas, Treatise on Coatings, Vol. 2, Characterization of Coatings:
       Physical Techniques (R. R. Myers, J. S. Long, Eds.), Marcel Dekker, New York, 1976.

[24]   H. Ahmad, M. Yaseen, Polym. Eng. Sci. *19*, 858 (1979).

[25]   R. F. Boyer, H. Keskkula, in Encyclopedia of Polymer Science and Technology,
       Vol. 13, John Wiley & Sons, New York, 1970.

[26]   W. Sliwka, Angew. Chem. Int. Ed. Engl. *14*, 538 (1975).

[27]   L. Broens, D. M. Koenhen, C. A. Smolders, Desalination *22*, 205 (1977).

[28]   A. Beerbower, D. A. Pattison, G. D. Staffin, Am. Soc. Lubric. Eng. Trans. *6*, 80
       (1963).

[29]   O. Olabisi, J. Appl. Polym. Sci. *22*, 1021 (1978).

[30]   D. J. David, T. F. Sincock, Polymer *33*, 4505 (1992).

[31]   J. Sevestre, Peint. Pigments Vernis. *42*, 838 (1966).

[32]   F. Higashide, K. Omata, Y. Nozowa, H. Yoshioka, J. Polym. Sci. Polym. Chem.
       15, 2019 (1977).

[33]   C. M. Hansen, J. Paint Technol. *39*, 505 (1967).

[34]   C. M. Blow, Polymer *14*, 309 (1973).

[35]   H. T. Dellicolli, in Controlled Release Pesticides (ACS Symp. Ser. 53),
       (H. B. Scher, Ed.), American Chemical Society, Washington D.C., 1977.

[36]   A. M. Thomas, Agric. Food Chem. *12*, 442 (1964).

[37]   S. A. Simon, W. L. Stone, P. B. Bennett, Biochim. Biophys. Acta *550*, 38 (1979).

[38]   P. Schmid, Physiol. Chem. Phys. *5*, 141 (1973).

[39]   A. S. Michaels, P. S. L. Wong, R. Prather, R. M. Gale, Am. Inst. Chem. Eng. J.
       21, 1073 (1975).

[40]   E. Squillante, T. E. Needham, H. Zia, Proc. Int. Symp. Controlled Release Bioact.
       Mater. *19th*, 495 (J. Kopecek, Ed.), Controlled Release Soc., Deerfield, Ill, 1992.

[41]   D. H. Kaelble, J. Moacanin, Med. Biol. Eng. Comput. *17*, 593 (1979).

[42]   M. J. Lever, K. W. Miller, W. D. M. Patton, E. B. Smith, Nature (London) *231*, 368 (1971).

[43]   A. Cammarata, S. J. Yau, K. S. Rogers, Pure Appl. Chem. *35*, 495 (1973).

[44]   C. Hansch, in Structure–Activity Relationships (C. J. Cavallito, Ed.), Pergamon Press, Oxford, 1973.

[45]   R. Osman, H. Weinstein, J. P. Green, in Computer-Assisted Drug Design (ACS Symp. Ser. 112) (E. C. Olson, R. E. Christoffersen, Eds.), American Chemical Society, Wahington D.C., 1972.

[46]   W. M. Meylan, P. H. Howard, in Techniques in Aquatic Toxicology (G. K. Ostrander, Ed.), CRC Press, Boca Raton, FL, 1996.

[47]   W. M. Meylan, P. H. Howard, J. Pharm. Sci. *84*, 83 (1995).

[48]   C. Hansch, A. J. Leo, Substituent Constants for Correlation Analysis in Chemistry and Biology, John Wiley & Sons, New York, 1979.

[49]   R. F. Rekker, H. M. de Kort, Eur. J. Med. Chem. *14*, 479 (1979).

[50]   G. J. Niemi, S. C. Basak, G. D. Veith, G. Grunwald, Environ. Toxicol. Chem. *11*, 893 (1992).

[51]   G. Klopman, J. Y. Li, S. Wang, M. Dimayuga, J. Chem. Inf. Comput. Sci. *34*, 752 (1994).

[52]   T. Suzuki, Y. Kudo, J. Computer-Aided Mol. Design *4*, 155 (1990).

[53]   A. K. Ghose, A. Pritchett, G. M. Crippen, J. Comput. Chem. *9*, 80 (1988).

[54]   N. Bodor, M. J. Huang, J. Pharm. Sci. *81*, 272 (1992).

[55]   P. Broto, G. Moreau, C. Vandycke, Eur. J. Med. Chem. *19*, 71 (1984).

[56]   M. J. Kamlet, in Progress in Physical Organic Chemistry *19* (R. W. Taft, Ed.), John Wiley & Sons, New York, 1993.

[57]   J. P. Hickey, in Techniques in Aquatic Toxicology (G. K. Ostrander, Ed.), CRC Press, Boca Raton, FL, 1996.

[58]   N. B. Godfrey, Chemtech, June 1972, 359.

9 Solubility, Crystallization

9.1 Solubility

The properties of active ingredients in formulations (pharmaceuticals, cosmetics, agro-chemicals, dyes, foods, drinks, and luxuries) depend largely on their state of aggregation. If a substance is formulated to reach a target quickly by diffusion, then the active ingredient should be contained in a solution of low viscosity. In microemulsions, emulsions, or vesicle preparations, material is transported more slowly. Matters are still worse if the active substance is in crystalline form: for one thing, the amount in solution is smaller, which reduces mass transport, and for another, the dissolution process itself may be rate-determining.

However, if the intention is the prevention or retardation of a chemical reaction such as hydrolysis, rearrangement, photolysis, or radical attack, then crystalline formulations are preferable, since the rate-determining step usually occurs in solution.

Molecules at the surface of the crystal often adopt a conformation unfavorable to attack. In addition, a comparison with the liquid state must include the crystallization energy or the enthalpy of fusion. Reactions at solid surfaces are therefore generally slower than those in solution. The high fastness to light of organic pigments is often due to their high fusion enthalpies, together with their high melting points. These cause their low solubility in the binding material so that decomposition, already energetically hindered, can occur practically only at the surface of the crystals.

It is not always necessary to use crystalline active substances for reasons of stability. In many cases, dissolved active substances can be protected from photochemical and free-radical attack by the addition of antioxidants and UV/light absorbers.

Regardless of whether the active substance is formulated as a solution, as a liquid, or as a solid, its solubility is important. Knowledge of the solvent, temperature, and structure dependence is always an advantage to a formulation chemist.

9.1.1 Thermodynamics of Solubility

The solubility of solids can be expressed by the following formula:

$$a = f \cdot x = \exp\left[-\frac{\Delta H_F}{R} \cdot \left(\frac{1}{T} - \frac{1}{T_F}\right)\right] \tag{9.1}$$

or:

$$\log x = -\frac{\Delta H_F}{2.303 \cdot R}\left(\frac{1}{T} - \frac{1}{T_F}\right) - \log f \tag{9.2}$$

a: activity of the dissolved substances
f: activity coefficient of the dissolved substance
x: solubility of the solid substance (as a mole fraction)
ΔH_F: enthalpy of fusion
R: gas constant
T: temperature of the solution
T_F: melting point

The first term in Equation 9.2 is the ideal solubility. The parameters for enthalpy of fusion ΔH_F and melting point T_F depend only on the solid and are normally easily determined experimentally by means of differential scanning calorimetry (DSC). The *solvent dependence of the solubility* is quantified by the *activity coefficient f*, which is different for each combination of solid and solvent. Enthalpy of fusion and melting point are usually measurable, so the solubility can be calculated for each temperature as long as the activity coefficient *f* for the appropriate solid–solvent combination is known. There has been no lack of attempts to develop thermodynamic models for the calculation of these activity coefficients from structure increments for solvents and the substances to be dissolved, for example UNIFAC, SUPERFAC, NRTL, ASOG, and MOSCED. All these models are based on mixing of liquids; that is, instead of a solid being dissolved directly, the substance could first be melted, then supercooled to temperature *T*, and finally mixed with the solvent. Both routes lead to the same solution (Figure 9.1).

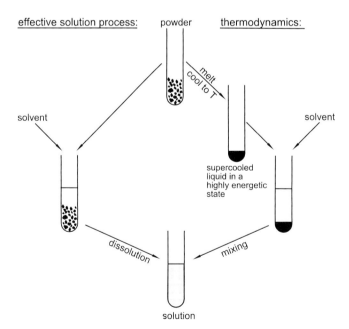

Figure 9.1. Two routes by which a solid can be dissolved: a) direct mixing of powder and solvent; b) melting of the powder followed by mixing of the molten substance and the solvent.

Equation 9.1 follows this theoretical cycle; it is the thermodynamic description of the mixing of two liquids, namely the hypothetically molten solid with the solvent. The reference state for the process is therefore the substance that was melted at the temperature T_F and cooled to the mixing temperature T. This hypothetical cooled liquid would be in a highly energetic, very unstable state and would interact with solvents with which it was mixed in a completely different manner to normal liquids. This is also the reason that *quantitative solubility models are only suitable for solids with low enthalpies of fusion, that is, those in which the hypothetical supercooled liquid differs little from normal liquids.* Even low molecular mass solid pharmaceuticals exhibit broad scattering in quantitative models [1], and still more extreme deviations are observed for poorly soluble organic pigments with high enthalpies of fusion and high melting points.

9.1.2 Temperature Dependence of Solubility

While it is difficult to quantify the solvent dependence of the solubility of a solid with simple models, Equation 9.1 can be applied to the temperature dependence with good results. Rewritten in logarithmic form it becomes Equation 9.3 (C = solubility, for example in g/L).

$$C = K_1 \cdot 10^{-\frac{K_2}{T}} \tag{9.3}$$

or, as logarithms:

$$\log C = \log K_1 - \frac{K_2}{T} \tag{9.4}$$

In Equation 9.4, log C is expressed as a function of $1/T$ in a linear equation with the constants log K_1 and K_2. As Figure 9.2 shows, the constants may be determined directly from the line of the graph, or by calculation from two solubilities determined experimentally at different temperatures, $C_1(T_1)$ and $C_2(T_2)$:

$$K_1 = C_1 \cdot 10^{\frac{T_2}{T_2 - T_1} \cdot \log\left(\frac{C_2}{C_1}\right)} \tag{9.5}$$

$$K_2 = \frac{T_2 - T_1}{T_1 \cdot T_2} \cdot \log\left(\frac{C_2}{C_1}\right) \tag{9.6}$$

Once the two constants have been determined, the solubility for any temperature can be calculated from Equation 9.3. However, it is prudent not to rely on just two solubility determinations for the calculation of the constants but rather to check with at least three values whether the line in the range in question really is straight. Protonation, association, complex formation and other processes in solution can cause deviations from a straight line, as can phase transitions in the solid.

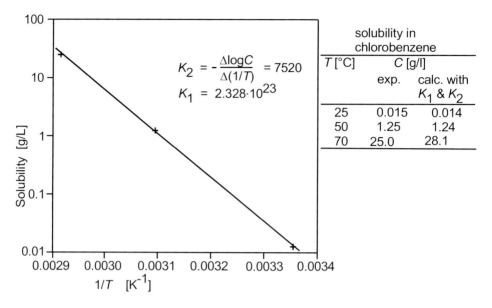

T [°C]	C [g/l]	
	exp.	calc. with K_1 & K_2
25	0.015	0.014
50	1.25	1.24
70	25.0	28.1

solubility in chlorobenzene

$$K_2 = -\frac{\Delta \log C}{\Delta(1/T)} = 7520$$

$$K_1 = 2.328 \cdot 10^{23}$$

Figure 9.2. Solubility measurements (the substance formula is shown in Figure 9.4): graphical determination of the constant K_2 from the gradient. K_1 was determined by substitution of the values for K_2 and T_2 in Equation 9.3.

It is worth reminding the reader that, roughly speaking, the temperature dependence is expressed in the constant K_2 and the solvent dependence in the constant K_1 of Equation 9.3 (Figure 9.3). Whether a substance be dissolved in water, oil, micelles, vesicles, or in a polymer, its relative temperature dependence is approximately the same, regardless of the substrate, unless a chemical change occurs or the dissolved substance is in equilibrium with other forms in solution.

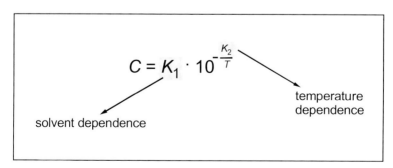

Figure 9.3. Solubility: separation of temperature and solvent dependence.

9.1.3 Phase Transitions in Solids

It is not unusual for a solid to experience a so-called phase transition. On warming, such solids convert into differently ordered states at certain temperatures, for which they require energy.

Figure 9.4 shows such a case, specifically, the calorimetric behavior of a dye with long hydrocarbon residues as substituents.

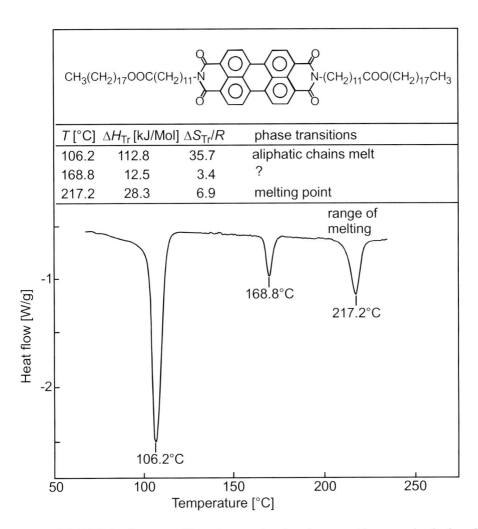

T [°C]	ΔH_{Tr} [kJ/Mol]	$\Delta S_{Tr}/R$	phase transitions
106.2	112.8	35.7	aliphatic chains melt
168.8	12.5	3.4	?
217.2	28.3	6.9	melting point

Figure 9.4. DSC plot for a crystalline substance showing phase transitions occuring in the solid. In the first phase transition, at 106.2°C, the HC chains, which make the largest contribution to the lattice energy, are "melted". This would increase the solubility by a factor of 15000 (if this is possible, depending on the solvent; calculated from Equation 9.1). The entropy of transition $\Delta S_{Tr}/R$ is of the same order of magnitude as that for similar hydrocarbons [2].

Besides the melting peak at 217.2°C in the DSC plot, two solid phase transitions are visible at 106.2 and 168.8°C. It is clear that the first phase transition requires much more energy than fusion itself. For the molecule shown here, the transition is a partial "melting" of the HC chains. Instead of being restricted to lattice oscillations, these can now rotate freely about their long axis in the crystal structure. More frequently, though, phase transitions involve changes in the lattice structure rather than partial melting of molecular residues.

After a phase transition has been completed, the enthalpy of fusion ΔH_F is reduced by the enthalpy of the phase transition ΔH_{Tr}. Accordingly, the solubility, expressed in Equation 9.1, changes suddenly. In the example, when the compound passed the first phase transition the solubility would increase by a factor of 15 000 (assuming that this were possible, depending on the solvent)!

Attempts to obtain large crystals of such a substance by cooling a supersaturated solution are destined to fail; when the solution temperature drops below the phase transition temperature many small crystals form suddenly. There may be other anomalies: crystallization in other modifications, another crystal habit, other crystal defects such as twinning or the inclusion of other molecules (solid solutions), etc.

9.1.4 Association in Solution

As has already been mentioned, equilibrium processes that may occur after dissolution of the substance must not be overlooked: protonation in protic solvents, tautomerization, association, and so on. All these processes increase the solubility, with different temperature dependences for different substances. This fact can result in distortion of the log C–$1/T$ curve.

Associates are in equilibrium with one another (M_1 monomer, M_2 dimer, ...):

$$M_1 + M_1 \Leftrightarrow M_2; \qquad M_2 + M_1 \Leftrightarrow M_3; \qquad \text{............} \tag{9.7}$$

The concentration of dissolved substance is given by Equation 9.8 (C_1: monomer concentration, ..., k_2: equilibrium constant for the dimer, ...):

$$C = C_1 + 2C_2 + 3C_3 + ... \quad = C_1 + 2k_2C_1^2 + 3k_3C_1^3 + ... \tag{9.8}$$

Substances with a tendency to form associates do not form them in all solvents. One substance may form them preferentially in alcohol, say, another in polar aprotic solvents such as dimethyl sulfoxide, a third in apolar solvents.

Figure 9.5 shows solubility determinations for two substances, one of which (a) associates strongly in DMF, while the other (b) is present in unassociated form (measuring method described later).

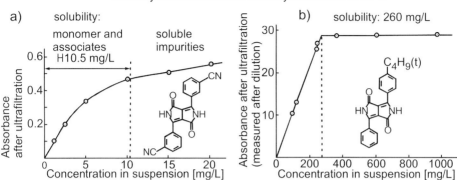

Figure 9.5. Solubility determination for finely divided powders: a) substance strongly associated in solution; b) no association observable.

9.1.5 Solvent Dependence of Solubility; Effect of Substituents

Thermodynamic models are based on the hypothetical melting and supercooling of the solid and its mixing with the solvent as already described. If the enthalpy of fusion is small and the melting point is low, the hypothetical molten solid is similar to the solvent and the various incremental models for mixing described can be applied with good results.

A simple model is based on solubility parameters for solvents (see Chapter 8). Equation 9.9 can be derived from Equation 9.1 [3].

$$\ln C_S = \frac{V_S}{V_L} + k_0 + \sum_n \left(k_{Dn} \cdot \left(\delta_D^L \right)^n + k_{Pn} \cdot \left(\delta_P^L \right)^n + k_{Hn} \cdot \left(\delta_H^L \right)^n \right) \qquad (9.9)$$

(C_S: solubility; V_S and V_L: molar volume of the solid S or solvent L; δ_D^L, δ_P^L, δ_H^L: solubility parameters of the solvent L, index D: dispersion term, P: polar term, H: hydrogen-bonding term; k_0, k_{Dn}, k_{Pn}, k_{Hn}: constants)

Using seven constants ($n = 2$ in Equation 9.9) and the solubility parameters of the solvents, Richardson et al. [4] found that calculated and experimentally determined solubilities for pharmaceutical products agreed well.

Less satisfactory results are obtained for substances with high melting points and high enthalpies of fusion. Calculated and experimentally determined values for a test substance with an enthalpy of fusion five times higher and a melting point 159° higher than those of the substance used by Richardson deviated considerably, even when 16 constants ($n = 5$) were used, as can be seen from Figure 9.6.

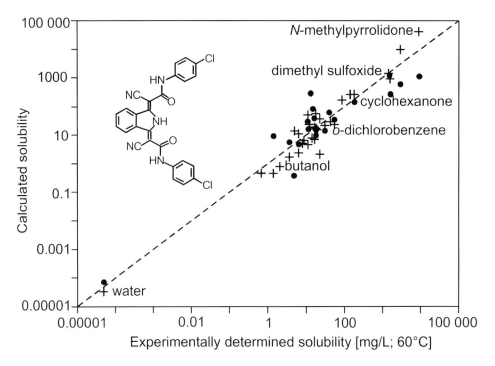

Figure 9.6. Solvent dependence of the solubility of a pigment; comparison of experimentally determined and calculated values. Two models: + Equation 9.9, *n* = 5 (to aid understanding, some of the points are labeled with the solvent); • Equation 9.10 (for those solvents for which solvatochromic parameters are known).

The model of solvent effects developed by Kamlet et al. [5] is based on the concept that a cavity must be created in the solvent in which a molecule separated from the mass of the solute can be placed and where it is subjected to forces of attraction. This model permits various properties that depend on the interactions between solute and solvent to be quantified by LSER (linear solvation energy relationship) equations, for example solubility:

$$\log C_S = k_1 + k_2 \cdot \left(\delta_H^L\right)^2 \cdot V_S + k_3 \cdot \pi_L^* \cdot \pi_S^* + k_4 \cdot \alpha_L \cdot \beta_S + k_5 \cdot \beta_L \cdot \alpha_S + k_6 \cdot (mp - T)$$

$$(9.10)$$

(C_S: solubility; $k_1 \dots k_6$: constants; δ_H^L: solubility parameter of the solvent for hydrogen bonding; V_S: molar volume of the dissolved substance; π^*, α, β: solvatochromic parameters for dipolar effects and for hydrogen-bond donor and acceptor properties of the solvent ($_L$) and the solute ($_S$); *mp*: melting point; *T*: solution temperature)

The term for conversion of the solid into the supercooled liquid depends on the enthalpy of fusion, the melting point, and the solution temperature, according to

Equation 9.2. However, Yalkowsky et al. [6] found that, for substances with melting points and enthalpies of fusion that are not extremely high, this term can be simplified (the last term in Equation 9.10). Yalkowsky obtained a value of 0.011 for the constant k_6; Kamlet's value is 0.0099 [7].

This model does not achieve a better agreement with reality than Equation 9.9 for the influence of the solvent, as can be seen from Figure 9.6. The π^*, α, and β values for the solvents were taken from reference [8].

Other semiempirical or empirical methods include those of Yalkowsky et al. [6], Yoshimoto et al. (incremental) [9], and Klopman et al. (incremental) [10]. In addition, the use of log P (the logarithm of the octanol–water distribution coefficient) is used to estimate the water solubility of pharmaceutical and agrochemical agents. Log P values can be calculated incrementally [11].

None of these methods is suitable, though, for the estimation of solid solubilities simply on the basis of their structural formulae. At the very least an estimate of the melting point is required too in order to take into account the conversion of the solid into the hypothetically supercooled liquid (the last term in Equation 9.10).

9.1.6 Determination of the Solubility

The rate of dissolution of a solid in a solvent can be represented by Equation 9.11 (C: concentration in solution; C_S: solubility; A: surface area of the solid; k: solution rate constant):

$$\frac{dC}{dt} = k \cdot A \cdot (C_S - C) \tag{9.11}$$

It follows from Equation 9.11 that:
- the larger the surface area of the solid (small particles, large quantity of solid), the faster the substance dissolves;
- sparingly soluble substances dissolve more slowly than readily soluble ones;
- towards the end of the dissolution process, before the solubility has been attained, the dissolution slows considerably;
- the constant k includes the viscosity of the solution, amongst other things, so the greater the viscosity, the slower the rate of dissolution.

One method of determining the solubility of a solid in a solvent involves raising the temperature until all the substance present is dissolved. This yields the solubility for that temperature. However, just before the substance dissolves completely the temperature increase must be very slow, since at this point the rate of dissolution is very small. Solubility values obtained by this procedure are generally too low, since the final state was attained dynamically because of the speed at which the temperature was raised and is not an equilibrium state. For poorly soluble substances in particular the error can be considerable. However, this method is useful when only a quick, rough determination of solubility is required for orientation purposes.

If more exact values are needed, then an excess of solid is mixed with the solvent and the dispersion stirred at constant temperature until equilibrium is reached (hours or days, depending on the substance). For complete certainty, the substance can be dissolved in excess at a slightly higher temperature and then the excess allowed to crystallize out at the desired temperature. The effective solubility then lies between the two values.

To determine the amount of solute in solution, the undissolved solid must then be separated from the solution. For large particles, sedimentation can be used; finely divided solids must be filtered in an apparatus equipped with a thermostat, preferably by ultrafiltration through filters with fine pores. Even then it is not impossible for particles of solid to pass through the filter and upset the solubility determination. This is a particular problem for fine, poorly soluble powders.

Often spectrometric methods are applied too, VIS, IR, NMR, or GC methods, and calibration curves used to determine the concentration in the saturated solution (or in a diluted solution if the saturation concentration is too high). If the dissolution temperature lies above the temperature at which the measurement is carried out, the solution must be diluted in any case to prevent the solute crystallizing out. If the method selected is specific to the substance to be determined, in other words if impurities are distinguishable as such, then exact values can be obtained.

Gravimetry is a general method suitable for relatively volatile solvents such as water. The method works if the solvent can be evaporated without loss or decomposition of the substance itself. If impurities are present, the solubility determination is carried out with a series of dispersions of increasing solid content. A graph like the one reproduced in Figure 9.7 then reveals the solubilities of the substance itself and the impurity.

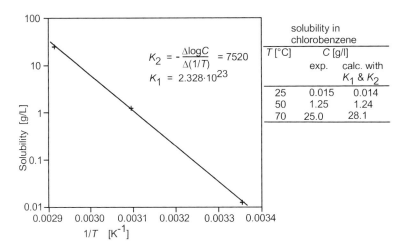

Figure 9.7. Gravimetric solubility determination for a series of suspensions of increasing concentration.

When the solubility is very low, a spectrometric method must be used instead of gravimetry. If it is impossible to distinguish between substance and impurity, the solubility determination will be correspondingly inexact, as is the case in Figure 9.5a. To avoid the deformation of the curve resulting from association shown there, the solution for measurement should be diluted to such a strength that association is completely suppressed.

9.2 Crystallization

Crystallization is an important technique in the chemist's armory, whether used in isolation, in purification, or in obtaining powders of specified size. The best-known example of purification by crystallization is probably the high purity of refined sugar compared with that of the raw product, but not only solids are isolated and purified by crystallization: liquids too are separated from by-products on a large scale by crystallization at low temperatures rather than by distillation. Crystallization has particular advantages in the case of azeotropic mixtures and when the two compounds have similar boiling points. Concentration of fruit juices by freeze drying and desalination of seawater are further examples of technically important crystallization processes concerning liquids.

To ensure that the crystalline powder obtained has the specified particle size and size range, nucleation, particle growth, and aggregation must be controllable in *industrial crystallization*. These processes occur simultaneously, whether in batch or continuous processes. A. D. Randolph and M. A. Larson worked out the theory behind the processes [12].

Aside from nucleation, the process of crystallization can be split into three stages:
1. transport of the substance to the crystal surface;
2. migration on the surface and incorporation into the crystal lattice;
3. loss of the heat of crystallization.
Any of these steps may be rate-determining. The slowest of these controls the kinetics of crystallization.

9.2.1 Methods of Crystallization

The many different designs of equipment for industrial crystallization can be classified into just a few types [13]: batch or continuous crystallizers with or without agitation or with controlled or uncontrolled supersaturation; removal of particles in a specified size range in a fluidized-bed step by obstruction of settling (classification); with or without circulation of the mother liquor or magma (mother liquor and crystals).

Crystallization by cooling: Uncontrolled cooling seldom gives the best crystals. In the initial phase, when cooling occurs very fast, too many nuclei form. More advantageously, the temperature change is controlled so that at the start of the process the cooling rate is much lower than at the end. Agitation during crystallization results in smaller but more even and also purer crystals than those obtained without agitation, when the magma settles out into agglomerates containing considerable amounts of the mother liquor. Figure 9.8a shows a crystallizer with stirrer and forced vertical circulation. Crystallizers are often also equipped with cooling/heating elements. These crystallizers can be used for both batch and continuous crystallization; in the latter case, the temperature is kept constant while supersaturated solution is fed into the crystallizer and crystals are removed.

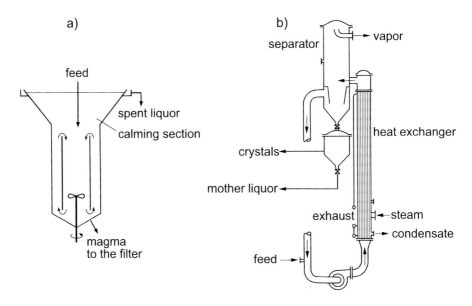

Figure 9.8. a) Principle of the agitator crystallizer with forced circulation; b) APV Kestner evaporation crystallizer (from reference [13]).

Crystallization in Evaporation and Vacuum Crystallizers: Some substances have solubilities that decrease little on cooling or even increase (certain salts); for these, supersaturation must be attained by evaporation of the solvent. An example is the APV Kestner evaporation crystallizer shown in Figure 9.8b, which is used principally for crystallizing aqueous solutions, for example salts, citric acid, and so on. In vacuum crystallizers, the hot solution is injected into the crystallizer at a temperature higher than its boiling point at the selected pressure, thus cooling the solution adiabatically and evaporating the solvent.

Precipitation: Precipitation differs from crystallization by cooling of a supersaturated solution or by evaporation of the solvent in that a second solution is added as an agent. This might be a poorer solvent, the addition of which decreases solubility, or it could be a substance that causes the precipitate to form by a rapid reaction, for example the precipitation of an acid by acidification of the solution, or the formation of a poorly soluble salt. This type of crystallization should not be confused with the precipitation of crystals from a slow chemical reaction in a homogeneous solution. As with the cooling of a supersaturated solution or the evaporation of solvent, in this case too supersaturation is maintained continuously in a single solution.

For precipitation, in contrast, the critical variable is the way the two solutions are mixed. Eddies form on mixing, their size dependent on the speed of mixing, and *each microeddy acts as a microreactor* in which nucleation and crystal growth proceed independently of the processes in surrounding eddies. The diffusion of the agent into these microreactors from the liquid around them is the rate-determining step, analogously to the case of rapid chemical reactions [14]. The more intense the mixing or agitation, the smaller the microeddies and the greater their surface area, so the more agent can diffuse in per unit time. The supersaturation is greater, and therefore the crystals formed are smaller.

9.2.2 Population Balance

When a crystallizer is working continuously it is useful to be able to monitor the nucleation and crystal growth. The dynamic equilibrium between nucleation, crystal growth, addition, and removal is expressed in the *population balance* of the crystals.

The population density n is defined as the number of crystals per unit volume with crystal lengths between L and $L+dL$.

$$n = \frac{dN}{dL} \tag{9.12}$$

The number of crystals per unit volume with lengths between L_1 and L_2 is therefore as defined in Equation 9.13:

$$\Delta N = \int_{L_1}^{L_2} n \, dL \tag{9.13}$$

In the dynamic equilibrium ΔN is constant; it is the population balance between the small growing crystals in the size range (L_2–L_1), the crystals growing beyond this range, and the particles in this range that are continually removed from the reactor. If we let Q be the rate of intake and removal, V the reactor volume, G the rate of growth of the particles (= dL/dt), and $\bar{n}$ the mean population density in ΔL, then for the time interval Δt and the particle size range ΔL (= L_2–L_1), Equation 9.14 applies:

$$n_1 \cdot G \cdot V \cdot \Delta t = n_2 \cdot G \cdot V \cdot \Delta t + Q \cdot \bar{n} \cdot \Delta L \cdot \Delta t \tag{9.14}$$

If the residence time of the solution supplied to the reactor is $T = V/Q$, then it follows from Equation 9.14 for $\Delta L \to 0$ that:

$$\frac{\mathrm{d}n}{\mathrm{d}L} = -\frac{n}{G \cdot T} \tag{9.15}$$

Integration yields the population density n, the particle size distribution per unit volume (n° is the population density of the nuclei):

$$n = n^{\circ} \cdot \exp\left(-\frac{L}{G \cdot T}\right) \tag{9.16}$$

The growth rate G can be determined from the population density distribution or the particle size distribution measured experimentally. The nucleation rate J is also of interest, and can be calculated from Equation 9.17:

$$J = \left.\frac{\mathrm{d}N}{\mathrm{d}t}\right|_{L=0} = \left.\frac{\mathrm{d}N}{\mathrm{d}L}\right|_{L=0} \cdot \frac{\mathrm{d}L}{\mathrm{d}t} = n^{\circ} \cdot G \tag{9.17}$$

G and n° can be determined by graphical means from a plot of $\log n$ against L, as shown in Figure 9.9.

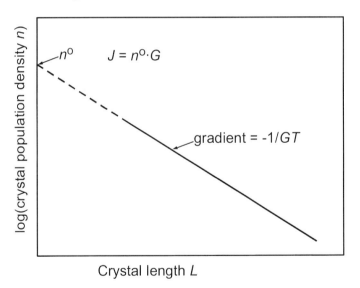

Figure 9.9. Determination of the growth rate G and the nucleation rate J from a population plot for continuous crystallization.

This MSMPR crystallization (mixed suspension, mixed product removal) is the simplest type of continuous process. Extended models also include nucleation by attrition, aggregation, and more complex process parameters [15]. Likewise, a similar model has been developed for batch processes [16].

9.2.3 Particle Size Distribution

The particle size distribution curve reproduced in Figure 9.9 resulted from a continuous process in an MSMPR crystallizer. For batch crystallization or precipitation, however, the particle size distribution is different. For crystalline powders from such processes, the particle size is usually characterized only in terms of crystal length. Length/width and length/breadth ratios are assumed to be constant for the entire population, so the particle size distribution can be expressed in terms of a normal distribution of the characteristic variable x (see reference [17] for bivariate particle size distributions).

$$f(x) = \left[\frac{1}{\sigma \cdot (2\pi)^{1/2}} \right] \cdot \exp\left[-\frac{(x-\bar{x})^2}{2\sigma^2} \right] \tag{9.18}$$

This function is symmetrical with respect to $\bar{x}$ and is normalized such that:

$$\int_{-\infty}^{\infty} f(x)\mathrm{d}x = 1 \tag{9.19}$$

The mean $\bar{x}$ is defined by Equation 9.20:

$$\bar{x} \equiv \int_{-\infty}^{\infty} x \cdot f(x)\mathrm{d}x \tag{9.20}$$

The breadth of the distribution is given by the standard deviation σ, which is defined as follows:

$$\sigma^2 \equiv \int_{-\infty}^{\infty} (x-\bar{x})^2 \cdot f(x)\mathrm{d}x \tag{9.21}$$

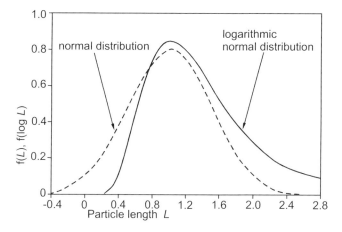

Figure 9.10. The logarithmic normal distribution represents the particle size distribution better than the normal distribution.

If the particle length L is selected as the characteristic variable ($x = L$), actual particle size distributions are reproduced only poorly, and in addition the formula also applies to negative particle lengths with no correspondence with reality (Figure 9.10). A better empirical fit is obtained if the logarithm of the particle length is chosen as the characteristic variable ($x = \log L$). This logarithmic normal distribution is no longer symmetrical with respect to particle length and starts at $L = 0$ ($\log L = -\infty$), as will be seen from Equation 9.10.

9.2.4 Mechanisms of Growth

The model described in Section 9.2.2 furnishes the numerical relationship between nucleation, crystal growth, and solution addition and removal. However, it provides no insight into the mechanisms involved. These mechanisms are examined below.

The degree of supersaturation is important both in nucleation and in crystal growth. Supersaturation is the difference between the concentration C in the solution and the equilibrium concentration C_e (in the presence of crystals) at the same temperature. The relative supersaturation S relates this difference to the equilibrium concentration (Equation 9.22).

$$S = \frac{C - C_E}{C_E} = \frac{C}{C_E} - 1 \tag{9.22}$$

Nucleation is the first step in crystallization. If molecules (or atoms or ions) in a supersaturated solution join up to form a nucleus, (volume) energy is liberated. At the same time, though, the surface of the nucleus forms, for which (surface) energy is required. For small nuclei this surface term outweighs the volume term. Small nuclei are therefore unstable and disintegrate. Only when a critical radius r_c is passed does the free enthalpy ΔG of the nucleus decrease when more molecules are added. Nuclei with a radius larger than r_c are therefore stable and can grow (Figure 9.11).

$$\Delta G = volume\ term + surface\ term = -\frac{4}{3} \cdot \pi r^3 \cdot \frac{RT}{V_M} \cdot \ln\frac{C}{C_E} + 4\pi r^2 \cdot \gamma \tag{9.23}$$

(V_M: molar volume; R: gas constant; T: temperature; γ: surface tension; r: nuclear radius).

Nucleus formation lowers the degree of supersaturation to the extent that no new, stable volume nuclei can be formed. Smaller crystals even redissolve, benefitting the larger ones, which grow. This process is called *Ostwald ripening*. Often the degree of supersaturation is such that although no more volume nuclei can form, surface nuclei can form on the faces of crystals, where they result in crystal growth.

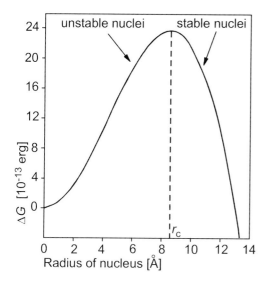

Figure 9.11. Change in the free enthalpy of nucleus formation related to the nuclear radius: water nuclei from the gas phase (from reference [18]).

A molecule (or atom or ion) that lands on a growing crystal surface cannot remain there, but either desorbs once more or diffuses across the surface to an edge, where it is incorporated into a kink in the growing layer (Figure 9.12). The number of kinks depends on the temperature. New kinks form continuously as a result of thermal motion, and molecules diffusing over the surface lodge here. The difference in crystal growth rates thus depends on how the edges are formed.

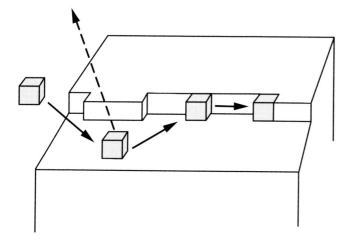

Figure 9.12. Crystal growth: edges and kinks.

An example of such a mechanism is two-dimensional nucleation, in which the growth rate depends on the rate of formation of the surface nuclei on the crystal faces. The nuclei grow much faster than they form. The growing islands spread over the crystal surface until they meet the edges of other islands; at the same time, new surface nuclei are forming on top of the islands. This growth mechanism is known as the *birth and spread model* (Figure 9.13). Occasionally so many surface nuclei form that they cannot grow far before encountering other growing surface nuclei (*polynuclear two-dimensional nucleation*). In the opposite case, a new surface nucleus forms only after the underlying layer is complete (*mononuclear model*).

According to the birth and spread model, the growth rate R is given approximately by Equation 9.24:

$$R = K_1 \cdot S^{5/6} \cdot \exp\left[-\frac{K_2 \cdot \gamma^2}{T^2 \cdot S} \right]$$

(9.24)

(R: growth rate normal to surface; C: concentration in solution; C_E: equilibrium concentration at the surface; S: supersaturation $= (C - C_E)/C_E$; γ: interfacial tension; T: temperature)

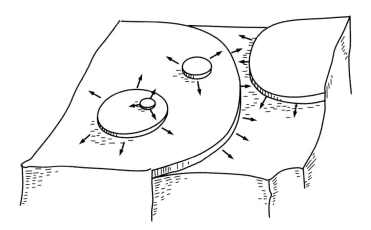

Figure 9.13. Birth and spread model. The formation of new nuclei is slower than their growth (from reference [19]).

If the degree of supersaturation is no longer adequate for the formation of surface nuclei, growth can still occur at crystal defects. Screw dislocations permit the docking of molecules at the spiral edge, so that the tip of the spiral turns at a constant rate of rotation and generates new edges on which molecules can settle until they meet another spiral (Figure 9.14a). This mechanism plays an important role in large crystals. Figure 9.14b shows various forms of screw dislocations in a crystal of NaCl.

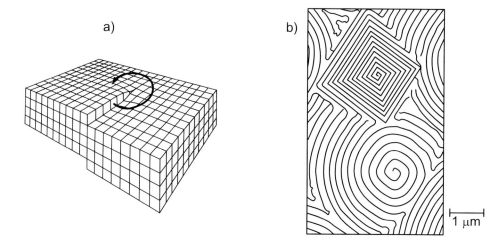

Figure 9.14. Crystal growth by screw dislocations. a) initiation of the screw dislocation at a slip plane; b) screw dislocations on the surface of a crystal of NaCl (source: [20]).

The theory behind this mechanism was explored by W. K. Burton et al. (the BFC model) [21]. The growth rate is given by Equation 9.25:

$$R = k_3 \cdot T \cdot S^2 \cdot \tanh\left(\frac{k_4}{T \cdot S}\right) \tag{9.25}$$

9.2.5 Ostwald Ripening

Small particles are more soluble than large ones, Equation 9.26. In a heterodisperse suspension material transport therefore occurs from the small to the large particles; the large ones grow at the expense of the small ones. This often undesirable process occurs in pastes and liquid formulations of powders, for example, if the substance concerned is too readily soluble in the liquid medium. The effect is particularly significant if a formulation stable for years is required; agrochemical flowables, for example, have reduced effectiveness if the particles are large, and pigment-containing paints have weaker color when the crystals are larger. On the other hand, the process is actively encouraged in the recrystallization of pigments to obtain powder of high hiding power.

The theory behind this process of Ostwald ripening, in which large particles grow at the expense of small ones, was investigated by Lifshitz and Slyozov [22] and Wagner [23]. The theory is known as the *LSW theory*. In both papers, the dependence of the solubility on the particle size is expressed by the Gibbs–Thomson Equation (9.26) or a simplification thereof.

$$C_R = C_\infty \cdot \exp\left[\frac{2 \cdot \gamma \cdot v_m}{R \cdot k \cdot T}\right] \tag{9.26}$$

(C_R: solubility of the particle with radius R; C_∞: solubility of an infinitely large particle; γ: interfacial tension; v_m: volume of a molecule; k: Boltzmann constant; T: temperature)

The time-dependent particle size distribution $f(R, t)$ is obtained by solving the continuity equation, Equation 9.27.

$$\frac{\partial f}{\partial t} = -\frac{\partial\left(f \cdot \dot{R}\right)}{\partial R} \tag{9.27}$$

After some rewriting, this equation yields a formula that relates the mean particle size $\overline{R}$ to the recrystallization time, Equation 9.28. $\overline{R}_0$ is the initial mean particle radius and m is a constant that depends on the rate-determining step. K is a constant. Instead of the radius R, it is of course possible to use particle length L or particle breadth B. The scaling exponent m is 1 for viscous flow, 2 for interfacial control, 3 for volume diffusion, and 4 for interfacial diffusion [24].

$$\overline{R}^m = \overline{R}_0^m + K \cdot t \tag{9.28}$$

In the most frequent form of Ostwald ripening, by volume diffusion, the growth in mean particle size is related to the cube root of time ($m = 3$). Very poorly soluble substances, however, for example some pigments in certain solvents, recrystallize with considerably weaker time dependence (higher m value). Thus m values of 5 for length and 8 for breadth were measured for recrystallization of the compound depicted in Figure 9.6 from o-dichlorobenzene at 60°C. The corresponding values for the better solvent N-methylpyrrolidone were comparable with those for control by volume diffusion, 3 ($m = 2.7$ for length, 3.5 for breadth). For the pigment shown in Figure 9.15, the m values were as high as 11 (length) and 10 (breadth), which are normal values for very poorly soluble substances.

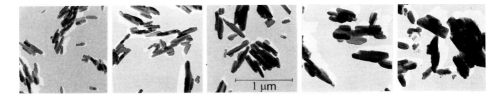

Figure 9.15. Ostwald ripening of N,N'-phenylenebis[4-[(2,5-dichlorophenylazo]-3-hydroxy-2-naphthalenecarboxamide] in o-xylene at 150°C (15 min; 1 h; 4 h; 24 h; 168 h).

In Equation 9.28, the constants K and m include, amongst other things, the dependence of recrystallization on the solubility in the solvent concerned. Figure 9.16

shows this dependence for the pigment depicted in Figure 9.6, in the same solvents. As can be seen, there is a *relationship between mean particle size and solubility almost independent of other solvent properties.*

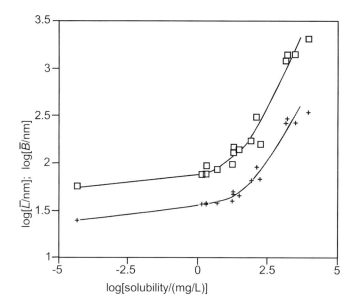

Figure 9.16. Mean particle length $\overline{L}$ and breadth $\overline{B}$ of the pigment from Figure 9.6 on recrystallization for 2 h at 60°C in various solvents as a function of solubility.

Sparingly soluble substances recrystallize only slowly. Since Ostwald ripening in bulk usually depends on the cube root or a still higher root of the time, large crystals are more easily obtained by increasing the temperature than by extending the time allowed for recrystallization.

References for Chapter 9:

[1] A. Martin, P. L. Wu, A. Adjei, A. Beerbower, J. M. Prausnitz, J. Pharm. Sci. *70*, 1260 (1981).
[2] A. A. Bondi, Physical Properties of Molecular Crystals, Liquids and Glasses, John Wiley & Sons, New York, 1968.
[3] A. Grubenmann, Dyes and Pigments *21*, 273 (1993).
[4] P. J. Richardson, D. F. McCafferty, A. D. Woolfson, Int. J. Pharmaceut. *78*, 189 (1992).
[5] M. J. Kamlet, R. M. Doherty, J.-L. M. Abboud, M. H. Abraham, R. W. Taft, Chemtech *16*, 566 (1986).
[6] S. H. Yalkowsky, S. C. Valvani, J. Pharm. Sci. *69*, 912 (1980).

[7] M. J. Kamlet, Progr. Phys. Org. Chem. *19*, 295 (1993).

[8] M. J. Kamlet, J.-L. M. Abboud, M. H. Abraham, R. W. Taft, J. Org. Chem. *48*, 2877 (1983).

[9] K. Wakita, M. Yoshimoto, S. Miyamoto, H. Watanabe, Chem. Pharm. Bull. *34*, 4663 (1986).

[10] G. Klopman, S. Wang, D. M. Balthasar, J. Chem. Inf. Comput. Sci. *32*, 474 (1992).

[11] S. H. Yalkowsky, S. C. Valvani, J. Chem. Eng. Data *24*, 127 (1979).

[12] A. D. Randolph, M. A. Larson, Theory of Particulate Processes, Academic Press, New York, London, 1971.

[13] J. W. Mullin, in Crystal Growth (B. R. Pamplin, Ed.), Pergamon Press, Oxford, New York, Toronto, Sydney, 1975.

[14] R. J. Ott, P. Rys, Helv. Chim. Acta *58*, 2074 (1975).

[15] M. J. Hounslow, AICHE Journal *36*(1), 106 (1990).

[16] M. J. Hounslow, R. L. Ryall, V. R. Marshall, AICHE Journal *34*(11), 1821 (1988).

[17] A. Grubenmann, Part. Charact. *3*, 179 (1986).

[18] A. W. Adamson, Physical Chemistry of Surfaces, 3rd ed., John Wiley & Sons, New York, 1976.

[19] M. Ohara, R. C. Reid, Modeling Crystal Growth Rates from Solution, Prentice-Hall, Englewood Cliffs, NY, 1973.

[20] M. Krohn, H. Bethge, in Current Topics in Material Science, Volume 2 (1976 Crystal Growth and Materials) (E. Kaldis, Ed.), North Holland, Amsterdam, New York, Oxford, 1977, p. 147.

[21] W. K. Burton, N. Cabrera, F. C. Frank, Phil. Trans. Roy. Soc. *A243*, 299 (1951).

[22] I. M. Lifshitz, V. V. Slyozov, J. Phys. Chem. Solids *19*, 35 (1961).

[23] C. Wagner, Z. Elektrochem. *65*, 581 (1961).

[24] H. Gleiter, in Physical Metallurgy (R. W. Cahn, Ed.), P. Haasen-Verlag.

10 Detergency

10.1 General Remarks and Basic Principles

Detergency is easily the most important area of application for surfactants. Cleaning and washing are very complex processes, which even today are not fully understood. Since there is a huge variety of types of dirt and of substrates, *there is no single mechanism for the process of detergency, but rather many depending on the substrate – textile, glass, metal, china.*

The process of detergency can be defined in general as the removal of liquid or solid dirt from a solid, the substrate, with the aid of a liquid, the cleaning bath. Dirt itself is defined, rather simplistically, as "material in the wrong place".

In the old days, soap was the usual cleaning material, or detergent, but it has the well-known drawbacks that it forms insoluble, inactive fatty acids in an acid environment and insoluble precipitates with Ca^{2+} and Mg^{2+} in hard water. Additives such as Na_2CO_3, phosphates, etc. can prevent these disadvantages. In the last 50 years, soap has been partly replaced by synthetic detergents which do not suffer from these negative effects. The most important of these are the *alkyl sulfates, alkyl aryl sulfonates*, and *nonionic poly(ethylene oxide) derivatives.*

10.2 Fundamental Phenomena in Detergency

We will limit this discussion to processes using aqueous cleaning media.

The following examples serve to illustrate the most important facet of the nature of the substrate:

a) processes for the removal of dirt (liquid and/or solid) from smooth surfaces, for example the degreasing of metal, glass, ceramics (including washing dishes), the cleaning of painted surfaces;
b) processes for the removal of dirt from porous or fibrous materials, for example the washing of raw wool and cotton, removal of spinning oils from spun fibers, degreasing of leather, laundry, desizing of textiles.

Each of these examples represents a complex process. It is not surprising that the individual processes take very different courses; for example, emulsification is a necessary stage in the removal of wool fat from the fibers of raw wool, but is totally unsuitable for the removal of a film of oil from a polished surface. We must therefore start by considering the fundamental phenomena of detergency, which always occur, and only after this shall we look at the subsidiary phenomena that have a role to play only in particular cleaning processes.

10.2.1 Mechanism of Detergency

A good cleaning agent or detergent must have the following properties:

1. good wetting power, so that the agent makes close contact with the surface to be cleaned;
2. the ability to remove the dirt into the bulk of the liquid or to assist this process;
3. the ability to solubilize or disperse the dirt once removed and to prevent it from redepositing on the cleaned surface and forming a residue.

10.2.1.1 Wetting

The best wetting agents are not necessarily the best detergents, and vice versa. For a homologous series of detergents such as soaps, alkyl sulfonates, and alkyl aryl sulfonates, the *optimum wetting is achieved by the C_8 tenside*, although the compounds with longer chains have stronger surface activity. The reason is that the smaller molecules diffuse to and adsorb at the surface faster.

The optimal detergent action belongs to the C_{14} tensides, but the C_8 compounds give the best all-round performance.

10.2.1.2 Removal of Dirt

Dirt is generally oily in nature and contains particles of dust, soot, and so on. Its removal relies on the replacement of one wetting substance with another, and can be considered in terms of the change in surface energy.

(a)

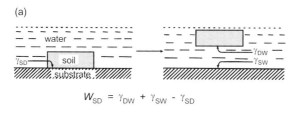

$$W_{SD} = \gamma_{DW} + \gamma_{SW} - \gamma_{SD}$$

(b)

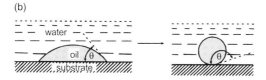

Figure 10.1. a) Work of adhesion between dirt particle and substrate. b) Increase of the contact angle of an oil drop by preferential lowering of the substrate/water interfacial tension γ_{SW}; from reference [1].

The work of adhesion between a particle of dirt and a solid surface is given by Equation 10.1 (see Figure 10.1a):

$$W_{SD} = \gamma_{DW} + \gamma_{SW} - \gamma_{SD} \tag{10.1}$$

The task of the detergent is to lower γ_{DW} and γ_{SW}, which decreases the adhesion W_{SD} and facilitates the removal of the dirt particle by mechanical agitation. *Nonionic detergents are usually less effective in the removal of dirt than anionic tensides.*

10.2.1.3 Liquid Soiling

If the dirt is liquid (oil or fat; Figure 10.1b), its removal is a problem of contact angles. The substrate is preferentially wetted by the tenside solution. The liquid dirt, which is initially present as a thin film spread over the substrate, is gathered up into droplets by the action of the detergent solution; these droplets can then be removed by the flow in the bath or by mechanical means.

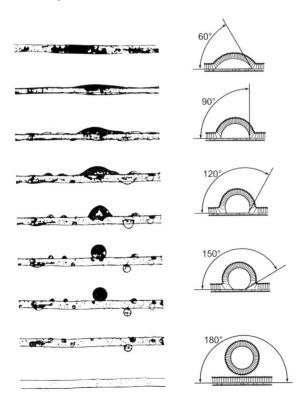

Figure 10.2. Removal of dirt by roll-up. The contact angle of the dirt is increased to 180°, so that it rolls off in balls; from reference [2].

The addition of the detergent increases the contact angle at the dirt/substrate/water interface (Figures 10.1 and 10.2), so that the dirt "rolls up" and off the substrate. Detergents that adsorb both at the substrate/water interface and at the dirt/water interface are the most effective. If the tenside adsorbs only at the dirt/water interface and lowers the interfacial tension, or if it dissolves in the oil (liquid dirt) and reduces the interfacial tension γ_{SD} of the oil relative to the substrate, dirt removal is rendered more difficult, since θ is reduced and roll-up prevented.

Tensides that adsorb at the air/water interface and reduce the surface tension of the water, or the cleaning bath, create foam, proving that *foam formation is not necessarily a sign of the detergent activity of the tenside!*

For example, nonionic detergents are usually very good cleaners of liquid dirt, but they do not foam. Since consumers tend to imagine the two properties are linked, nonfoaming detergents are not always easily accepted.

There are many investigations and publications concerning the influence of foam on the efficacy of detergents. Most conclude that *foam is no indicator of the efficacy of detergents.* It is known that the formation of foam in washing machines is disadvantageous. However, foam can have its uses, for example as a place for the dirt to gather. It can encourage the removal of oil by absorbing this from surfaces into the Plateau border of the lamellae. Foam can also support the removal of dirt from fibers by lending a certain rigidity to the system, which increases the effect of mechanical agitation.

Since foaming can have considerable negative effects in watercourses and drains, the prevention and destruction of foam is important. Foam-breaking is covered in Section 4.6.

10.2.1.4 Prevention of Redeposition of Dirt

The process of detergency is depicted in Figure 10.3.

To prevent the dirt particles from redepositing on the substrate once they have been removed, they must be stabilized in the cleaning bath by colloid-chemical means. The prevention can be effected by means of electrical charge and steric barriers resulting from the adsorption of tenside molecules from the cleaning bath by both the dirt particles and the cleaned material. Tensides or inorganic ions with the same charge sign (mostly anions) raise the electrical potential of the Stern layer and thus prevent agglomeration of the particles.

The most effective detergents for this purpose, however, are nonionic tensides with poly(ethylene oxide) chains, which are highly hydrated and oriented towards the aqueous phase (the cleaning bath). These create a steric barrier that causes mechanical and entropic repulsion (due to decrease in the entropy of the POE chains). Likewise, the addition to the cleaning bath of special components that adsorb onto the dirt or the substrate can prevent agglomeration of the dirt particles by electrical or steric repulsion, for example sodium carboxymethylcellulose or other polymers.

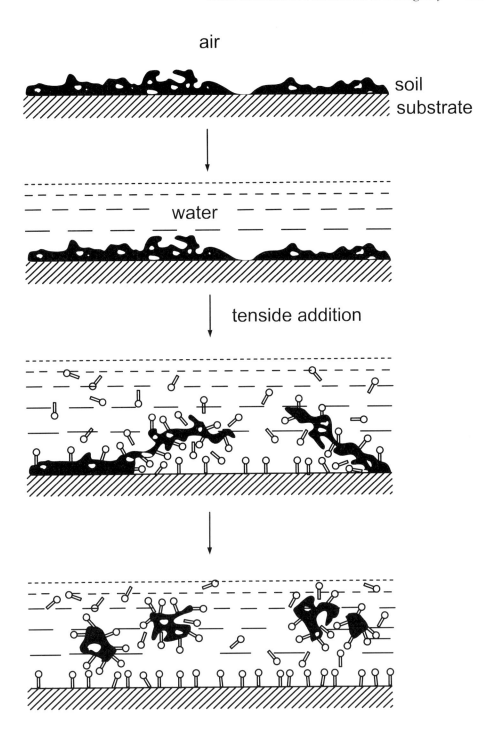

Figure 10.3. Loosening and removal of dirt [1, 3].

10.3 Special Phenomena in Detergency

Unlike the general phenomena dealt with in Section 10.2, the following phenomena are relevant only in particular cases.

10.3.1 Solubilization in Micelles

Solubilization in tenside micelles is important for the removal of *small quantities of oily soiling* from substrates. Whether from hard or textile surfaces, this removal becomes significant only once the *CMC* has been exceeded for nonionic tensides, and indeed even for some anionic tensides with a low *CMC*. Maximum effectiveness is reached only at a multiple of the *CMC*. The degree of solubilization of the oily dirt depends on the chemical structure of the tenside, on its concentration in the bath, and on the temperature.

At low tenside concentrations, the dirt is solubilized into more or less spherical micelles. Only a small amount of oil can be solubilized in this manner, whereas at high concentrations (10–100 × CMC) solubilization resembles formation of a microemulsion. The high tenside concentration permits large quantities of greasy dirt to be held in solution.

For ionic surfactants, the concentration used is not much larger than the *CMC*. *Solubilization is therefore seldom sufficient to remove all oily dirt.* For nonionic surfactants, the amount solubilized depends principally on the temperature of the bath, relative to the cloud point of the tenside. *Solubilization of grease increases sharply close to the cloud point.*

There are various views concerning soiling removal by micelles. Since the most successful detergents form micelles, some hold the opinion that micelles must be directly involved in the detergency, solubilizing grease. However, *the detergent action is dependent on the concentration of unassociated tenside and is practically unaffected by the presence of micelles, since only monomeric tenside molecules adsorb at the interfaces.* The micelles therefore act, at best, as a reservoir from which nonassociated tensides adsorbed from solution are replenished. It seems that the molecular properties of tensides with good detergency also favor the formation of micelles, so if anything these should be seen as competition rather than as an aid to dirt removal.

10.3.2 Emulsification

For emulsification, the interfacial tension between the oil droplets and the cleaning bath must be very low, so that emulsification may occur with very little mechanical work. Adsorption of the tenside at the dirt/bath interface therefore plays a significant role. *The suitability of the cleaning bath for emulsification of the oily dirt is insufficient to prevent all redeposition of soiling on the substrate.* If emulsified oil droplets collide with the substrate, some will stick and adopt the equilibrium contact angle. *This contrasts with solubilization, in which the oily dirt is completely removed from the substrate.*

Dispersion of the dirt particles in the cleaning bath alone does not amount to effective cleaning. There is no correlation between the detergency and the dispersing power of a cleaning bath. Tensides that are excellent dispersants are often bad detergents, and vice versa. On the other hand, in the case of anionic and nonionic tensides there is a correlation between increased adsorption on the substrate and on dirt; furthermore, in the case of nonionic tensides, there is also a correlation between the solubilization of greasy dirt and the detergent effect.

Finally, *cationic tensides have little detergent effect, since most soiling and most substrates are negatively charged in aqueous media at neutral or alkaline pH.* Adsorption of the positively charged tenside ions on substrate and dirt reduces their negative electrical potential, hinders the removal of dirt, and encourages its redeposition.

10.4 Detergent Additives, Builders

In addition to surfactants, detergent formulations contain a wide variety of other substances. These are intended not only to eliminate the negative effects of multiply charged cations on the detergency, but also to increase the detergent effectiveness of the tensides. These substances are called *builders*. They are chiefly inorganic salts used at fairly high concentrations. Some organic polymers are also used at lower concentrations to prevent redeposition of dirt.

The principal functions of builders are as follows:

1. Sequestration of Ca^{2+} and Mg^{2+} ions (by formation of soluble, nonadsorbing complexes). Sodium and potassium polyphosphates are used for this purpose, especially sodium tripolyphosphate $Na_5P_3O_{10}$, with a binding power for Ca^{2+} of 158 mg CaO/g. Even better binding power is possessed by nitrilotriacetic acid (285 mg CaO/g). Sodium silicate, carbonate, and hydroxide salts, which precipitate multiply charged cations as their insoluble salts, are also used.

2. Deflocculation and dispersion of dirt particles by adsorption of the builder on the particles and by increasing their negative potential and thus their mutual repulsion. Polyphosphate ions, with their multiple negative charge, are particularly well suited to this task.

3. Creation of an alkaline environment and buffering. High pH increases the negative potential of dirt and substrate and increases detergency. Buffering is necessary to prevent the lowering of the pH and thus the surface potentials of the dirt and the substrate. Sodium carbonate is particularly effective.

Some builders have special purposes: sodium silicates to prevent corrosion of the aluminum parts of washing machines and of the glaze on china, sodium carboxymethylcellulose at low concentration (up to 2%) to prevent redeposition of dirt on fibers. Builders also help to create mild alkaline conditions, which aid cleaning.

10.5 Laundry Detergents

As has already been mentioned, the processes of washing are complex. Various chemical and physical processes occur together. Both water-soluble soiling and, for example, adsorbates of pigments, grease, carbohydrates, proteins, and natural and synthetic dyes must be removed from the surface.

To remove insoluble particulate soiling, the adhesion energy must be overcome, so mechanical energy must be supplied in the first step of the washing process; this is assisted by adsorption of tenside. At this stage the composition of the washing liquid controls the complete lifting of the dirt particle in the next step.

The process of lifting dirt is the reverse of flocculation, and depends heavily on electrostatic interactions. Therefore, adsorption of ionogenic tensides or multiply charged ions such as alkali metal phosphates or silicates (builders) must be employed to ensure that substrate and soil have the same charge sign, so that their separation is electrostatically favored.

After lifting, dirt is present in the washing liquid in dispersed form, and the adsorbed tenside layers should delay or prevent the flocculation of solid particles and the coalescence of emulsified droplets. Macromolecular additives such as carboxymethyl-cellulose have a similar effect by steric hindrance, preventing flocculation and coalescence, which would otherwise result in redeposition of the dirt and graying of the laundry. Macromolecular additives thus also act as graying inhibitors.

Many forms of soiling have to be chemically changed before they can be removed, whether by redox processes by means of bleaching agents (combined with bleach activators and stabilizers), for example the natural dyes present in tea, wine, and fruit juice, or by enzymatic degradation of denatured protein adsorbates. Complex formation and ion exchange are other important processes.

Greasy soiling is mostly liquid at wash temperatures greater than 40°C and spreads over the substrate surface in more or less unbroken layers. These oily layers need to be emulsified during the washing process; the "roll-up" already mentioned is important here.

Other additives to laundry detergents are optical brighteners, corrosion inhibitors, fragrances, and colorants.

Table 10.1 gives the approximate composition of an all-purpose laundry detergent, which is used at a loading of 6–10 g/L water. In the USA and Japan, usual loadings are 1–1.5 g/L, so the composition is altered accordingly.

Environmental concerns have brought numerous low-phosphate or phosphate-free detergents onto the market, and the trend continues. For the same reason the biodegradability of tensides is now taken into consideration.

Besides all-purpose detergents, detergents specifically designed for use at 60°, with delicates, wool, or net curtains, for small amounts of hand- and machine-washing, and for prewashing are also available. In addition, liquid detergents are sold, which usually contain solubilizers such as ethanol and propylene glycol, and sometimes brightening cationic tensides, but no bleaches like perborate and only rarely builders. Likewise there is no perborate in detergents for hand-washing, though these have very high tenside contents. Cationic tensides, which are found in certain liquid detergents, are also present in fabric conditioners which, analogously to hair conditioners (see Chapter 11), are intended to leave a coating on the cloth and thus achieve a soft feel on washing, the desired fleeciness, and prevention of the accumulation of static electricity.

Table 10.1. Approximate composition of an all-purpose laundry detergent (from reference [4]).

Agent	Example	Amount [%]
anionic tensides	alkyl benzenesulfonate	5–10
nonionic tensides	fatty alcohol polyglycol ether	1–5
foam inhibitors	soaps, silicon oils	1–5
complex formers	sodium triphosphate	10–40
ion-exchange resins	Zeolite 4A	0–30
bleach	sodium perborate	15–35
bleach activators	tetraacetyl ethylenediamine	0–4
stabilizers	ethylenediaminetetraacetate, magnesium silicate	0.2–2.0
graying inhibitors	carboxymethylcellulose, other cellulose ethers	0.5–2.0
enzymes	proteases	0.3–1.0
optical brighteners	stilbenedisulfonic acid and bis(styryl)biphenyl derivatives	0.1–0.3
corrosion inhibitors	sodium silicate	2–7
perfumes		0.05–0.3
colors		0–0.001
extender	sodium sulfate	2–20

Due to the great turnover of more than 500 kg washing per hour and unit, detergents for use in laundries have a different composition from those intended for domestic use. In addition, since the hardness is removed from the washing water in advance, the detergent mixture need not contain so much phosphate. Soap is still used on a large scale. Textiles soiled with grease can be cleaned with fat-dissolving detergents, which contain solvents. Because the types of soiling vary, the washing processes often have to be adapted accordingly in order to enhance the efficacy of the washing liquid.

Washing powders are produced largely by spray drying, as described in Chapter 6. The constituents are stirred with water to form a slurry, which is atomized at the top of the drying tower under a pressure of up to 80 bar. Hot air evaporates the water in the droplets, creating a powder of hollow spheres [4, 6, 7]. In the case of flow in the same direction, the slurry enters the tower at the hottest zone. The water evaporates quickly, expanding the particles and creating a loose, low-density powder. Because of the short dwell time in the hot zone, this process is particularly suitable for gentle drying of delicate laundry detergent with high tenside content. However, the principal method of production is the countercurrent process. Since in this case the drying begins at low temperature and high air humidity, thick-walled beads are created. In addition, the low rate of descent favors agglomeration, so more compact, coarser powders are formed; see also reference [8].

References for Chapter 10:

[1]   D. J. Shaw, Introduction to Colloid and Surface Chemistry, Butterworths, London, 1966.

[2]   Ciba–Geigy Rundschau 1971/2: N. Bigler, Die Tenside (Ciba-Geigy AG, Basel, Div. Farbstoffe, Ed.).

[3]   L. M. Kushner, J. I. Hoffman, Synthetic Detergents, Sci. Amer., Oct. 1951, p. 26.

[4]   Ullmanns Encyklopädie der technischen Chemie, 4th ed., Vol. *24*, Verlag Chemie, Weinheim, Deerfield Beach, FL, Basel, 1983, p.108.

[5]   H. J. Lehmann, Chem. unserer Zeit *7*, 82 (1973).

[6]   K. Henning, Chem. Lab. Betr. *27*, 46, 81 (1976).

[7]   Winnaker–Küchler 4th ed., Vol. *7*, Carl Hanser Verlag, München, p. 84.

[8]   P. Berth, M. J. Schwuger, Chemische Aspekte beim Waschen und Reinigen, Tenside Detergents *16*, 175 (1979).

11 Cosmetics

11.1 Skin as the Substrate for Cosmetics

Cosmetics are substances or preparations intended for use externally or in the buccal cavity for cleaning and care and to modify the appearance or odor of the user. They differ from medicaments in that the latter are administered to the body for the purpose of curing, reducing, preventing, or diagnosing illness, suffering, or physical damage. Although most cosmetics are clearly not medicaments in terms of the above definitions, sometimes the distinction is difficult; for example, skincare preparations can influence the secretion of sebum, antiperspirants can inhibit sweating, and sunscreens prevent damage to the skin.

We distinguish between the *active substance* and the *base* of a cosmetic preparation. In moisturizers, for instance, the base consists of water, polyols, lipids, emulsifiers, consistency regulators, humectants, and small amounts of preservatives and scent. Active substances could be vitamins, plant extracts, extracts of animal protein, and other substances. Since the metabolic mechanisms of the skin are complex, however, the cream base can have a larger effect than the active substances added, depending on skin type. Colloid chemistry is therefore important for cosmetics, as it defines the properties of the lipid/water system. In addition all substances that contact the skin can influence it; even water affects the cornified layer (Figure 11.1), making it swell.

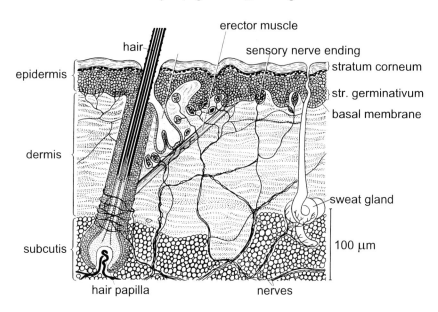

Figure 11.1. Cross-section through the skin (from reference [1]).

The skin, as the body's interface with the external environment, has many functions. It acts as a barrier to harm and has a role in the exchange of substances with the outside world. It receives stimuli and also has a signaling function (blushing). Skin consists of two distinct layers, the dermis and the epidermis (Figure 11.1). The major function of the epidermis is to form the stratum corneum, the cornified layer, as a barrier against external influences. It is an outer layer of several levels, the top ones containing loose, cornified cells that slough off, while replacements are created continuously in the lower levels by cell replication (in the stratum germinativum, the germinal layer). The basal cells lining the dermal–epidermal junction are called serrated basal cells and serve mainly to anchor the epidermis to the dermis. The transition from dermis to epidermis is marked by numerous papillae containing nerve endings and blood vessels projecting into the epidermis. These are arranged in rows and can be seen in the epidermis in the ridged structure observable on the fingertips, for example. The epidermis projects into the dermis, too, in the form of cutaneous glands surrounded by a network of dermal blood vessels, which secrete various substances such as sweat or sebum. Hairs are epidermal as well, though the follicles are anchored deep in the dermis. However, the dermis forms papillae to nourish the growing hair. In addition there are sebaceous glands situated beside the hair follicle in order to lubricate the hair.

The secretions of the glands form a protective hydro-lipid film on the skin. If this is removed by a solvent such as alcohol, that is, if the skin is degreased, the water from the swollen cornified layer (stratum corneum) evaporates, and the skin "dries out".

Washing with detergent also removes the hydro-lipid mantle along with dirt – indeed, this occurs to a lesser degree even when water alone is used – so the skin can then dry out because of evaporation of water from the unprotected cornified layer. Although tenside is adsorbed onto the cornified layer during washing, this film is insufficient to act as a barrier to evaporation. Soaps, though, do increase the swelling of the cornified layer for long enough for the hydro-lipid mantle to be regenerated by sebaceous secretion. Alkaline detergents thus cause less drying of the skin than neutral ones and are therefore better for dry skin than neutral washing liquids.

The mildest cleaning is definitely that achieved by means of cleansing milks, emulsions in which the lipid/water ratio strongly favors the lipid; this way, a lipid layer can form on the skin. However, in terms of effective cleaning, these emulsions are inferior to pure tenside solutions, even if the emulsifier is present in a large excess.

In contrast to cleansing agents, lipid emulsions are intended to form a lipid layer that prevents evaporation. Almost complete cover can be achieved with a film of an anhydrous lipogel, such as Vaseline, over the skin. Channels form in this because of perspiration, though, and water can evaporate through these without difficulty.

Special w/o emulsions are better; they distribute not only fat but also water and emulsifier onto the skin. This too covers up the skin, if incompletely. The water released by the cornified layer is emulsified in the layer of oil and remains in equilibrium with the latter.

Skincare products must take these intricate water absorption and desorption mechanisms into account for different skin types. Preparations for dry skins must contain larger proportions of lipids, and cleansers must be formulated so as not to increase the already excessive drying effect [2].

11.2 The Effects of Tensides on the Skin

For any discussion of the effect of tensides on the cutaneous barrier, we must consider not only the lipids but also the proteins of the epidermis, which result from cell differentiation. The basal cells of the germinal layer (stratum germinativum) differentiate via the spiny layer (stratum spinosum) and granular layer (stratum granulosum) into the flattened, cornified cells of the cornified layer. Figure 11.2 shows the details of the transition from the granular layer to the cornified layer.

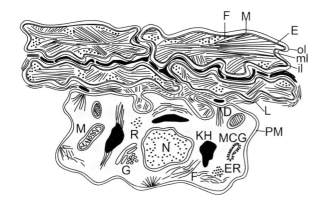

Figure 11.2. Sketch of the cornified layer and a differentiated granular cell in the initial stage of transformation. Cornified cells are filled with intermediate filaments (F) and an amorphous matrix (M). The exterior (E) consists of the thin outer (ol) and middle (ml) lamellae and a thickened inner lamella (il). The intercellular space is filled with parallel lamellae 2 nm thick (L) which originated in the membrane-covering granulaes (MCG). The granular cell contains a nucleus (N), bundles of intermediate filaments (F), keratohyalin granules (KH), mitochondria (M), Golgi bodies (G), endoplasmic reticulum (ER), and free ribosomes (R). The granular cell is wrapped in a triple layer of plasma membrane (PM), which itself adheres to the cornified cells by desmosomes (D) [3].

Over the 27 days it takes for the differentiation cycle to be completed, various internal autolytic processes create the keratinized matrix for proteolipid components, and also water-soluble nitrogenous substances in the intercellular spaces of the cornified layer. In the process of keratinization in which the basal cells in the germinal layer become cells in the cornified layer, however, the composition of the lipid components changes as well. The polar lipids like glycosphingolipids and phospholipids present in the deeper levels of the epidermis become nonpolar lipids in the differentiation process, so the keratinized layer contains lipids such as sterol esters, free fatty acids, ceramides, and triglycerides (Table 11.1).

The lipids in the hydro-lipid film come from a different source, principally from the secretions of the sebaceous glands, which have the following composition: squalene (10%), wax esters (22%), triglycerides (25%), di- and monoglycerides (10%), sterol esters (2.5%), and sterols (1.5%), amongst others.

Table 11.1. Lipid components in various layers of the epidermis (in %) [4].

Lipid	Basal/Spiny	Granular	Cornified
phospholipids	63	25	0
glucosyl ceramides	7	10	0
ceramides	0	15	50
cholesterol	10	21	25
free fatty acids	7	17	15
various	13	12	10

The following facts are known concerning interactions between tensides, lipids, and keratin [4]: lipids are not extracted if the tenside concentration is less than the *CMC*. Above the *CMC*, however, a small quantity of lipids is dissolved out of the cornified layer, around 4–7% of the total present. These lipids are free fatty acids, cholesterol, and cholesterol esters, but not the principal lipid type present, ceramides. Tensides interact with the lipid phase and with keratin; if the contact time is short, they are absorbed only by the lipid phase. If they are allowed to act on the skin for longer, an equilibrium develops between the tenside in water and in the lipid phase, and the tenside bound to the keratin. In addition, even the composition of the lipids in the cornified layer can change as a result of a disturbance in the lipid synthesis. The tenside adsorption is greatest for 12C alkyl chains, even in the case of nonionizable ethoxylates with degrees of ethoxylation of 4–5. As has already been said, the adsorption of anionic tensides increases the swelling of the cornified layer, which can be explained in terms of specific interactions with the head groups. The quarternary structure of keratin can be revealed by tensides (see Chapter 13). Tenside molecules interact in a similar manner with soluble proteins in the intercellular spaces; these are extracted into the aqueous phase.

All these interactions with pure tenside solution leave the skin chapped and scaly, and prone to irritation. On the other hand, they have the advantageous effect of increasing skin permeability to active substances.

11.3 Cosmetic Preparations

As was explained in Sections 11.1 and 11.2, cosmetics can affect the physiology of the skin. Both desirable and harmful effects are possible. These problems must be considered carefully when formulating cosmetics. Various technical aspects of formulation are important: emulsions and microemulsions, liposome systems, dispersions of solids, solutions, rheology, diffusion, and reactivity are a few of the key words related to this topic. A knowledge of colloid chemistry, as covered in previous chapters of this book, is therefore a great help in the development of new formulations. Emulsions, whether oil/water or water/oil, are particularly important in cosmetic formulation. Given the large number of raw materials, there are many combinations that can make up optimized products for particular applications [5], such as:

Cosmetics for the skin: moisturizers like day creams, night creams, moisturizing milks, after-sun lotions, cleansing emulsions, hand and body lotions, skin protection preparations, facial toners and lotions, sunscreen emulsions and preparations for use with sunbeds, body oils, foot lotion emulsions, skincare products for babies, facial foundation emulsions, make-up pencils, concealer, eyeliner, and eyeshadow, lipsticks, face masks, skin bleaching creams, depilatory products, powder, combined cream/powder products, compact powders, aerosol powders, antiperspirant–deodorants, deodorant sticks, insect-repellent sticks, bath salts and bath cubes, bath oils, foam bath, toothpaste, tooth powder, mouthwash, and denture care products.

Cosmetics for the hair: shampoos and cleaning products, antidandruff shampoos, conditioners, setting lotions, styling creams and fixatives, hair oils, pomades, brilliantine, hair tonics, hairsprays, dyes, cream dyes, toners, gel colors, decolorizers, bleaches, perming agents, straighteners.

Nail care: nail polish, varnishes, nail varnish remover, cuticle softener/remover, nail hardener, bleach spray, dry nail-varnish spray, lengthener, cream.

Shaving aids: shaving soaps, shaving creams, shaving foams, pre-shave treatments, aftershave lotions, shaving powders.

Notes on aspects of the application and the action of these cosmetics, and many recipes, can be found in reference [2].

11.4 Emulsions in Cosmetics

The general information necessary for the development of emulsified formulations has already been provided in the chapter on emulsions. In this section, the aspects specific to the cosmetics sector will be treated.

11.4.1 Emulsion Types

One of the hardest tasks in the field of cosmetics is the development of a stable w/o emulsion that is safe for the skin and meets the modern requirements in terms of consistency, namely, is soft, smooth, and easily spread. Furthermore, the development of an emulsion which is stable to storage is often a compromise; emulsions which have good skin compatibility tend to be unstable at higher temperature ranges, whereas very stable emulsions can occasionally cause soreness.

W/o emulsions correspond closely to the physiological conditions of the skin, since this is the form adopted by the skin's own fatty exudate. As was mentioned in Section 11.1, these emulsions form a thin film on the skin surface and thus control the dehydration of the cornified layer.

However, o/w emulsions are more popular, because when these emulsions are applied they have an initial cooling effect owing to evaporation of the water. Adjusted to the correct pH, with ionic tensides in the alkaline range, they plump the skin up and can prevent excessive hardening of the epidermis. In addition the lipids are presented to the skin in very finely dispersed form, an advantage of which particular use is made in day creams. As body lotions for application to large areas of damp skin, o/w emulsions are more suitable. They are less likely to block pores, and do not make the skin look so shiny as do w/o emulsions, which increases their acceptance by consumers. Increasingly, instead of tenside solutions adjusted to an alkaline pH, neutral, nonionic emulsifiers are being used.

Multiple emulsions are those in which a disperse phase is contained within another disperse phase. A w/o/w emulsion contains water droplets dispersed in oil drops dispersed in water. At least two emulsifiers must be used, for instance a sorbitan ester with long hydrocarbon residues and a polyethoxylated sorbitan ester. In the cosmetic sector, products containing multiple emulsions include refreshing and moisturizing emulsions. Their advantage over w/o emulsions is that, like o/w emulsions, they feel more pleasant and less oily in use. For sufficient stability, however, they must contain thickeners. Coalescence and creaming are the instabilities to be prevented, just as for normal emulsions. A further problem is diffusion of the water from the dispersed water droplets into the outer aqueous phase.

11.4.2 Hydrocolloids as Protective Colloids and Consistency Regulators

Good emulsions need cotensides to keep the emulsifier film firm (see Chapter 2). But other additives, such as protective colloids and consistency regulators, are also necessary to optimize the emulsion properties, for example its rheological properties. Gel formation, with its influence on the initial yield strength and thixotropy, has a favorable effect on the long-term stability of emulsions.

As gel-forming thickeners, hydrocolloids are able to interact with the emulsion droplets by gathering around the droplets, complex-style, and reinforcing their stabilizing layer. The strength of the interaction depends not only on the emulsifier system used and the hydrocolloid, but also on the processing conditions.

Both organic and inorganic products are used. Inorganic compounds such as bentonite (for example Veegum) form thixotropic gels. They can be used when the fatty and hydrocarbon bases of the emulsion are not readily compatible and are also useful in moisturizing emulsions because of their water-retaining action.

Organic hydrogel formers [6] such as the anionic polyelectrolytes sodium alginate or carboxymethylcellulose dry on the skin to form protective films. Polyacrylic acids such as Carbopol and Rohagit and their salts also belong to this group.

Nonionogenic hydrogel formers such as methylcellulose and hydroxypropyl-methylcellulose have the advantage that they can be used together with cationic ingredients if necessary, unlike anionic polyelectrolytes. A special mention must be made of hydroxyethylcellulose, which is an important gel former in antiperspirants and roll-on deodorants.

11.4.3 Phospholipids and Proteins

These substances are treated in detail in Chapter 13. The lecithin/water system is depicted in Figure 13.9, and Section 13.3 covers proteins.

Lecithin is a lipophilic emulsifier for w/o emulsions, but can also be used in liposomes (see Chapter 3). Hydroxylated lecithin, on the other hand, is suitable for use in o/w emulsions.

Casein, which can be swollen and dispersed and contains calcium ions, becomes water-soluble when alkali is added and adopts a loose structure. Like heteropolymers, it adsorbs on emulsion droplets, unlike lipoproteins and phosphoproteins, which adsorb as granular particles.

11.4.4 Basic Composition of Cosmetic Emulsions

Creams are semi-solid, unctuous masses, unlike liquid emulsions, which resemble milk. Some pharmaceutical ointments are also emulsions, though with a much lower water content. Some, though, consist only of fatty components to which the active substances have been added. This type is equivalent to cosmetic skin oils and fats.

The following ingredients are part of a cosmetic emulsion:

– the oil phase, to which belong the emulsifier system, consistency regulators, oil-soluble preservatives, and oil-soluble antioxidants as well as the actual oil components;
– the aqueous phase, which may make up 85–95% of the emulsion, containing water-soluble preservatives and any humectants and thickeners;
– the remaining ingredients such as active substances, perfume oils, colorants.

11.4.5 Skincare Emulsions

According to K. Schrader [2], cosmetic emulsions can be made up as follows:

Moisturizing emulsions, which moisten the cornified layer and are intended to keep it moist for some time, can be w/o or o/w creams (cold cream, emollient cream, day cream, night cream, vanishing cream, ...). Moisturizing lotions with high water content and body milks are also used. Their bases may contain the components listed in Table 11.2:

The aim is to develop nonsticky, easily spread emulsions that are pleasant to use; this is not easy, particularly in the case of w/o emulsions. Researchers are therefore now attempting to develop w/o emulsions that produce the sensation of an o/w formulation on the skin, thus uniting the dermatologically more valuable properties of the w/o formulation with the cosmetic properties of the o/w emulsion. Waxes used to give consistency, for example, often leave a sticky feeling on the skin and must therefore be chosen carefully.

Various active substances and other materials are added to the emulsion base on the lines of Table 11.2, for example collagen, animal protein derivatives, hydrolysed milk protein, various amino acids, urea, sorbitol, and royal jelly. Individual recipes may be found in reference [2].

Table 11.2. Base of moisturizing emulsions [2].

Ingredient	[%]
water	20–90
polyol	1–5
various oil components, consistency regulators, and fats	10–80
emulsifiers	2–5
humectants	0–5
preservatives and fragrances	as required

Cleansing emulsions contain more emulsifier and oil, preferably paraffin oil, in comparison to moisturizing products. As was explained in Section 11.1, the emulsifier attacks not only dirt but also the hydro-lipid mantle. In order to minimize this problem – otherwise the products applied after cleansing, such as day or night creams, will not be sufficient to compensate for the loss – mild tensides such as protein derivatives can be used to wet the dirt (cf. Section 10.2.1.2) in addition to the emulsifiers. Higher polyol (e.g. propylene glycol or butylene glycol) content can also be used. Typically, cleansing milks contain 2–8% emulsifier, 20–50% oil, and 5–10% polyols besides preservatives and water.

Table 11.3. Example of an o/w-type cleansing milk [2].

Ingredient	[%]
oil phase:	
glyceryl stearate, HLB 11	8.0
stearyl heptanoate (PCL solid)	3.0
isopropyl myristate	15.0
cetyl alcohol	4.0
paraffin oil	15.0
methylsilicone oil 100	0.5
aqueous phase:	
chlorhexidine digluconate 20% (microbicide)	1.0
glycerin	2.0
allantoin (keratolytic agent for skin regeneration)	0.2
water	to 100
fragrance phase: perfume oil	0.2

In the example shown in Table 11.3 the aqueous phase is mixed and emulsified with the oil phase at 75°C. After 5 min further stirring, the emulsion is cooled to 40°C and fragrance is added, and finally it is run through a homogenizer at 30°C.

Hand and body lotions are formulated much like facial milk. Body lotion especially is often applied to damp skin and must be spreadable, so o/w emulsions are proven for this application. Care must be taken that the tenside content is not too high, as otherwise residues on the skin can be absorbed by towels. Such residues can be problematic with hand lotions too: the oil must be absorbed as quickly as possible, but this leaves a dull feeling on the skin. For this reason the development of good emulsions for the hands is not easy, and indeed is only possible with the aid of a combination of additives.

Face masks are used for a period of 5–30 min and act as cleansers or as sources of moisture or some active substance. The emulsions used are very similar to those in o/w day creams. Cream masks resemble foundation make-up, and scrubbing masks contain mostly poly(vinyl alcohol) as a gel former, alcohol, softener, wetting agent, fragrance, and water.

Foot lotions are emulsions designed to refresh the feet, inhibit sweat, soften the cornified layer, calm itching, absorb moisture, and prevent fungal infections. They should be formulated with ingredients that dry the skin out more than ingredients in other preparations, such as triethanolamine lauryl sulfate, magnesium lauryl sulfate, and sparingly ethoxylated fatty alcohols. Aluminum salts have proven to be useful antiperspirants.

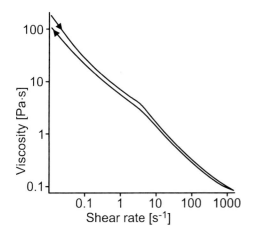

Figure 11.3. Viscosity curve for baby lotion [6].

Baby care products must be specially formulated with the particular sensitivity of babies' skins in mind, since these react almost like mucosa. The ease with which they absorb substances means that there is a danger of absorption of toxic compounds and a greater susceptibility to bacteria such as streptococci and staphylococci. Bacterial production of ammonia from urine (*Bacterium ammoniagens*), especially at neutral pH, combined with friction can lead to diaper rash, dermatitis, which is particularly pronounced between the thighs. Products for babies must protect against wetness with a thick film of grease that adheres firmly to the skin and has a drying effect, and should create a slightly acid environment on the skin. To obtain a stable, thick layer, finely divided solids such as zinc oxide are usually added to the emulsion. The lipophilic base is often lanolin, a derivative of lanolin, vaseline, or paraffin oil. Figure 11.3 shows a typical viscosity curve for baby lotion.

11.4.6 Skin-Protection Emulsions, Foundation and Make-Up, Deodorants

The selection of emulsifiers and their concentration is particularly important for *sunscreens*. Solar radiation can induce changes in the skin if emulsifiers, especially unsuitable ones, are present in excessive amounts. Emulsifiers that have little negative impact on the skin should therefore be chosen, and used in the lowest possible quantities, always bearing in mind the requirements of stability in storage. Simple formulations rather than wildly heterogeneous compositions usually seem to be better tolerated by the skin. In addition to UV absorbers, sunscreens may contain compounds that protect the skin, such as allantoin, insect repellents, or local anesthetics such as lidocaine. Generally, o/w emulsions are more popular. The intensive irradiation experienced in tanning studios, however, requires w/o emulsions with very high proportions of fat.

Antiperspirants and deodorants have different functions. The former, common in the United States, act both to inhibit sweating and to deodorize. The deodorants that are more popular in Europe, on the other hand, inhibit only odor.

Human sweat is almost odorless when it is secreted by the apocrine glands; only after it has been decomposed by bacteria on the skin does the unpleasant odor develop. There are several possible methods of combating this smell: masking with perfume oils, oxidation of the odoriferous compounds with peroxides, adsorption by finely dispersed ion exchange resins, inhibition of the skin's bacterial flora (the basis of most deodorants), or the action of tensides, especially appropriate ammonium compounds.

Antiperspirants contain astringent substances that precipitate proteins irreversibly and thus inhibit perspiration.

The general composition of antiperspirant or deodorant preparations is along the lines of 60–80% water, 5% polyol, 5–15% lipids (stearic acid, mineral oil, beeswax), 2–5% emulsifiers (polysorbate 40, sorbitan oleate), 10% antiperspirant (aluminum chlorohydrate), 0.1% antimicrobial, 0.5% perfume oil.

Facial foundations include preparations for the face, neck, and decolleté, and blush (rouge). Cream foundation and cream blush contain 30–60% pigment and 40–60% cream base (standard emulsion), amongst other things. In contrast, fluid foundation and fluid blush are pigment dispersions containing thixotropic agents.

Skin-protection products can be classified according to their protective action: against chemical attack such as that by acids and alkalis; against dirt and dust, tar and lubricating oils; against physical attack such as that by UV or heat; against mechanical damage; lubricants and products for use in massage; insect repellents. They should not have any effect on the tissue but simply form a flexible film on the skin. Preparations that protect actively against dirt and dust, for example, contain about 20% solid salts of fatty acids such as zinc stearate. A protective effect against organic solvents is achieved with combinations of soaps, starches, and emulsifiers like sodium lauryl glutamate, glyceryl monostearate, and stearic acid, while for protection against aqueous solutions semisolid lipids and silicone oils are utilized.

11.5 Microemulsions and Liposomes in Cosmetics

The general aspects of microemulsions and liposomes are treated in Chapter 3. As was described there, microemulsions form spontaneously and are thermodynamically stable, unlike emulsions. Microemulsions containing ionic tensides require them in high amounts (15–20%). Somewhat lower concentrations are necessary for nonionic tensides, for instance 5% for certain poly(ethylene glycol) ethers.

Microemulsions are interesting for the cosmetics industry not only because of their thermodynamic stability but also because of their transparency. In addition, the small particle size (100–500 nm) means that they penetrate the skin particularly well. Gel sunscreens, perfume gels, skin cleansing and skincare gels are all built on this base [7]. The high tenside content means the possibility of skin reactions cannot be ruled out, a problem that can be magnified by the small particle sizes.

To obtain a clear, transparent gel, it is important that the temperature be held above 90°C when the water and oil phases are mixed. Air must not be incorporated during production, as it will be unable to escape. To manufacture a clear gel by the emulsification of a mineral oil, for example, the chain length of the oil must be less than that of the emulsifier. On the other hand, long-chain fatty alcohols may be added to increase the viscosity; hydrocolloids are unsuitable, as they reduce the stability of the microemulsions.

Liposomes, which occupy the size range of 10 nm–100 μm, have shells of one or more bilayers of amphiphilic lipids (see Figure 3.5). These lipids may be naturally occurring compounds such as phospholipids, glycolipids, or sphingolipids, or they may be synthetic.

As explained in Section 11.2, liposomes will not be able to penetrate the cornified layer. However, it is possible for them to increase the moisture content of the hydro-lipid mantle by forming mixed phases with the lipids of the cornified layer; after all, the polar residues of lecithin are surrounded by up to 23 water molecules. Active substances like cortisol, though, are enriched by 5–10% in the epidermis and dermis compared with emulsions [8]. It is therefore assumed that the presence of liposomes encourages the penetration of active substances into the skin.

11.6 Solutions

Skin oils and fats used as solvents for active substances are intended to soften, smooth, and protect the skin, and to counteract the defatting effect of washing. They can also serve as lubricants for massage. Combined with emulsifiers, they are used to remove dirt from the skin, such as pigments left from make-up. Other applications include hydrophobic protection for swimming, sunscreens, and skincare products with additives such as essential oils, extracts of medicinal herbs, or vitamins.

Facial toners too are all manufactured as solutions. These are used to remove residues from cleansing and prepare the skin for subsequent application of a day or night skincare product. The formulations must be such that they do not dry the skin out unnecessarily, otherwise it will be very difficult for the following steps in the skincare regime to return the skin to its original state.

Body rubs based on alcohols are designed to refresh the skin after physical exercise or after bathing. They contain refatting, easily spread substances such as ethoxylated lanolin alcohols or polyol fatty acid esters, as well as mild disinfectants. Plant extracts or other active substances are often also included.

Preshave and aftershave lotions differ from toners and body rubs in having a higher alcohol content and special active ingredients. Like toners, shaving lotions are mildly acidic, with a pH of 5.5–6.5. Astringent aluminum salts are added, and often bactericides too to help prevent inflammation.

11.7 Bath and Shower Products

Cosmetics for the bath, designed for healthy users, are intended to stimulate the metabolism, improve the bather's feeling of wellbeing, and treat or cure mild skin disorders. Above all, though, they should reinforce the cleansing action of the water and, by means of essential oils, cause a feeling of wellbeing. The stimulation of the metabolism is also largely due to the presence of essential oils, but not from the tannic plant extracts included as astringents.

In addition, the product should be formulated to avoid leaving a dirty tidemark around the bathtub and to facilitate cleaning of the tub. Thick, viscous products create an impression of concentrated content and are therefore psychologically preferable to runny formulations. Since the tensides used are almost always fatty alcohol ether sulfates, the viscosity can be adjusted by addition of NaCl (Figure 11.4). One bath requires about 6 g of washing agent, which is contained in 10–30 g of the formulated product.

Detergents adsorb onto the skin from diluted solution and, in the case of alkyl aryl sulfonates, lend it a feeling experienced as one of unpleasant stickiness. Irritation related to detergents can be reduced by the use of suitable combinations. Ether sulfates are better than alkyl sulfates or sulfonates in terms of compatibility with mucosa. Solutions containing more than 20% washing agents seldom support bacterial or fungal growth, but more dilute solutions need to contain preservatives.

Bath salts are based on various salts such as sodium sulfate, sodium chloride, sodium tripolyphosphate, sodium hexametaphosphate, trisodium phosphate, sodium metaborate, or sodium thiosulfate, or solid anionic tensides. The other ingredients, such as color, perfume oils, and tensides, are either sprayed onto the base or mixed with it in a drum. For *bath cubes*, on the other hand, sodium hydrogencarbonate is employed as the base, since it releases carbon dioxide in acid solution with, say, tartaric or adipic acid. Oxygen bath cubes contain sodium perborate, which liberates oxygen in the presence of catalysts such as manganese salts. It is a good idea to mix starch, anhydrous sodium sulfate, or aerosil into the tablets to control moisture content.

Bath oils and *foam baths* have characteristics that vary depending on the content of emulsifier or tenside. Pure oils, with fragrance mixed in, float on the bathwater and form a film of oil on the body when the user gets out of the bath. Products containing a small amount of emulsifier yield milky or cloudy baths and leave no dirty ring, or only a slight one. The addition of detergents as foam formers reduces the skincare action of the product.

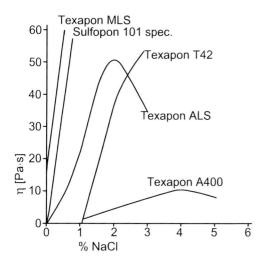

Figure 11.4. Increasing viscosity by addition of salt to solutions of alkyl sulfates and amphoteric tenside (Dehyton K); tenside ratio 3:1; total concentration 15% (source: Henkel [4]).

11.8 Gels

Fat-free gels form the base for plant extracts such as cucumber juice, melon juice, tomato juice, and apple juice, as well as honey or glucose syrup. As well as water and, usually, alcohol, they contain large amounts of glycerin (up to 20%) and various gel-formers such as gelatin, agar-agar, alginates, carragheen, pectins, tragacanth, wheat starch, bentonite, Veegum, colloidal kaolin, methylcellulose, polyvinylpyrrolidone, poly(vinyl alcohol), or

Carbopol. Gels based on vegetable mucilage are seldom compatible with alcohol. Exceptions are quince mucilage and pectin gels, which do not flocculate when ethanol is added.

All vegetable gel-formers are mildly acidic and so reinforce the protective effect of the skin's natural mantle. Examples of the various gel products on the market are lemon hand gel, perfume gel, hairstyling gel, firming gel, antiseptic gel for handwashing, foam bath gel, sunscreen gel, aftersun gel for sunburn, protective gel for the skin, and protective gel for babies.

11.9 Pencils and Sticks

Stick formulations have gained popularity at the expense of the aerosol. Not only lipsticks, eyebrow pencils, foundation sticks, but also scent sticks, insect repellent, and deodorant sticks are successfully marketed. The base material of a stick is either a wax or grease mass (usually the case for make-up), or a shape-stable gel (for example for scent sticks and deodorants). The base should not soften below 50°C; if it transfers well onto skin, it can soften at this temperature.

Deodorant sticks are based on soaps such as sodium stearate, which can form solid gels mixed with 10–100 times their own quantity of alcohol. Alcohol-free formulations contain polyols instead. The cooling and refreshing effect of the stick formulation is amplified by the addition of menthol and essential oils.

Lipsticks with a pure fat base have high gloss and excellent hiding power, but tend to come off the skin too easily. More permanent lipsticks contain hydrophilic solvents such as glycols or tetrahydrofurfuryl alcohol in addition. The raw materials for lipstick base include: ozocerite (absorbs oil well, prevents crystallization), ceresin wax (absorbs oil well, microcrystalline), Vaseline (forms water-impermeable films), paraffin (thixotropic, often crystalline), beeswax (good permanence, increases resistance to fracture), myristyl myristate (improves transfer onto skin, soft effects), cetyl lactate and myristyl lactate (nonsticky, form an emulsion with moisture on the lips), carnauba wax (binds oil, increases melting point, hardness, and surface luster), lanolin derivatives (various properties, sticky or nonsticky, hydrophilic or hydrophobic), oleyl alcohol (oil component), and isopropyl myristate (decreases stickiness).

11.10 Powders, Cream Powders

Powders are solid particles in the 100–200 μm size range. Powders smaller than 10 μm block pores and cause inflammatory reactions to foreign bodies; particles that are too large, on the other hand, feel rough. Powders form the base for a variety of active substances and can also be used to dry and cool the skin and for its mechanical protection. Colloidal silicon dioxide, magnesium carbonate, and starch are additives used to increase the drying effect, while starches or stearates improve the cooling effect. Powders have to adhere well to the skin, so starches or possibly lubricating fatty components are added. In addition they must be physiologically and chemically inert. Their sliding properties are increased by addition of talc. The most important powder bases are silicates (kaolin, aerosil, talcum), carbonates (magnesium carbonate, calcium carbonate), oxides (zinc oxide, titanium dioxide), stearates (zinc stearate, magnesium stearate, aluminum stearate), starches, and protein decomposition products.

Baby powders consist of very absorbent ingredients together with small amounts of antiseptic. The grains must not be too hard and must not have sharp edges or points, lest they injure the skin. The powder base should be only sparingly soluble in both water and oils.

In cream/powder products, available in pots or cast into tubes, the powder is incorporated into a base. For anhydrous products, this is a mass of oil, fat, or wax with highly thixotropic properties. The pressure with which it is applied releases the solid particles, which adhere to the skin. The waxes most frequently used are beeswax and microwaxes; less often, carnauba wax and candelilla wax. As well as the anhydrous formulations, there are cover cream emulsions of the w/o type and the o/w type which contain powder, pigments, and pearl-gloss pigments.

11.11 Oral and Dental Hygiene Products

Although teeth feel like bones, they are actually – like nails and hair – outgrowths of the same embryonic layer that produces the skin. In cross-section (Figure 11.5), a human tooth can be seen to contain a pulpy interior surrounded by dentin and enamel. The enamel is coated with a surface skin, or pellicle.

The pellicle, which is hardly affected by toothbrushing, can regenerate quickly by the adsorption of proteins from the saliva onto the hydroxyapatite of the tooth enamel. It consists largely of glycoproteins and presents a surface for bacterial colonization, the source of plaque. Plaque is the coating of living and dead microorganisms in a polysaccharide–glycoprotein matrix that adheres firmly to our teeth. It consists of 80% water, and the remainder is 60–70% bacterial mass. Incorporation of calcium ions turns plaque into tartar; addition of crystallization inhibitors to toothpaste can help limit this process.

Van der Waals and electrostatic interactions (Chapter 5) are responsible for the first step in deposition of bacteria on the pellicle. In the second step, specific adhesive sites on the bacteria develop connections to receptors in the pellicle. The reinforcement and crosslinking of these bonds uses specific polysaccharides synthesized by oral bacteria from saccharose. In fresh plaque, streptococci predominate, alongside other bacteria such as Neisseria and Actinomyces spp. As the plaque matures, the bacterial population alters from gram-positive streptococci to more filamentous, gram-negative species and fuso-bacteria, and then later to spirillae and spirochetes.

Bacterial decomposition produces acids in the plaque, which attack the crystal structure of tooth enamel and cause carious lesions through demineralization. Subsequently the dentin may also be destroyed. The toxins which arise as another consequence of the decomposition by bacteria can inflame the gums. Thus careful oral and dental hygiene to prevent the formation of deposits on the teeth is obviously worthwhile. The mechanical cleaning achieved with a toothbrush is supported by the use of oral and dental care products containing active substances intended to prevent the formation of deposits. Antimicrobial substances added to inhibit the bacteria responsible for creating plaque include various bisbiguanides and heavy metal ions, or certain quaternary ammonium compounds. The use of enzymes to destroy the matrix is also under discussion. However, the products must not have a nonspecific antiseptic effect, as this would destroy beneficial oral flora too.

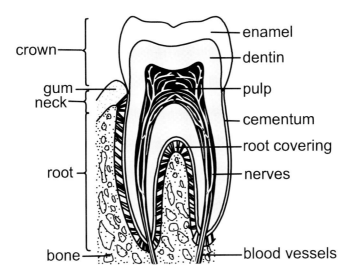

Figure 11.5. Section through a molar [2].

Oral hygiene products can be classified as disinfectant, astringent, and scented mouthwashes. They contain alcohols in aqueous solution as well as the appropriate active substances and flavorings.

Tooth powders consist of abrasive grains of calcium carbonate (prepared chalk), soap powder, or fatty alcohol sulfates. They are flavored and usually dyed pink. Since they are more abrasive than toothpastes, they can cause damage.

Liquid tooth cleaners, such as Pearl Drops, are toothpastes formulated as liquids, and usually more abrasive.

Toothpastes support the mechanical cleaning action of the toothbrush and refresh the mouth. A typical toothpaste contains the following ingredients (Table 11.4):

Table 11.4. Composition of toothpaste [2].

Ingredient	[%]
polishing mass	15–55
humectant	10–30
binder	0.5–2
preservative	up to 0.5
tenside	0.5–2
flavor	0.5–1
active ingredients	as required
water	to 100

The polishing agent is pure precipitated calcium carbonate that contains no abrasive impurities such as silicon or aluminum oxides, so that the enamel will not be damaged. Once this has happened, dentin is abraded twenty-five times faster, and the root elements thirty-five times faster, than the enamel. Calcium phosphate and the sparingly soluble salt sodium metaphosphate are other polishing masses used together with fluoride-containing agents; recently, synthetic zeolites and aluminum hydroxide have also been used.

Glycerin, sorbitol, xylitol, and poly(ethylene glycol) are used as humectants, and hydrocolloids such as carboxymethylcellulose, methylcellulose, hydroxyethylcellulose, carragheen, tragacanth, alginates, poly(acrylic acid) salts, and montmorillonite are used as binders. The most common preservatives are *p*-hydroxybenzoic acid and sodium benzoate, and the tenside of choice is sodium lauryl sulfate [9]. This is present not simply as a foaming agent but also to assist, as a wetting agent, the penetration of the toothpaste into cracks and gaps. As far as active substances are concerned, fluoro compounds play the dominant role; a maximum of 0.15% active fluoride ions is permitted.

11.12 Shaving Aids

Shaving aids are intended to prepare the skin before, and treat it after, dry or wet shaving. Products are shaving soaps, shaving creams, shaving foams, pre- and aftershave lotions, all products in which the fragrance is very important.

Shaving soaps, available in stick form, differ from toilet soaps in their particularly high potassium soap content, as this improves their foaming properties. They often also contain humectants like glycerin or polyols, and other additives such as foam stabilizers (e.g. amine oxides, alginates).

Foaming shaving creams, sold in tubes, must not separate or decompose even after long storage. These, too, are potassium soaps containing 5–25% glycerin. They are given an opalescent appearance by the addition of 2–4% stearin (glyceryl tristearate).

Nonfoaming shaving creams are recommended for dry skins. They are applied without the aid of a shaving brush and have a less degreasing effect. These creams are mostly superfatted o/w emulsions, which contain for example lanolin, cetyl alcohol, Vaseline, or paraffin as the superfatting agent and monoglyceride esters or ethoxylated sorbitan esters as the emulsifiers.

Shaving foams are liquid shaving soaps packaged and foamed by a gas like aerosols.

Preshaves have a high alcohol content in order to degrease the skin, so that the razor can better grip the stubble. They also contain lubricants to aid spreading of the oil layer.

Aftershaves are intended to neutralize the alkaline environment created on the epidermis by shaving, for example with citric or tartaric acid. They should restore the biological buffer of the skin's mantle, have an astringent and refreshing action, and encourage the healing of small injuries, for which purpose appropriate active substances are added to the aqueous–alcoholic solution.

11.13 Hair Cosmetics

11.13.1 Scalp and Hair

The sites of action for hair treatments are the surface of the hair shaft (the cuticle), its interior (the cortex), and, especially for greasy hair, the cornified layer of the skin and its sebaceous glands (see Figure 11.1). The cuticle changes greatly from the scalp to the tip of a hair, from nearly perfectly smooth to jagged and irregular, because of wear. The mean diameter of a hair is 70 μm, and the mean growth rate in the anagenic phase is 0.35 mm/day.

The *cuticle* (about 2.2 μm thick) consists on average of six layers of cuticular cells, with a mean thickness of 0.37 μm. This coating is clearly visible in Figure 11.6. Each cell has two layers, which form the actual cornified substance, the keratin, of the cuticle. The inner, endocuticular layer contains all the nonkeratinous components of the cell such as incompletely or nondecomposed organelles and membranous structures, mitochondria, secretion channels, and granules. The individual scales of the cuticle are bound together by a sort of mechanically stable "cement"; the individual scales remain firmly glued together even when the cuticle is worn.

The *cortex*, the bundle of fibers on the interior, consists of spindle-shaped cells that form a system of bound microfibrils and "cement". In between are located the granules of pigment that give hair its color.

The follicle (see Figure 11.1) is formed by the skin. In the thickened, onion-shaped end adjacent to the hair papilla, the keratinocytes multiply by cell division. The resulting pressure of growth pushes cells upwards, where they differentiate and contribute to hair growth. Hair falls out in the normal course of the growth cycle when a new hair has developed underneath it in the follicle, except in the case of hair loss related to illness, when no new hair is present.

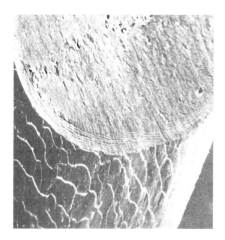

Figure 11.6. Cuticle and cortex in a human hair [10].

The surface of hair is hydrophobic and difficult to wet with water. Hair itself, though, is hygroscopic and can absorb up to 27% water, swelling as it does so. The swelling is dependent on the pH of the medium, however, and is least at the isoionic point, where the greatest number of salt bonds is present in the hair.

One problem in haircare is the production of sebum, which is strongest on the scalp. Sebaceous glands are holocrine glands, that is, their secretion results from the destruction of the cells themselves. The rate of formation of sebum therefore depends on the number and size of the cells undergoing fatty degeneration. In the subsequent process, in which hair becomes greasy, the sebum is transferred from the scalp to the hair and renders it greasy and stringy, especially in the case of long hair.

The scalp sheds tiny particles of skin from the cornified layer as a normal physio-logical process. These are not usually a problem, but in the case of dandruff large agglomerates of corneocytes are sloughed off at once and these are cosmetically unpleasant.

11.13.2 Haircare Preparations

Shampoos are intended exclusively for the hair and scalp. The principal ingredient of any shampoo is a tenside. The product must foam well and have a good detergent effect with respect to grease and dirt. Other criteria are manageability, hold, and volume, and good dermatological, toxicological, environmental and economic properties.

Not only dirt but also microorganisms like bacteria, yeasts, and fungi are removed by hairwashing. Since both hair and most dirt particles are negatively charged in water – hair only adopts a positive charge at pH values of less than 3 – anionic tensides are the obvious choice for detergents. Furthermore, anionic tensides tend to disrupt the bonds between SH groups (important for styling hair), unlike nonionic tensides. Because of their opposing charge, cationic tensides do not wash well; at most they can be used as conditioning agents. Cationic polyelectrolytes are other conditioning agents, which increase the weight of the hair, and fatty alcohols and fatty acid derivatives are used to counteract excessive defatting of the hair on washing. The general composition of shampoos is listed in Table 11.5 [11].

Table 11.5 General composition of shampoos.

Ingredient	[%]
water	50–70
tensides (preferably anionic)	7–15
consistency regulators (nonionogenic)	0.5–2
foam stabilizers (amine oxides, amphoteric substances)	3–5
superfatting agent (fatty acid or fatty alcohol ethoxylates)	0.5–1
active compounds (e.g. cationic conditioner)	~ 2
opalescent or pearl-gloss agent (e.g. ethylene glycol distearate)	~ 1
preservative (e.g. *p*-hydroxybenzoic acid ester)	~ 0.2
dye	~ 0.005

Hair treatments for use after shampooing have become modern necessities. They are intended to hold a hairstyle in place, increase shine, improve the feel of the hair, and above all make the hair easier to comb, whether wet or dry. Hair treatments can be divided according to their composition in to those formulated to be rinsed out after application and those that remain in the hair.

Conditioners, which are rinsed out, are taking an increasing share of the market. They are designed chiefly to give ease of styling, and besides water they contain cationic and hydrophobic substances like fatty alcohols, emulsifiers, acids to adjust the pH, thickeners, active substances such as hydrolysates of proteins or anti-dandruff compounds, humectants, dyes, and perfume oils.

Hair masks are used to treat damaged hair and hair beds. The principal ingredients of the creamy products are cationic compounds which coat the hair; often the coatings are not removed by shampoo and continue to improve manageability even after 3–4 washes.

Hair fixatives are available in a wide variety of forms such as lotions, gels, and foams. Polymeric film-formers, amongst the principal constituents, hold the hairstyle in place and decrease the frequency of washing. Copolymers are used, of the type vinylpyrrolidone/vinyl acetate, vinyl acetate/crotonic acid, methyl vinyl ether/maleic anhydride. Hair fixatives also contain softening agents which increase the flexibility of the polymers and thus reduce their tendency to flake off. Hairsprays are a special form of hair fixative.

Unlike hair fixatives, styling creams are not very popular; they have been replaced in importance by gel and foam products. *Hair oils*, *pomades*, and *brilliantines* may be compared to skin oils since, like these, they are solutions in fat or oil. *Hair lotions* belong to the same category; they contain active substances, for example to counteract dandruff or hair loss, in aqueous or alcoholic solution.

The same criteria apply to hair dyes as to textile dyes. Just as for textiles (see Chapter 15), dye that resists washing and is lightfast must be formed inside the hair shaft for permanent color. Temporary colorings also exist, which are washed out on the first shampooing, as do semipermanent toners, which fade slowly with washing and exposure to light. It should be borne in mind that, while the product should color the hair, it must leave the scalp unchanged.

Direct dyes require no further treatment to achieve the color desired; they coat the hair but cannot usually penetrate it because of the size of their molecules. They are therefore easy to remove and are intended as semipermanent rather than permanent dyes. Typical substances are nitro- and anthraquinone dyes as well as azo and quinonimine compounds with a quaternary acyclic ammonium group.

Lasting and intense color is achieved by the oxidative development of soluble intermediates absorbed into the hair shaft. In their application, these *oxidation dyes* thus resemble vat dyes for textiles, which diffuse into the material as the soluble leuco form and are converted there into insoluble pigment by oxidation, or *developing dyes*, which are converted to their final colored form in the fiber. Figure 11.7 shows the principle of oxidative coupling by means of examples.

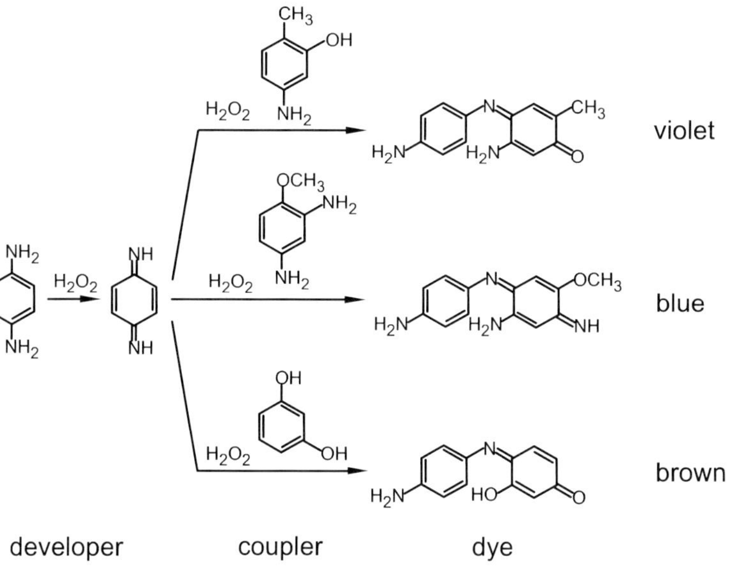

Figure 11.7. Oxidative coupling in hair dyes; analogous to [12].

To dye hair, the coupler and developer are applied to the hair together with the oxidant in a suitable formulation – cream, gel, or liquid – and allowed to act for 20 30 min. It is important that the diffusion speeds of the coupler and the developer do not differ greatly, since otherwise oxidative self-coupling can yield undesired colors. The composition of an oxidation dye is described in Table 11.6. The oxidant is mixed into such formulations just before use; it is usually hydrogen peroxide. This oxidant has the desirable property of destroying and bleaching the melanin in the hair, so permitting even dyeing.

Table 11.6. Composition of an oxidation dye.

Ingredient	Function
dye intermediates	yield dyes after reaction
stabilizers/reducing agents	stabilization during storage
alkaline pH adjusters, e.g. ethanolamine	swelling of the keratin, oxidation accelerator
carriers	creation of suitable form for application
complex formers	prevention of effects due to heavy metals
fragrances	masking of the amine smell

Other products include *toners* such as toning shampoos or gel hair colors, which are available in a wide variety of tones to refresh color or cover up undyed hair between dyeing sessions, to even out hair color, to tone graying hair, and so on. As well as color, such toning products may also contain emulsifiers, thickeners, cationic conditioning agents, or organic acids.

Permanent waving of the hair (perming) is related to dyeing in that, in both cases, chemical reactions are carried out in the hair. As will be described in Chapter 13, hydrogen bonds, ionic salt bonds, and disulfide bonds bind polypeptide chains together and lend stability to their structure. Applied to the hair, this means that that these bonds must first be broken and then, after reshaping, reformed to give permanence to the new style. After a thorough wash, the hair is styled with curlers; the action of alkali, usually ammonia, makes the hair swell and disrupts ionic and hydrogen bonds, while reduction with thioglycolic acid breaks the disulfide bonds so that the hair can be shaped. Subsequently the disulfide bonds are reinstated by oxidation with hydrogen peroxide, and acidification restores the hydrogen bonds and salt bonds.

The aqueous alkaline thioglycol solution (the *cold wave*) may also contain tensides and oils. The acidic *fixing solution* or *fixing emulsion* often has cationic polymers added to it as conditioning agents.

11.14 Bases and Auxiliaries

Chapter 2 summarizes the topic of emulsifiers, so no further remarks on the subject will be made here. Other base and auxiliary substances used specifically in cosmetics, though, will be dealt with in this section [13].

Alcohols, like ethanol, *n*-propanol, or isopropanol are the principal solvents other than water for liquid products. In addition they are employed in emulsions, aerosols, and sticks.

Natural fats and oils, triglycerides of various fatty acids, are closely chemically related to human fat and therefore seldom cause any skin reaction. Since they are easily oxidized (become rancid), emulsions of such triglycerides require the addition of oil-soluble antioxidants (see Section 12.4). Amongst the many natural fats and oils used in cosmetics, the following are common: *soya oil* (bath oils easily transferred to the skin), *peanut oil* (high-quality skincare products, emulsions, hair oils, skin treatment oils, bath oils), *olive oil* (high-quality oil for skincare preparations and bath products), *sunflower oil* (skin oils and emulsions), *avocado oil* (high-quality, biologically effective skincare products; in creams, liquid emulsions, hair oils, and skin treatment oils), *castor oil* (lipsticks, eyelash products, superfatting agents for alcoholic hair tonics and brilliantines), *pork fat* (pharmaceutical and cosmetic ointments and creams), *spermaceti oil* (ointment base for skincare products), *wheatgerm oil* (high-quality oil for treatment of injuries and for skincare), *evening primrose oil* (skincare for dry, rough or chapped skin; also has antiinflammatory properties and stimulates circulation), *jojoba oil* (high-quality emulsions and skin oils).

Synthetic oils are becoming increasingly important because of their superior stability. Mid-length triglycerides are used, mostly of capric and caprylic acid, as well as PCL liquid (a mixture of three branched fatty acid esters), myristyl myristate, liquid lanolin, isopropyl fatty acid esters (palmitate, myristate, isostearate, oleate), liquid dicarboxylic acids, and others.

Paraffin oils, which, unlike many natural lipids, barely penetrate the skin, are used for products that should have only a surface effect, such as massage oils, baby oils, and sunscreens. They are also used in combination with so-called fatty film looseners, can reinforce some emulsions, and have good compatibility with the skin.

Vaseline is a mixture of solid and liquid straight-chain, branched, and cyclic, and also some unsaturated aliphatic hydrocarbons. They form three-dimensional networks in which the liquid hydrocarbons are trapped. Vaseline is used in pharmaceutical and cosmetic ointments, in products aimed at a surface effect, and in make-up.

Although *soft and hard paraffins* are recommended for use as consistency regulators, they feel unpleasant on the skin and can be substituted with ozocerite (or ceresin wax), which has a lower melting point.

Microcrystalline waxes, which consist of a mixture of straight-chain paraffins and branched isoparaffins, soften at 60–70°C and are used as consistency regulators in emulsions and stick formulations.

Waxes generally have a low melting point and contain long-chain compounds. According to their chemical class, they are known as hydrocarbon, alcoholic, ketone, ester, or amide waxes. Vegetable waxes include *beeswax* (sensitive to oxidation; consistency regulator in creams and ointments, stabilizer in cold creams), *carnauba wax* (derived from the Brazilian wax palm, used for make-up), *candelilla wax* (from a Mexican spurge, used in stick formulations such as lipsticks, sunscreen sticks, insect repellent sticks).

As a supplement to the information given in Chapter 2 concerning auxiliaries for emulsions, we should mention:

Lanolin, used as a cream base with excellent skin compatibility and high water-absorbing ability and as a stabilizer and refatting agent, is an unctuous mixture of various esters of higher alcohols and small quantities of free alcohols, fatty acids, and hydrocarbons. It is easily oxidized, so antioxidants must be added.

Sterols such as *sterolane* (manufactured from wool wax, as emulsifiers for skin-compatible w/o emulsions), or *cholesterol* (the most important animal sterol in pharmaceuticals and cosmetics), are used in w/o emulsions as cotensides.

Cetyl and *stearyl alcohols* are employed as consistency regulators and cotensides.

Soaps of singly, doubly, and triply charged cations (Na, K, Ca, Mg, Al; also, triethanolamine) can be used, for example, as emulsifiers in day creams; the multiply charged soaps are also utilized as gel-formers in w/o emulsions.

Esters and *ethers of polyvalent alcohols* are used in large amounts nowadays as nonionic w/o and o/w emulsifiers in creams. Examples include sorbitan and glycerin esters and ethers, combined with poly(ethylene glycol), fatty alcohols, and fatty acids. The large number of possible combinations makes it possible to tailor emulsifiers of any desired HLB value.

References for Chapter 11:

[1]   G. Czihak, H. Langer, H. Ziegler, Biologie, Springer-Verlag, Berlin, Heidelberg, New York, 1981.
[2]   K. Schrader, Grundlagen und Rezepturen der Kosmetika, 2nd ed., Hüthig Verlag, Heidelberg, 1989.
[3]   A. G. Matoldsy, The Skin of Vertebrates, in: R. I. C. Spearman, P. A. Riley (Eds.), Linnean Society Symp. Ser., Number 9, Academic Press, London, 1980.
[4]   W. Abraham, in Surfactants in Cosmetics, 2nd ed., Surface Sci. Ser. 68, (M. M. Rieger, L. D. Rhein, Eds.), Marcel Dekker, New York.
[5]   F. Greiter, Moderne Kosmetik, Hüthig Verlag, Heidelberg, 1985.
[6]   D. Laba, in Rheological Properties of Cosmetics and Toiletries (D. Laba, Ed.), Marcel Dekker, New York, 1993, p. 403.
[7]   J. Chester, Soap. Parf. Cosmet. *40*, 393 (1967).
[8]   W. Wohlrab, J. Lasch, R. Laub, C. M. Taube, K. Wellner, in Liposome Dermatics (O. Braun-Falco, H. C. Korting, H. I. Maibach, Eds.), Springer-Verlag, Berlin, 1992.
[9]   K. Schrader, in F. Greiter, Aktuelle Technologien in der Kosmetik, Hüthig Verlag, Heidelberg, 1987.
[10] Das Haar und seine Struktur, 2nd ed., Wella AG, Darmstadt, 1991.
[11] A. L. L. Hunting, Encyclopedia of Shampoo Ingredients, Micelle Press, 1983.
[12] N. Maak, In Between Congress Joint Conference (IFSCC, Deutsche Gesellschaft der Kosmetik-Chemiker e.V., Ed.), Verlag für chemische Industrie Ziolkowsky KG, Augsburg, 1987.
[13] H. P. Fiedler, Lexikon der Hilfsstoffe für Pharmazie, Kosmetik und angrenzende Gebiete, Editio Contor, Aulendorf, 1981.

12 Pharmaceutical Technology

Product identity, efficacy, and purity are the important criteria for medicaments. Painstaking investigation of the product quality is necessary with respect to the following points:

– the content of active substance must be the same in each dose
– there must be no undeclared substances present
– the efficacy and therapeutic properties must remain unchanged at least until the expiry date.

The initial form of the active substance is a crystalline or amorphous powder, or a liquid. However, the therapeutic efficacy and the therapeutic index LD_{50}/ED_{50} (the lethal dose for 50% of patients divided by the effective dose for 50% of patients) depend not only on the active substance but also on its presentation. The formulation of the active substance is therefore important. Depending on the formulation, the properties of an active substance can change drastically, both in storage and on use. Thus, for example, an acid-sensitive active substance must be formulated so that it can pass through the stomach undamaged.

12.1 Absorption of the Active Substance

Pharmaceutical substances are often taken orally because of the ease of this form of administration. This means that the amount of the active substance passing into the bloodstream must be carefully judged.

– There is normally a relationship between the concentration of the active substance in the body and the therapeutic response, that is, the greater the dose, the stronger the reaction.
– The therapeutically effective concentration must be reached rapidly at the target so that the pharmacological action can begin.
– Depending on the desired effect, the action must be designed to be short- or long-term.

The factors that play a role in oral administration are physicochemical and physiological variables as well as the variable type of presentation. Pharmacologically active substances are usually relatively small organic molecules that penetrate membranes well. Newer medicaments based on polypeptides, in contrast, with higher molecular masses and greater sensitivity to hydrolysis, are seldom suitable for oral administration. Efforts are being made, however, to modify such molecules so that some degree of absorption is possible in the gastrointestinal tract without enzymatic degradation.

12.1.1 Anatomy and Physiology of the Gastrointestinal Tract

The primary functions of the gastrointestinal tract are secretion, digestion, and absorption. It also protects the body from irritants, removing them by means of diarrhea and vomiting. Stress factors can initiate these mechanisms, but also influence the absorption in the tissues.

Numerous glands open into the first section of the tract, the buccal cavity, such as the sublingual gland, the submandibular gland, and the parotid gland. The gullet (esophagus) passes from the pharynx through the diaphragm to the stomach (the area around the entrance to the stomach is called the cardia). The major section of the stomach is the fundus. The exit, the pylorus, on the right-hand side of the stomach, leads into the small intestine, the first section of which is the duodenum. As the intestine descends, the bile duct of the liver joins it, and then a duct from the pancreas, the principal source of digestive enzymes. The subsequent sections of the small intestine are the jejunum and the ileum. In the lower abdomen the small intestine gives way to the large intestine, which has a blind end in the form of the cecum and appendix. The final section of the large intestine is the rectum, which is closed off from the exterior by the anal sphincter muscle (Figure 12.1).

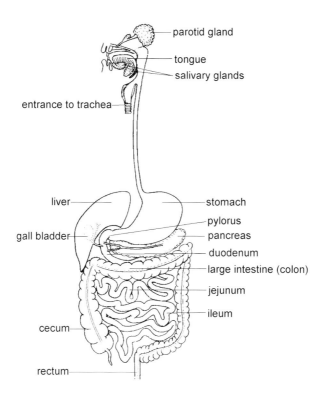

Figure 12.1. Gastrointestinal tract (from reference [1]).

The special shape of the surface of the small intestine is perfectly suited to digestion. Wrinkles in the mucous membrane (Kerckring folds) and fingerlike projections (villi) with surfaces covered in their turn with microvilli about 1 μm long increase the available surface area inside the small intestine by a factor of 600.

The mucosa of the small intestine can be split into three layers: the muscular layer at the bottom (muscularis mucosa), 3–10 cells thick, which divides the mucosa from the submucosa; the lamina propria between the muscularis mucosa and the intestinal epithelium, and the surface epithelium together form the structure of the villi. The lamina propria contains various types of cells, blood and lymphatic vessels, and nerves. Molecules to be absorbed must pass through this layer (Figure 12.2).

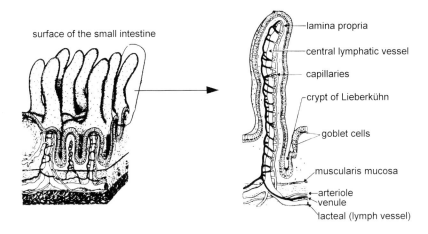

Figure 12.2. Anatomy of the villi in the mucous membranes of the small intestine (from reference [2]).

The chemical breakdown of foods and their absorption occur in the intestine. The food is broken down hydrolytically under the catalytic action of hydrolase digestive enzymes which are produced not only in the large glands attached to the intestine but also in the epithelial cells of the intestinal wall itself and the small glands embedded in it. In intracellular digestion, dissolved substances and tiny particles are ingested by the cells of the alimentary canal and broken down there. Larger particles and molecules that can be broken down quickly are digested extracellularly in the lumen of the intestine or of the intestinal glands by enzymes secreted into this lumen.

Various enzymes are involved in digestion, according to the different categories of food.

Proteins are split at the peptide bonds by proteases. Endopeptidases attack specific amino acids in the interior of long-chain proteins. Other peptidases fulfil other specific digestive functions. Trypsins act extracellularly in an alkaline environment; they are produced as inactive trypsinogens in the pancreas and activated by another enzyme secreted by the intestinal wall. In the highly acid environment of the stomach, pepsin produced by the fundus glands contributes to the digestion of proteins.

Carbohydrates are broken down by attack at the glycosidic C–O–C bonds by carbohydrases. Polysaccharides with hexoses joined by α-glycosidic bonds (for example starches, glycogen) are digested by amylases. Glycosidases often act specifically on one type of glycosidic bond.

Lipids are broken down by the different esterases. The esterases which are most effective in the breakdown of long-chain glycerin esters are known as lipases. Other esterases preferentially split esters of short-chain fatty acids. Phosphatides are decomposed by specific phospholipases and, less specifically, by phosphodiesterases. The breakdown of the insoluble lipids is facilitated by the presence of bile, which acts as an emulsifier.

Ribonucleases and deoxyribonucleases break down nucleic acids. Mononucleotides are split into nucleosides and orthophosphate by nonspecific phosphomonoesterases.

12.1.2 Bioavailability, Transport of Active Substances, and Absorption

Bioavailability is the percentage of active substance that arrives in the blood compared with the dose administered. Oral administration in particular exposes the active substance to a wide range of influences that may alter it. Intestinal flora, enzymes, and food and drink are among the factors that can affect the chemical consistency, the gastrointestinal transport rate, or the absorption rate.

On being swallowed, peroral medicaments travel into the intestine via the stomach, passing through very different anatomical and physiological environments on the way. The pH value, for instance, changes from 1–3 in the stomach to 5–7 in the duodenum and 7–8 in the ileum. The specific surface area too alters dramatically from the stomach to the small intestine, where absorption occurs. Before entering the intestine, therefore, pharmaceutical formulations are exposed to high proton concentration; hydrolysis, flocculation, and precipitation may occur at the low pH prevailing in the stomach. But it is not only the acid that has an effect; the entire contents of the stomach matter. The time, quantity, and type of food consumed determine the speed at which medicaments enter and, indeed, leave the intestine and the state in which they do so. Thus, it is hardly surprising that the available concentration of a medicament in the blood can vary greatly depending on the gastric contents when it was administered. As can be seen from Figure 12.3, a full stomach delays the arrival of active substances in the blood, as all the contents of the stomach remain there longer. Accordingly, under such conditions acid-sensitive substances can be more severely degraded than would be the case in an empty stomach, and are therefore only available for absorption from the intestine in small quantities. This is illustrated with the example of the acid-sensitive, water-soluble antibiotic erythromycin in Figure 12.3.

Bioavailability is not always impaired by the consumption of food. Many acid-resistant substances arrive in the intestine later than they would have done had the stomach been empty, but nevertheless unharmed, and the entire dose can be absorbed.

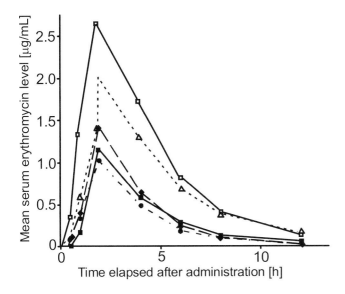

Figure 12.3. Concentration of active substance in the serum after administration of 500 mg erythromycin stearate in 20 mL water Δ, 250 mL water □, 250 mL water after a meal with high carbohydrate content ■, 250 mL water after a meal with high fat content ◆, 250 mL water after a meal with high protein content ● (from reference [3]).

The time at which a medicament is administered is therefore important in the case of the oral route. The period that has elapsed between the last meal and the administration can affect both the bioavailability and the blood concentration of an active substance heavily.

12.1.3 Distribution, Storage, and Elimination of Medicaments in the Body

Active substances in solution in the vicinity of the intestinal epithelium can diffuse through the membrane, as long as no substances are present to interfere. If they are not metabolized in the cells of the intestinal wall, active substances can attain the circulation. Once a substance has passed the outer layer of the mucous membrane and reached the lamina propria, it can pass into the bloodstream or (to a much lesser extent) the lymphatic system (Figure 12.2). Lipophilic substances with high o/w distribution coefficients ($>\sim$10 000) are an exception; these concentrate in the lymphatic vessels.

From the phenomenological point of view, the oil/water distribution coefficient $K_{o/w}$, the pK_a value, and the molecular volume are the determining values in the passive absorption of a substance. In particular, the distribution between the aqueous phase and the lipophilic membranes is important [4].

Pharmaceutical substances are distributed throughout the body largely by the plasma in blood vessels, which makes up about 4% of body mass. After passing out of the capillaries into the intercellular fluid in the interstitial volume (about 16–20% of body mass), the active substances easily diffuse to individual cells. The space occupied by the cerebrospinal fluid around the central nervous system is an exception. The *blood–brain barrier*, the unbroken epithelium that surrounds this space, only allows certain lipophilic substances through.

In plasma, the active substance may bind to plasma proteins; this can hinder or even prevent the exit of the substance from the circulation into the tissue fluid. Such complexes can therefore act as a depot for the active substance, since they are in equilibrium with the unbound, soluble substance, which can diffuse out of the blood.

Lipophilic active substances can also be stored in fatty and muscular tissues, lengthening the duration of the active substance release. Sometimes this is disadvantageous (anesthetics, impaired driving ability).

The purpose of degradation of active substances is to render them water-soluble so that they can be excreted by the kidneys. Enzymes in the liver are chiefly responsible for this metabolism. Hereditary enzymatic differences can result in different rates of metabolism in different population groups, so that the effects and side-effects of an active substance may differ too.

About 90% of an active substance or its metabolites is excreted through the kidneys. Certain lipophilic molecules are also excreted via the liver, the bile, and thence the intestines. This form of excretion is independent of kidney function; the existence of the route is exploited in the treatment of infections of the bile duct.

Besides the kidneys and the liver, glands such as sweat or mammary glands, and even the breath, can act as routes for excretion in small amounts.

12.2 General Remarks on Drug Formulations and Delivery

Active substances, that is, substances that have a pharmacological effect in the organism, may be of *natural origin*, such as extracts from plants or from animal glands, hormones, sera, or minerals. *Semisynthetic* active substances also have a natural origin, unlike *synthetic* ones; they are chemically modified once obtained.

Auxiliaries help to lend shape to the active substance and to preserve it, or to regulate its action. They should not cause any undesirable side effects, and in particular should not impair the action of the medicament or make it less well tolerated.

Drug dosage forms consist of active substance and auxiliaries formulated into powders, tablets, suppositories, ointments, dressings, or tinctures. If they are packaged and sold under a trade name, then they are described as *pharmaceutical specialties*.

The choice of packaging is not trivial. It must protect the medicament from moisture, light, dust, and microbial attack, and also act as a dosage aid. Examples are blister packs for tablets, sealed foil packaging for suppositories to maintain their shape, and bottles with droppers and with negligible alkali release for liquids.

Pharmaceutical specialties should be marked with an expiry date. In the absence of such a date, it may be assumed that the product can be used without problems for five years from the date of manufacture, provided it has been stored away from moisture at temperatures lower than 25°C. Unstable medicaments should be stored under refrigeration.

Depending on the route of administration – oral, sublingual, rectal, cutaneous, percutaneous, parenteral – the drug can act systemically or locally. In the case of *systemic* action, the active substance reaches the target organ in the bloodstream after absorption. In the case of *local* action, the point of administration and the point of action are the same; no transport of the substance need occur.

Drugs are most frequently administered *orally*, that is, they have to be swallowed. This mode of administration is unsuitable if the active substance is destroyed by the digestive juices produced by the pancreas, or if a very even concentration of the substance in the blood is necessary or the substance is not easily absorbed. In such a case, the substance is better absorbed on an empty stomach. Similarly, the absorption of lipophilic active substances is improved if they are taken with a meal rich in fats. Defecation influences the process too. Diarrhea reduces the duration of contact in the intestine and in many cases the extent of the absorption of active substances, whereas constipation increases it. The *first-pass effect* must not be forgotten: drugs are absorbed mostly from the small intestine, where they pass through the portal vein to the liver. From there, they are distributed to all organs. However, certain active substances are metabolized to a large degree in this first pass through the liver, and only small quantities reach their targets.

Sublingual administration, in which the medicament is held under the tongue, results in rapid absorption through the mucosa of the tongue and avoids the effects of digestion and the first-pass effect.

Rectal administration is indicated in cases of vomiting or unconsciousness, as well as for small children. The action of the drug may be systemic (for example, analgesics) or local (for example, to treat hemorrhoids). The extent of absorption depends on how full the rectum is. The largest amount of the active substance (80%) passes directly into the circulation, the remaining 20% travels through the liver, so the first-pass effect is insignificant.

Intact skin presents a substantial barrier to *cutaneous* or *percutaneous* administration, in which active substances are applied to the skin or the mucous membranes (see Figures 11.1 and 12.2). It must be borne in mind that when the skin is damaged the absorption may be unexpectedly (usually) high. This form of administration is suitable for lipophilic drugs, whether with systemic (percutaneous administration) or local (cutaneous administration) action.

Parenteral administration involves *injection* or *infusion* of a solution of the active substance. Absorption and passage through the liver are thus circumvented, so the active substance is rapidly distributed throughout the body without the first-pass effect. To avoid damage to the walls of the vein, solutions for intravenous administration should be as close to isotonic as possible (have the same osmotic pressure as blood), and have a pH similar to that of the blood. Intramuscular and subcutaneous administration have slower but longer-lasting effects than intravenous administration. In this case, suspensions and oily solutions can be administered to act as drug depots and release drugs over a period of, in some cases, months. If the concentration and pH of the solution injected vary greatly from those of the blood, injections can be painful and may cause tissue necrosis.

12.3 Drug Dosage Forms

Drug dosage forms can be classified by consistency as solid, semisolid, and liquid formulations [5]. Solid formulations include powders, granules, microcapsules, capsules, tablets, dragées, suppositories, and decoctions. Ointments, creams, gels, and pastes are semisolid formulations, and solutions, emulsions of low viscosity, and suspensions make up the class of liquid formulations. Products that are applied to the skin occupy the border between pharmaceuticals and cosmetics, as described in Chapter 11.

12.3.1 Powder

Powder preparations comprise loose, dry particles in a size range of 40–300 μm. They may contain auxiliaries as well as the active substance.

Powders intended for oral administration can be dispensed as *loose powders* or, in the case of substances with high activity, in *measured portions*. There are also powders for *parenteral* administration; before use, these are dissolved in sterile liquid to furnish a solution for injection.

Powders for topical use on skin or mucous membranes, or for treatment of wounds by application to injured tissue, increase the surface area of the skin and therefore have a cooling effect. They also possess drying and antiinflammatory properties. The particles should be less than 100 μm in diameter, and powders for use on open wounds must be sterile. The powder base may be inorganic, for example talc, zinc oxide, or clay; less commonly used substances include magnesium oxide, magnesium carbonate, and titanium dioxide. Since these substances are alien to the body, some patients may not tolerate them. Organic powder bases are better suited to the treatment of open wounds in this respect, but, unlike inorganic substances, they provide a good substrate for bacterial growth and often cannot easily be sterilized because of their sensitivity to heat. Starches such as maize and wheat starch, and occasionally potato starch, are used as bases for powders. Since they tend to clog, nonswelling starch derivatives are preferable. Lactose is biocompatible and completely absorbable, but its adhesive properties are not entirely satisfactory. Other additives can make a considerable difference to the adhesion and the flow of a powder, since in many cases these two properties are inadequate. Aerosil (very finely dispersed silicon dioxide) added in a proportion of 0.5–5% raises both the powder's adsorption ability and its absorbency. Addition of metal soaps (5%) improves flow and adhesion and increases the cooling effect of the powder (see also Section 11.10).

12.3.2 Granules

Granules and agglomerates are described in detail in Section 6.2. Some additional aspects related specifically to pharmaceutical technology will be covered here.

Granules are solid agglomerates of powder particles and usually flow better than powders. They have a homogeneous composition and, unlike powder mixtures, they cannot separate out even if they contain components of differing densities. There are two reasons for granulation: granules are easier to administer than powder, and their homogeneous composition makes granules a suitable preliminary stage in tablet-making.

Products such as coated granules, enteric-coated (gastric-acid-resistant) granules, and granules with modified release of the active substance or effervescent granules have special characteristics. *Coated granules* are coated with one or more layers applied as a solution or suspension in a volatile solvent. *Enteric-coated granules* are protected by layers of cellulose acetate phthalate and anionic copolymers of methacrylic acid and its esters, so that the granule only disintegrates when it reaches the intestine. *Effervescent granules* contain acids and hydrogencarbonates or carbonates so that, when the granules are mixed with water, carbon dioxide is liberated.

There is more than one granulation procedure:
– wet granulation: build-up granulation, granulation by division;
– dry granulation.

Build-up granules are formed by the joining of powder particles, for instance by fluidized-bed granulation or by dish granulation. These largely automated processes produce the granules in a single stage and with a narrow particle size distribution.

In *fluidized-bed granulation*, the particles are held in suspension by a current of air flowing upwards and sprayed with a liquid. Particles colliding with one another stick together and dry in the stream of air. When they reach the desired size, they are removed from the chamber. Grain size and characteristics of the granules are easily controlled by adjustment of the process parameters.

Dish granulation yields even, round granules with a smooth surface. Powder is fed continuously into a slanted rotating dish and sprayed with liquid. Rotation and drying with a jet of warm air furnish the even pellets, which on reaching a certain minimum size fall from the dish automatically. These granules are often packed in hard gelatin capsules for administration.

Granulation by division involves the comminution of a mass of wet substance. The powder is first aggregated while wet, for example by warming if one of the components has a low melting point. *Sinter granules* are formed, which must be processed while warm to prevent setting. If, on the other hand, a liquid in which some of the components partially dissolve is used to moisten the powder, for example water, sugar syrup, or alcohol, then *crust granules* form owing to crystallization of the dissolved substances on drying. *Agglutinated granules* are those that are moistened with gelatin solution, starch paste, or solutions of other polymers like polyvinylpyrrolidone or alginates. This type of granule is more stable than crust granules, since the polymers used lend a certain elasticity to the aggregates.

The amount of liquid must be judged such that the mass neither flows nor crumbles. The damp dough is dispersed by shaking or pressing through a sieve. The resulting *shaken granules* consist of near-spherical granules that flow well and have a high bulk density, unlike *press granules*, which are made up of long, rod-shaped grains. A perforated disk may be used rather than a sieve, but the resulting granules are still rod-shaped, with a relatively smooth surface.

The granules are dried at around 30–40°C, but a certain amount of moisture should remain if necessary for further processing. Any residual powder or loose aggregations of granule grains must be removed by sieving or comminuted, respectively. This process is called *equalization*.

Dry granulation is used for moisture- or heat-sensitive substances in particular, and for other powders because of its relative ease and rapidity. The powder is pressed into briquettes which are then broken or ground and the grains classified to obtain the required size. The grains are of course irregular. Those that are too large are further broken up, those that are too small are returned to the briquetting stage. The basics are covered in Chapter 6.

12.3.3 Microcapsules, Nanoparticles, Liposomes

Microcapsules are solid particles or liquid droplets coated with gelatin or another polymer and able to flow. They vary from micrometers to millimeters in size. Depending on their size, the coating makes up 2–30% of the total material. Microencapsulation turns liquids into "powders" that can be processed further in their particulate form with solids. The coating provides protection from oxygen, moisture, and other damaging influences, fixes volatile substances, and masks unpleasant tastes and smells. In mixtures, micro-encapsulation can prevent any interaction between active substances. In addition, careful selection of the coating material can furnish drug formulations with delayed action.

Microcapsules are used in instant decoctions, peroral suspensions, and hard gelatin capsules. *Coacervation* is the most common encapsulating process; the substance to be encapsulated is dispersed or emulsified with the aid of the coating material (e.g. gelatin) dissolved in water, which acts as a dispersant. A change in pH or temperature, or addition of alcohol or electrolyte, alters the solubility of the coating material to the extent that it separates out in the form of tiny droplets onto the dispersed or emulsified particles. Once the coating has hardened by cooling or chemical cross-linking, the microcapsules can be filtered or centrifuged off, washed, and dried.

The Wurster process, the centrifugal process, and electrostatic microencapsulation are other processes. In the *Wurster process* (Section 6.2.8.5), solid particles to be encapsulated are sprayed with the coating material in a fluidized bed. The *centrifugal process* can be used to microencapsulate both solids and liquids. The substance is applied to a rapidly rotating disk and is flung outwards by centrifugal forces, passing through a curtain of falling coating solution. In *electrostatic microencapsulation*, the particles are coated with aerosol particles of encapsulating material bearing an opposing charge.

More recent developments are concerned with the production of *nanoparticles* [6], which have diameters of <1 μm. These tiny particles can be administered intravenously as stable dispersions and are suitable for controlled-release formulations. Target organs can be, for example, the liver, the spleen, or the lungs [7, 8]. There are various methods for manufacturing nanoparticles; the most important is emulsion polymerization. Other methods rely on interfacial polymerization, or on the dehydration of naturally occurring proteins. Numerous different carriers are manufactured and used; these include poly(alkyl cyanoacrylate)s such as poly(butyl cyanoacrylate), poly(methyl methacrylate), albumin, and gelatin. The active substances can be encapsulated, dissolved, adsorbed, or chemically bound [9]. The penetration of the nanoparticles into the body depends on their size, their composition, and their surface charge [10]. Thus, it was found that polycyanoacrylate-based nanoparticles accumulate in certain tumors [11]. So far, nanoparticles have been used mostly for the encapsulation of cytotoxic substances [12, 13], but also for bioactive peptides and proteins [14].

Liposomes too are used in medicine. The hollow spheres made up of phospholipid bilayers may contain the medicament in the aqueous interior or the lipophilic membrane, depending on the nature of the drug (cf. Figure 3.5). Liposomes have found uses in parenteral administration [15, 16], inhalation treatment [17], percutaneous administration [18–20], ophthalmics [21], cancer treatment [22], and controlled-release formulations [23, 24].

12.3.4 Capsules

Capsules are single doses of a drug in solid form, in varying shapes and sizes, and with a hard or soft coating. As well as capsules to be taken orally, capsules for rectal and vaginal use exist. Capsules for oral use are easier to consume than powders and hide any unpleasant taste or smell; the active substance is released within a few minutes of consumption. The formulation of active substances for use in capsules is particularly gentle, so even heat- and moisture-sensitive substances can be processed.

Capsules for oral administration are classified as follows:

- hard capsules
- soft capsules
- capsules resistant to gastric acids
- capsules in which the substance release has been modified.

Capsules made with starch or with wafers, once very common, are now increasingly being replaced by gelatin capsules because of the moisture sensitivity, mechanical weakness, and unsatisfactory disintegration properties of the former. In comparison, gelatin capsules are mechanically stable, afford better protection against oxygen and moisture, and are easier to swallow, since saliva renders the coating slippery. Gelatin capsules can be colored transparent or opaque with water-soluble dyes or pigments, both so that the medicament is recognizable and to protect the active substance from light. Layers of enteric coatings protect the capsule contents from aggressive gastric acids so that the capsule does not dissolve until it reaches the small intestine; this technique also protects the stomach itself from active substances that are not well tolerated. As for enteric granules, the coatings used are made of cellulose acetate phthalate and polymers containing acid groups.

Hard gelatin capsules are manufactured from gelatin without addition of plasticizers. They are produced by dipping a matrix into a hot solution of gelatin and drying at about 30°C, and are easily removed from the template, as long as this has first been coated with a mold-parting agent. In the Colton machine, developed some 50 years ago, the dipping time is about 12 seconds. After dipping in the gelatin solution, the coated templates are rotated 2½ times around their axes and pointed upwards, subjected to a jet of cold air to set them, and dried. The complete drying cycle lasts about 45 min. After removal from the template, the moisture content of the capsules is 15–18%. In subsequent stages, this is adjusted according to the intended end use, and defective capsules with cracks, streaks, traces of grease, or bubbles are removed. Finally the finished capsules are marked in an offset rotary printing process at a rate of up to three-quarters of a million capsules per hour.

The moisture content of the capsules is critical. If it is less than 12%, the capsules are fragile; if it is greater than 15%, they are too soft. Capsules are best stored at relative humidity levels of 40–60%. Filled capsules may become fragile if the contents absorb water from the gelatin, so the filling material should be preconditioned at an appropriate humidity. Very high moisture levels, or traces of aldehyde, cause the gelatin to cross-link during storage. Such capsules dissolve only slowly and no longer meet their specification.

The capacity of the capsules is standardized; they are available in the sizes listed in Table 12.1.

Table 12.1. Type and capacity of standard hard gelatin capsules.

No.	000	00	0	1	2	3	4	5
capacity [cm³]	1.37	0.95	0.68	0.50	0.37	0.30	0.21	0.13

Snap-fit® capsules are especially stable; there are ridges on both halves of the capsule that snap together when one piece is fitted into the other (Figure 12.4).

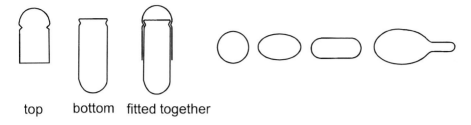

top   bottom   fitted together

Figure 12.4. *Snap-fit*® safety closure capsules (simplified) and soft gelatin capsules.

Soft gelatin capsules have a thicker shell than hard capsules. Plasticizers such as glycerin or sorbitol lend them elastic properties. They are used in preference to contain liquid or semisolid preparations, except for aqueous solutions or pastes, which would dissolve or swell the gelatin shell. Solids must first be granulated or dissolved or dispersed in nonaqueous liquids.

In industrial production, soft gelatin capsules are most commonly punched out. In the *Scherer process*, two ribbons of gelatin are fed over two opposing rollers with matching declivities in the shape of half-capsules. The ribbons are pressed against the rollers by a wedge containing a feed pipe, and filled. The capsule halves are then joined together, stamped out of the ribbon, washed, and dried at 20°C and 20% relative humidity (Figure 12.5). Up to 100 000 capsules per hour can be manufactured in a single step. This procedure is used to manufacture not only oral but also rectal and vaginal capsules.

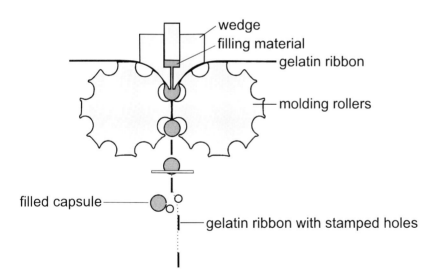

Figure 12.5. The Scherer process for capsule manufacture; from reference [25].

In the *dropping process*, another method for the manufacture of soft gelatin capsules, a double-walled dropper is used to extrude filling from an inner tube and gelatin solution from the surrounding cavity. The resulting coated drops fall into paraffin oil kept at 4°C, where they set to form seamless, spherical capsules without any inclusion of air. However, this procedure is now seldom used.

Hard and soft gelatin capsules must be designed to disintegrate within half an hour in water at 37°C. Enteric capsules must be able to resist 0.1 N hydrochloric acid for two hours while disintegrating within one hour in a phosphate buffer solution with a pH of 6.8.

12.3.5 Tablets, Dragées

Tablets are solid drug dosage forms of various shapes, usually made of powdered or granulated medicaments mixed with fillers, binders, disintegrating agents, and lubricants or other additives by compression. Agglomerates are more easily turned into tablets than is powder. Stabilizers, colors, or flavoring agents are acceptable additives. However, all additives must be physiologically harmless and must not negatively influence the activity of the tablet contents. Auxiliaries used in tablet-making include the substances listed below (Table 12.2):

Table 12.2. Auxiliaries used in tablet-making.

Type	Example
filler	lactose, saccharose
binder	starches, gelatin, sugar, cellulose ethers, polymers such as polyvinylpyrrolidone
disintegrating agents	starches, starch ethers
lubricants and mold-separating agents	talcum, stearates, silicones
flow regulators	talcum, aerosil

Simple circular tablets with biplanar or biconvex cross-sections and spherical tablets are the most common owing to their mechanical stability. This can be further increased in the case of biplanar tablets by the presence of beveled edges, and in the case of biconvex tablets with a truncated rim. Ellipsoidal, oval, heart and ring shapes can also be manufactured, as can shapes with corners (Figure 12.6).

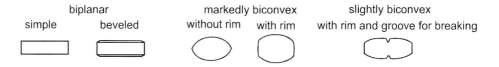

Figure 12.6. Cross-sections of various tablet shapes.

Tablets for peroral administration are swallowed whole or chewed, or allowed to disintegrate in liquid before consumption, and arrive in the intestine after contact with the strongly acidic gastric juices, as has already been described. The dose in the tablets must be accurate to within a tolerance of 5%, and in a disintegration test the tablet must fall apart within 15 min.

The press power during manufacture and the additives contained in a tablet have a considerable influence on its rate of disintegration. The disintegration rate can be decreased or increased by subjecting the tablets to a breaking or a sintering stage after compaction. Lubricants too can have varying effects on the disintegration rate: tablets containing hydrophilic sodium lauryl sulfate dissolve faster than those without it, whereas the hydrophobic lubricant magnesium stearate acts as an excipient, that is, retards disintegration. Elevation of the content of disintegrating agents such as starch accelerates the dissolution; these agents swell in water and thus speed the disintegration. It should also be remembered that there are other factors, namely, the crystal modification and habit, and the particle size, which can have a considerable influence on the rate of solution and therefore on the efficacy [26].

In addition to simple tablets, tablets with special characteristics are manufactured.

Multilayer tablets were designed to minimize the contact between incompatible active substances. *Mantle tablets* contain a core surrounded by a mantle with different pharmacological properties, perhaps a different active substance or a different initial dose, while the core, which disintegrates later, contains the follow-up dose.

Effervescent tablets are intended for fast action. They contain hydrogencarbonates and acids such as citric or tartaric acid, which, in combination, release carbon dioxide on dissolution. Packaging for effervescent tablets includes a desiccating plug in the lid to prevent moisture in the air liberating CO_2 from the tablets.

Retard tablets, with prolonged action, are manufactured from, for example, a number of different granules with staggered release times for the active substance they contain. Some types are designed to allow an interval of 12 hours between taking tablets, others 24 h.

Besides tablets for oral ingestion there are other forms, including *lozenges* and *chewable tablets* that have a local effect in the mouth and throat or are absorbed through the mucous membranes of the mouth. *Buccal* (held in the cheek cavities) and *sublingual* (held under the tongue) tablets have also been designed for the latter purpose. *Implantation tablets* are small, sterile tablets for implantation in tissue; *ocular tablets* are inserted into the conjunctiva. *Vaginal tablets* are usually ovoid in shape. *Soluble tablets* are used to prepare solutions for internal, external, or parenteral use.

Many tablets are coated; in the case of *dragées*, the coating has many layers, usually of sugar, on top of markedly biconvex tablets. Tablets with a very thin coating are called *film-coated* or *lacquered tablets*. The coatings used should be poorly or slowly water-soluble polymers such as ethylcellulose, shellac, or poly(vinyl alcohol). As in the case of coated granules, the coating protects the tablet from atmospheric humidity, oxygen, light, and gastric acids, and hides unpleasant smells or tastes. Coated tablets take longer to disintegrate than the uncoated variety. Dragées require 60 min to disintegrate, film-coated tablets 30 min. There are special requirements for the residence times for tablets with an enteric coating: they must withstand gastric juices for at least 2 h, while releasing the active substance within 60 min once they reach the intestine.

Dragées are coated in rotating chambers one layer at a time (Figure 12.7). To avoid difficulties in coating edges, tablets with a markedly biconvex shape should be used. Two procedures, cold and warm, are distinguished on the basis of the drying temperature. The process, which can last several hours or even days, involves the application of up to 50 layers and normally has five different stages: sealing, coating, smoothing, coloring, and polishing.

Usually, the sealing stage commences with application of a moisture barrier, a thin film of cellulose acetate phthalate (CAP), poly(vinyl acetate phthalate) (PVAP), or shellac stabilized with polyvinylpyrrolidone; subsequently 3–8 layers of concentrated sugar solution (50–65%) containing a small amount of gum arabic or gelatin are applied. To prevent the dragées adhering to one another, each completed layer is dusted with a powder made of a mixture of various solids such as talcum, calcium carbonate, kaolin, starch, and aerosil, along with gum arabic. In the actual coating process, the coating syrup and the powder, or a suspension of the powder, are applied and dried with infrared lamps or warm air repeatedly until the mass of the coating has reached 30–50% of that of the core; this may require 30 or more layers. The unfinished dragées often have a raw surface, so they are further coated with a special smoothing syrup containing no solids, and then colored with syrup to which 1–5% dye or pigment has been added. Finally they are polished in a polishing vessel to achieve the desired surface gloss. Carnauba wax and combinations of beeswax, paraffin, spermaceti, cocoa butter, etc., are used to impart gloss, either as solids or as solutions or emulsions.

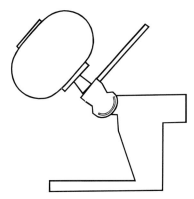

Figure 12.7. Sketch of an apparatus for coating dragées.

12.3.6 Suppositories

Suppositories are solid and usually torpedo-shaped drug dosage forms which melt or dissolve at body temperature and are introduced into the rectum. Suppositories for adults weigh 2–3 g, those for children 1 g. Synthetic solid fats are used as the base; they are obtained as a mixture of mono-, di-, and triglycerides from esterification of fatty acids with glycerin, and melt over the range 32–35°C. The mono- and diglycerides present act as emulsifiers and increase the water absorption capacity and the absorption of the active substance. Cocoa butter, once in common use for this purpose, easily becomes rancid owing to its unsaturated fatty acids and is therefore scarcely used any more. Occasionally water-soluble macrogols (poly(ethylene glycol)s) and elastic glycerin–gelatin masses are used as bases for suppositories. However, since these hydrophilic products absorb water from the rectal mucosa, they can stimulate the urge to defecate, which is counterproductive.

It is often necessary to add auxiliaries in order to improve the characteristics of the suppository. Consistency can be altered with beeswax or cetyl palmitate, for suppositories which are too soft, or with liquid paraffin or castor oil if they are too hard. Aerosil, bentonite, or glycerin monostearate increase viscosity and thus slow sedimentation and flocculation of particles of the active substance, especially when its concentration is low. Emulsifiers such as lecithin may also be necessary, for example to help the combination of aqueous solutions with the fatty base, and perhaps to increase bioavailability. Difficulties with dispersion and stabilization arise when powders containing the active substance are mixed into a hydrophilic base, and these have to be dealt with, as described in other chapters. Polysorbates, for instance, are used as emulsifiers or dispersants.

The same bases and methods of manufacture are employed for *vaginal suppositories*. They may be spherical or ovoid, and are used primarily for local treatments.

12.3.7 Ointments, Creams, Pastes, Gels

Ointments, semisolid drug formulations, are intended for application to healthy, diseased, or injured skin. Masking ointments are meant to protect healthy skin from damage, or to soften the skin. Ointments for injuries and corticosteroid ointments, which penetrate the upper layers of the skin, have a local curative effect. In the case of certain ointments, penetration is deeper; for example, in the case of nitroglycerin ointments the active substance has to enter the systemic circulation to achieve the desired effect on the heart. Details of interactions with the skin may be found in Chapter 11.

In a narrower sense, ointments are formulations that cannot be fitted into the categories of creams, pastes, or gels, such as single-phase, spreadable drug formulations. Hydrophobic ointments are based on paraffin, Vaseline, vegetable oils and fats, and waxes. They can only absorb small quantities of water. Hydrophilic ointment bases are made from macrogols, poly(ethylene glycol)s, which are miscible with or soluble in water.

Unlike single-phase ointments, creams are o/w or w/o emulsions. The line between pharmaceutical and cosmetic ointments and creams is fuzzy. Masking ointments, sunscreens, penetrating ointments, absorption ointments, and cooling creams are classified as one or the other according to their content.

Gels contain a structure former in the liquid phase to control the rheological properties. Individual aspects of the topics of rheology and gel formers are treated in the chapters on rheology, cosmetics, and food formulations. Active substances may be present in either dissolved or dispersed form.

Further details on pastes, too, will be found in the appropriate chapters. Pastes are highly concentrated suspensions containing up to 50% powder in a base which may be a liquid like glycerin, a cream, or a gel.

12.3.8 Liquid Formulations

In the field of pharmaceuticals, the word "solution" usually refers to an aqueous solution. Other solvents used are ethanol and oils. Besides those for external use like boracic acid solution and spirit of camphor, there are solutions for internal use, for example those available in ampules for drinking, like B-group vitamins.

Tinctures (literally, "colored liquids") and extracts are prepared by extraction of drugs (the vegetable or, rarely, animal starting materials for medicinal formulations). An alcohol/water mixture is the usual medium of extraction. However, extracts can be used in inspissated form as dry extracts, as well as in the form of liquid extracts.

Other liquid formulations are spirits, suspensions, and mucilages. Spirits are ethanolic preparations obtained by distillation of drugs with ethanol. Liquid suspensions used in place of solid formulations of medicaments are particularly important in pediatrics. As well as dispersion stabilizers, they often contain thickeners to delay sedimentation. Mucilages are, more or less, very viscous solutions made from mucilaginous drugs or synthetic swelling agents; they include appetite suppressants and mild purgatives.

As for the auxiliaries such as fats, waxes, emulsifiers, and gel formers used in liquid and semisolid drug formulations, further details can be found in Chapters 2, 11, and 13.

12.3.9 Special Forms, Control of Invasion

Of the various processes which affect the transport of active substances from the point of application to the point of action (*invasion*), like release, uptake, and distribution of the active substance, the release is the easiest to influence by technical means. The idea is to optimize the release of the active substance so as to control the start and the duration of its action.

The interplay between invasion and *evasion* (processes that result in an irreversible decline in the level of the active substance in the blood) governs the time course of the concentration in the blood, as shown for a single oral dose of a medicament in Figure 12.8.

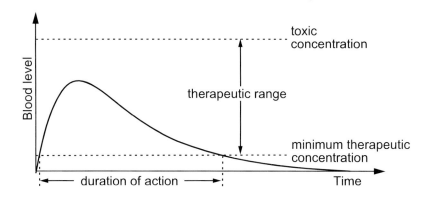

Figure 12.8. Concentration of active substance in the blood after a single oral dose.

The blood level of the active substance must be maintained within the therapeutic range: below the minimum therapeutic concentration, the desired effect is not achieved, but above the lower limit of the toxic concentration, unacceptable side effects may occur.

When repeat doses of a medicament are administered, the blood level plot depends not only on the dose but also on the interval between doses. If these two parameters are properly selected, the blood concentration can be kept within the therapeutic range (Figure 12.9).

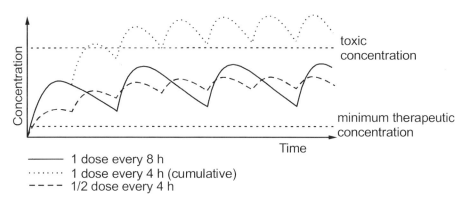

———— 1 dose every 8 h

········· 1 dose every 4 h (cumulative)

– – – – 1/2 dose every 4 h

Figure 12.9. Blood concentration curve for multiple dosage (reproduced from reference [27]).

Besides the dose and the dose interval, the invasion rate affects the course of the level of active substance in the blood; it can be altered by acceleration or retardation of the release of the active substance. Slower invasion results in a shallower blood-level curve and longer action. The effect is achieved with retard formulations.

There is a type of capsule which swells into a gel mass in the stomach and is lighter than the gastric juices, called the *hydrodynamically balanced system* (*HBS*). Regardless of the volume of the stomach contents, these capsules float on the surface, remaining in the stomach and releasing active substances for up to 6 h.

In the *facilitated absorption system* (*FAS*), a tablet with a special composition, the active substance is embedded in a "solution facilitator" such as an emulsifier, ensuring a high concentration at the site of absorption in the small intestine even in the case of poorly soluble substances. These tablets must be designed to resist the gastric juices.

An *OROS* (*oral osmotic system*) contains a core of medicament surrounded by a semipermeable membrane in which a small exit hole has been made by laser. Water diffuses through the membrane and dissolves the active substance, which is released at a constant rate through the opening because of the osmotic pressure and the difference in chemical potential.

Like the OROS, the *gastrointestinal therapeutic system* (*GITS*) involves a semipermeable surrounding membrane with a small opening made by laser. The interior of the GITS consists of two layers, the first of an auxiliary substance that swells in water and is thus osmotically active, and the second of the active substance suspended in a highly viscous gel, and therefore possessing a large surface area. The active substance is released in the same manner as in an OROS, but with higher osmotic pressure and a greater area of dispersed grains.

Transdermal therapeutic systems (*TTS*) resemble sticking plasters in appearance and, like these, are stuck to the skin, where they provide a continuous transdermal supply of active substance into the blood vessels and the circulation. They are particularly suited for use with substances that are severely degraded in the liver if taken orally (first pass effect). Since this degradation does not occur with a TTS, the dose can be kept a lot lower and the side effects are smaller accordingly. A TTS has a structure of several layers (Figure 12.10); viewed from the top, they consist of a cover impermeable to water, a reservoir of the active substance, an optional control membrane to regulate the release, an adhesive layer, and a protective strip removed before application.

A method currently under development, *iontophoresis* [28], serves to transport larger molecules like proteins through the skin. The proteins are ionized and transported actively through the skin or the pores by application of an electrical potential by means of an Ag/AgCl electrode system.

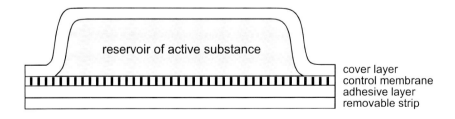

reservoir of active substance

cover layer
control membrane
adhesive layer
removable strip

Figure 12.10. Transdermal therapeutic system (TTS).

12.4 Preservatives and Antioxidants

It is often difficult to protect liquid formulations adequately against microbial attack. Emulsions, for example, are particularly susceptible to fungi and yeast. Quite apart from the pathological aspect, this has other negative consequences including a decline in the stability of the emulsion. Ingredients such as polypeptides, carbohydrates, or lecithin provide an ideal source of nutrients for various microorganisms. Their metabolic byproducts can have a great variety of effects. One obvious one is caused by CO_2, which can burst bottles when produced in sufficient quantities. Substances such as quaternary ammonium compounds, benzoic acid, or phenylmercuric nitrate are employed as antimicrobials. Table 12.3 lists various preservatives used in pharmaceuticals and cosmetics.

Table 12.3. Preservatives and their efficacy against bacteria, yeasts, and fungi [29].

Antimicrobial	Concn. [%]	Optimal pH range	Efficacy			
			Gram +	Gram –	Yeasts	Fungi
phenol	0.3	2–4–(8)	+	+	(+)	(+)
cresol	0.3	2–4–(8)	+	+	+	+
p-chloro-m-cresol	0.02	2–4–(8)	+	+	+	+
phenylethyl alcohol	1.0	2–4–(7)	+	+	(+)	(+)
chlorobutanol	0.5	2–4	+	+	(+)	(+)
benzyl alcohol	1.0	2–4–(7)	+	(+)	(+)	(+)
PHB methyl ester	0.18	2–7–(9)	+	+	0	0
PHB propyl ester	0.02	2–7–(9)	+	+	(+)	0
PHB methyl + propyl ester	0.2	2–7–(9)	+	+	(+)	(+)
PHB methyl + propyl ester + benzyl alcohol	0.2 + 0.5	2–7	+	+	+	+
sorbic acid	0.2	2–3–(5)	+	+	+	+
benzoic acid	0.1	2–3–(5)	+	+	+	+
phenyl mercuric nitrate	0.001	7–10	+	+	+	+
Merthiolat® thiomersal	0.02	2–7–(9)	+	+	+	+
Thiocide	0.01	2–7–(9)	+	+	+	+
Cialite	0.01	2–7–(9)	+	+	+	+
benzalkonium chloride	0.01	(3)–5–8–(10)	+	+	+	+

Legend: x–y effective;   (x)–y and x–(y) some effect;   + effective;   (+) some effect; 0 ineffective.   PHB: p-hydroxybenzoic acid.

It is often also necessary to add antioxidants in order to capture free radicals and to interrupt the cycle of oxidation by hydroperoxides. Radicals readily react with oxygen to form peroxy radicals, which in turn form hydroperoxides by abstracting hydrogen atoms from organic substances. Thermal decomposition or the effect of UV radiation breaks these hydroperoxides down into two radicals, which bind to oxygen in a chain reaction and thus can further attack the unprotected material. Antioxidants interrupt this chain reaction, for instance by forming unreactive, stable radicals. One example of such a cycle is depicted in Figure 12.11. The chain can also be broken by the reaction of two radicals with each other, but this occurs only to a small extent.

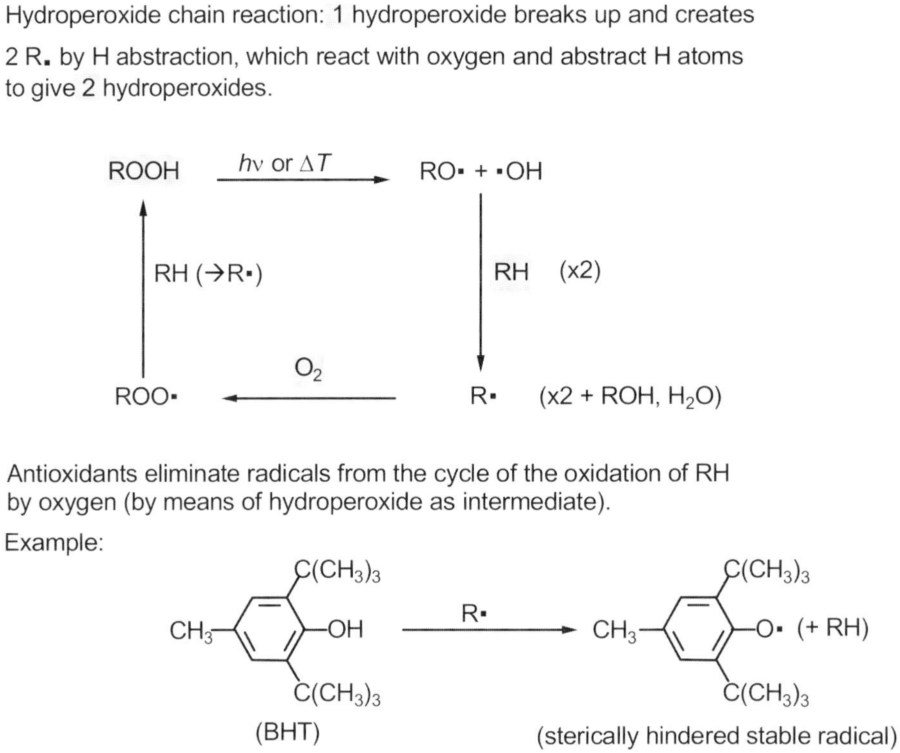

Hydroperoxide chain reaction: 1 hydroperoxide breaks up and creates

2 R. by H abstraction, which react with oxygen and abstract H atoms to give 2 hydroperoxides.

$$ROOH \xrightarrow{\textit{hv} \text{ or } \Delta T} RO\cdot + \cdot OH$$

$$RH (\rightarrow R\cdot) \qquad \qquad RH \quad (x2)$$

$$ROO\cdot \xleftarrow{\qquad O_2 \qquad} R\cdot \quad (x2 + ROH, H_2O)$$

Antioxidants eliminate radicals from the cycle of the oxidation of RH by oxygen (by means of hydroperoxide as intermediate).

Example:

$$\underset{\text{(BHT)}}{\text{BHT structure}} \xrightarrow{R\cdot} \underset{\text{(sterically hindered stable radical)}}{\text{radical structure}}$$

(structures: 2,6-di-tert-butyl-4-methylphenol with $C(CH_3)_3$ groups, CH_3, $-OH$ → $-O\cdot$ $(+ RH)$)

Figure 12.11. Oxidation of organic substances, represented by the hydroperoxide cycle.

In multiple-phase systems like emulsions, it makes a difference whether a water-soluble or an oil-soluble antioxidant is utilized. The antioxidant must be present in the same phase as the substance that requires the protection; in special cases, where both phases contain substances to be protected, a combination of water- and oil-soluble antioxidants is necessary.

Antioxidants for aqueous solutions include dithionite and the salts of the sulfurous acid. Since the corresponding acids have an unpleasant smell, these compounds are not really acceptable for oral uses. Other water-soluble antioxidants are ascorbic acid and cysteine hydrochloride. Examples of oil-soluble antioxidants are propyl gallate, tocopherols, ascorbyl palmitate, and di-*tert*-butyl hydroxytoluene (BHT). Addition of antioxidants to preparations containing fats is especially necessary in order to prevent the onset of rancidity.

Antioxidants are required in relatively small amounts: for example, ascorbyl palmitate 0.01–0.2%; tocopherols 0.001–0.5%; propyl gallate 0.001–0.02%; BHT 0.001–0.02%; sodium sulfite 0.05–0.3%; cysteine hydrochloride 0.01–0.1%; ascorbic acid 0.01–0.1%.

References for Chapter 12:

[1]   G. Czihak, H. Langer, H. Ziegler, Biologie, 3rd ed., Springer, Berlin, Heidelberg, New York, 1981.
[2]   J. W. Hole, Human Anatomy and Physiology, 2nd ed., Wm. C. Brown Co., Dubuque, IA, 1981.
[3]   P. G. Welling, H. Huang, P. F. Hewitt, L. L. Lyons, J. Pharm. Sci. *67*, 764 (1978).
[4]   M. Mayersohn, in Modern Pharmaceutics, 3rd ed. (S. G. Banker, C. T. Rhodes, Eds.), Marcel Dekker, New York, Basel, Hong Kong, 1996.
[5]   H. Sucker, F. Fuchs, P. Speiser, Pharmazeutische Technologie, Georg Thieme Verlag, Stuttgart, 1978.
[6]   J. H. Fendler, Nanoparticles and Nanostructured Films, Wiley-VCH, Weinheim, 1997.
[7]   R. C. Oppenhem, in Drug Delivery Systems (R. L. Juliano, Ed.), Oxford University Press, New York, 1982.
[8]   J. Kreuter, Pharm. Acta Helv. *58*, 217 (1983).
[9]   J. Kreuter, W. Liehl, J. Pharm. Sci. *70*, 367 (1981).
[10] S. D. Troster, U. Mueller, J. Kreuter, Int. J. Pharm. *61*, 85 (1991).
[11] E. M. Gipps, R. Arhady, J. Kreuter, P. Groscurth, P. P. Speiser, J. Pharm. Sci. *75*, 256 (1986).
[12] J. Kreuter, H. R. Harmann, Oncology, *40*, 363 (1983).
[13] J. Kreuter, in Drug Targeting (P. Buri, A. Gumma, Eds.), Elsevier, Amsterdam, 1985.
[14] J. C. Gautier, J. L. Grangier, A. Barbier, P. Dupont, D. Dussossoy, G. Pastor, P. Couvreur, J. Controlled Release *20*, 67 (1992).
[15] H. Sasaki, T. Kakutani, M. Hashida et al., J. Pharm. Pharmacol. *37*, 461 (1985).
[16] J. P. Sculier, A. Coune, C. Brassine et al., J. Clin. Oncol. *4*, 789 (1986).
[17] M. Ausborn, B. V. Wichert, M. T. Carvajal et al., Proc. Int. Symp. Control Rel. Bioact. Mater. *18*, 371 (1991).
[18] M. Jacob, G. P. Martin, C. Mariott, J. Pharm. Pharmacol. *40*, 829 (1988).
[19] V. Masini, F. Bonte, A. Meybeck, J. Wepierre, J. Pharm. Sci. *82*, 17 (1993).
[20] M. Mezei, V. Gulasekharam, J. Pharm. Pharmacol. *34*, 473 (1982).
[21] E. Hirnle, P. Hirnle, J. K. Wright, J. Microencapsulation *8*, 391 (1991).
[22] Y. Watanabe, T. Osawa, Chem. Pharm. Bull. *35*, 740 (1987).
[23] S. E. Tabibi, R. Mathur, D. F. H. Wallach, 83rd Annual Meeting of AACR, San Diego, CA, 1992.

[24]   V. M. Knepp, R. S. Hinz, F. C. Szoka, R. H. Guy, J. Controlled Release *5*, 211 (1988).

[25]   Ullmanns Encyklopädie der technischen Chemie, 4th ed., Vol. *18*, Verlag Chemie, Weinheim, 1979, p. 151 (Pharmazeutische Technologie).

[26]   E. M. Rudnic, M. K. Kottke, in Modern Parmaceutics, 3rd ed. (S. G. Banker, C. T. Rhodes, Eds.), Marcel Dekker, Inc., New York, Basel, Hong Kong, 1996.

[27]   J. Friedland, Arzneiformenlehre für pharmazeutisch-technische Assistenten, 3rd ed., Georg Thieme Verlag, Stuttgart, New York, 1992.

[28]   D. Parasrampuria, J. Parasrampuria, J. Clin. Pharm. Ther. *16*, 7 (1991).

[29]   K. H. Wallhäußer, Pharm. Ind. *36*, 716 (1974).

13 Food Formulations

13.1 Some Important Principles for the Formulation of Foodstuffs

1. To create a structure in the material, starches in the form of pastes, sugar, or sometimes fats can be used as binders. An excellent structure former is the wheat gluten found in bread dough; bread, pasta, and baked goods owe their mechanical properties to gluten, properties which are important both in that they render the goods transportable and for the characteristics of the goods when bitten, chewed, and swallowed. Mechanical properties like the modulus of elasticity and the bending strength of wheat dough or gluten depend on the moisture content. The bending strength is $60–110 \, \text{kg/cm}^2$, depending on the material, at a moisture content of 20%, but increases to $310–420 \, \text{kg/cm}^2$ when the proportion of water present is only 10%.

2. For aroma and flavor, sugar, maltodextrins, and fats must be present in the formulation.

3. Additives are used for a number of reasons: *colorings* to improve appearance; *antioxidants* to prevent rancidity in oils and fats and discoloration of fruits and vegetables; *preservatives* to block attack by bacteria and molds; *emulsifiers* to combine oil and water; *gelling agents* and *thickeners* as stabilizers (viscosity, creaminess); *anti-caking agents* to prevent the formation of lumps in powders; *acids*, *bases*, and *salts* in baking powder and processed cheese; *flavor enhancers* (e.g. glutamate); preparations containing *enzymes* (e.g. yeast); *surface improvers* such as waxes, oils, or paraffins; *sweeteners* (saccharin, etc.).

4. The wettability and solubility of powders can be adjusted as required by processing, for example by agglomeration or by freeze drying.

5. Some products may not be used under some circumstances: alcohols or pork products for religious reasons, cows' milk (lactose) because it causes digestive problems for many people.

A formulated product must be stabilized. The possibilities are: sterilization, pasteurization, dehydration, and freezing.

Various procedures give food products their shape: molding (stock cubes), extrusion of pastes (pasta, flour "rice"), spraying (milk, coffee). Fermentation, baking, and boiling are also involved in some forming processes; freezing, too, counts as part of the process in freeze drying.

There are many different types of dryers used in processing foods, each with additional functions. *Roll dryers* sterilize, form flocs, and caramelize. They dry to a very low final moisture content in a single continuous process. *Tunnel dryers* dry large quantities of loose material continuously in air or vacuum; *fluidized-bed dryers*, combined with *shaking troughs*, are used for the continuous drying of large amounts of powder or granules; *spray dryers* are utilized in the processing of very large quantities of pumpable liquids; *freeze dryers* are employed for products expensive enough to justify the high costs of constructing and running the plant.

Finally, packaging machines must not be forgotten; these have to be set up as appropriate for different requirements.

Provided with this armory of machines, the food industry can formulate and process practically anything desired into commercial products. Nonetheless, each product requires adjustments of the machinery to suit its particular characteristics.

It is worth describing here a new method of agglomeration for very fine, dry powders, agglomeration in a steam jet (Figure 13.1), which clumps the powder particles together into larger balls resembling a bunch of grapes. The particles moistened by the steam are made to collide with each other by turbulence. Solution bridges containing maltodextrin, sugar, or the like as binders, can only form if the particles remain in contact for a certain length of time. After drying, these bridges must be sufficiently strong to withstand sifting, packaging, and transportation. The process consists of permitting steam to expand from 2 atm to 1 atm from a special nozzle and adding a current of air and one of powder to this jet of steam (300 m/s). The turbulence occurring in the mixed flow on exit from the nozzle effects clumping of the particles. If the jet diverges by 8°, the turbulence dies out within about 3 m from the nozzle. The laminar flow in the lower 17 m of a 20 m tower serves simply to dry the damp agglomerates.

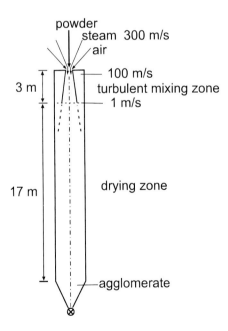

Figure 13.1. Sketch of steam-jet agglomeration.

13.2 Food Colloids

Food formulations are complex colloidal systems which may contain, amongst other substances, water and oil, fat crystals, protein aggregates, salts, soluble carbohydrates, and starch granules or gels. Their organoleptic properties are adjusted to agree with current fashions by means of structure-forming additives, flavorings, and colorings.

Emulsifiers present a special challenge as regards their safety and the stability of emulsions. Compounds used include mono- and diglycerides, sorbitan fatty acid esters, polyoxyethylene sorbitan esters, and phospholipids. The trend is towards the use of the most natural additives possible, such as proteins and phospholipids. These emulsifiers behave differently from the usual types of emulsifiers used in other areas of industry. Thus the tendency of food emulsifiers to form lamellar phases and vesicles is much more marked than it is for alkyl sulfonates, for example. Phospholipids form monolayers and multilayers around neutral droplets of lipids, or interact with adsorbed or dissolved proteins.

The raw materials of a food to be formulated often contain substances that interfere with the action of emulsifiers and can lead to instability such as flocculation or gelling. On the other hand, many of the proteins already present can act as good emulsifiers. Stability to temperature changes requires particular attention. Many foods are subjected to heat treatment like pasteurization (15 s at 72°C), sterilization (10 min at 120°C), or a UHT process (4 s at 140°C), which can denature proteins and alter texture and flavor characteristics.

Polysaccharides, a large group of varied macromolecular carbohydrates, are important auxiliaries. The food industry could not manage without them as auxiliaries used to change the structural characteristics of processed foods. As in the case of proteins, the thermal behavior of polysaccharides must be taken into account. In this case, degradation to give shorter macromolecules can impair their effectiveness as steric stabilizers, their rheology, and the formation of cross-links.

13.3 Proteins

Proteins vary enormously in size, from 50 to 2800 amino acid units. They possess characteristic structures, though these can be altered by denaturing. Often protein-containing raw materials only become suitable for use as surface-active additives once they have been denatured.

In emulsions, the substituents of the amino acids interact with the oil while the hydrophilic peptide groups prefer to be located in water. For structural reasons the denatured, unfolded proteins do not pack together tightly as adsorbates, and they reduce the interfacial tension less than do low molecular mass emulsifiers (Table 13.1).

Plant and animal proteins, utilized as foods or as emulsifiers, have very complex three-dimensional structures in solution [2]. In the emulsification process, these structures (secondary, tertiary, and quaternary structures) often have to be destroyed by addition of energy, a change in pH, or some other alteration in order to force the protein into a conformation suitable for adsorption and stabilization.

Table 13.1. Reduction of the interfacial tension at the o/w interface (from reference [1]).

Molecule	$\gamma_0 - \gamma\,[\text{mN/m}]$	Oil phase
β-casein	25	tetradecane
β-lactoglobulin	21	tetradecane
gelatin	15	tetradecane
phosvitin	12	tetradecane
lysozyme	17	toluene
glycerin monostearate	28	sunflower oil
$C_{12}E_2$ tenside	35	tetradecane

As is well known, proteins contain variable numbers of amino acids (though the number is constant for any particular protein) joined into a chain by peptide bonds. The sequence of amino acids in a polypeptide chain is known as the *primary structure* of the protein.

In solution, these polypeptide chains do not adopt random conformations but rather a predetermined structure, the *secondary structure*, which depends on the free rotation about C–C and C–N bonds. However, this rotation is somewhat restricted owing to the partial double-bond character of the peptide bond. The secondary structure adopted by a polypeptide chain depends on the bonds present in addition to peptide bonds, especially on hydrogen bonds between neighboring sections of polypeptide chain (Figure 13.2a).

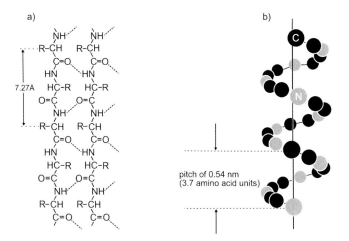

Figure 13.2. a) Hydrogen bonding *between neighboring* peptide chains results in β-folded sheet structures; b) formation of an α-helix by hydrogen bonding *within* the peptide chain.

The hydrogen bonds between spatially adjacent carbonyl and amide groups are not particularly strong (8.4–42 kJ/mol), but they can have a considerable influence on the chain conformation because of their number. Two or several polypeptide chains can form a local plane with a typical *folded-sheet structure* known as a β-sheet.

After peptide bonds and hydrogen bonds, the disulfide bond between cysteine residues is the most important type of bond. It is formed by dehydrogenation. Heteropolar ionic bonds between acid and basic groups in the physiological pH range also play a significant role, with bond energies of 42–84 kJ/mol.

As well as the folded sheet structure, the helix is a frequent type of chain conformation. It is formed by hydrogen bonds *within* a single peptide chain and is referred to as an α-helix. There are 3.7 amino acids for each turn along the chain, which has a pitch of 5.44 Å (Figure 13.2b).

In most proteins the α-helices are dextrorotatory. An example of an exception is collagen, the basic material of gelatin, in which they are levorotatory.

In proteins about 31% of the amino acids are present in the shape of dextrorotatory α-helices and 28% as β-sheet structures. Contrary to popular belief, the lengths of these two structures in proteins are relatively small. They are limited by the diameter of the protein globule, so that α-helices contain 10–15 amino acid units, β-sheets 3–10 units. The arrangement of helices and folded sheets, which are joined by lengths of disordered chain, is represented schematically in Figure 13.3.

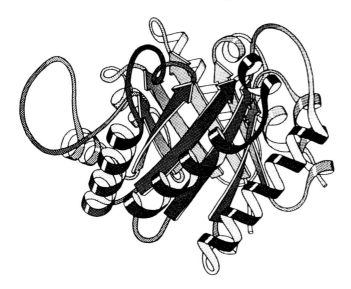

Figure 13.3. Schematic depiction of a polypeptide chain with α-helices and twisted β-sheets joined by lengths of unorganized amino-acid sequences (from reference [2]).

The three-dimensional shape of the protein molecule is known as its *tertiary structure*; such a structure is visible in the contours of the polypeptide chain in Figure 13.3. Roughly spherical tertiary structures occur in lipoproteins with high lipid content. *Globular proteins* deviate from this spherical shape; in solution, they adopt the form of a rotated ellipsoid with an axis ratio of 2:1. Fibrillar proteins depart still further from the spherical; their very elongated ellipsoids have an axis ratio of 30:1.

Several polypeptide chains may assemble to form a functional protein unit without any covalent bonds holding them together. This combined unit, which has a defined structure and consists of a defined number of polypeptide chains, the subunits, is called the *quaternary structure*. A quaternary structure is usually present if a protein has a relative molecular mass greater than 100 000.

Figure 13.4. Collagen: dextrorotatory superhelix (quaternary structure) consisiting of three levorotatory α-helices like those shown in Figure 13.2b.

The complexity of structural relationships is illustrated well by the example of collagen. Collagen, the starting material for the production of gelatin, comprises 30% glycine and 15–30% proline and hydroxyproline. The polypeptide chains have a levorotatory helical structure and a mean molecular mass of 95 000. Three such chains are twisted together into dextrorotatory superhelices 280 nm in length (Figure 13.4), which in turn are collected together in larger fibrils. When the water-soluble molecules are heated above 40°C, the superhelical structure is destroyed. The polypeptide chains now exist as free α-helices, although these are in equilibrium with the random coil structure (Figure 13.5).

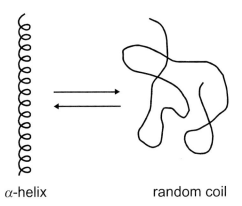

α-helix   random coil

Figure 13.5. Soluble collagen, gelatin: in solution the α-helix is in equilibrium with the random coil.

It is these random coils that possess interfacial activity. In the raw material collagen, obtained from cartilage, bones, connective tissue, etc., the polypeptide chains are usually cross-linked; the cross-links must be broken by the action of acids or bases, and of the resulting fractions, those with $M > 30\,000$ are defined as gelatin. On cooling, the polypeptide coils gel owing to the formation of hydrogen bonds and local refolding.

The refolding of random coils back into the tertiary structure is a slow process. In the case of small proteins, it takes seconds; for large molecules, minutes. The process is important in gelation, for example of gelatin. Here, the individual coils join up by forming collagen-like links or, in some cases, depending on the protein, β-sheet links, thus limiting their freedom of motion (Figure 13.6).

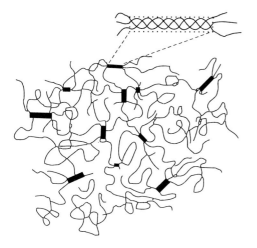

Figure 13.6. Gel formation in gelatin.

Folded, structured proteins are compact not only in their crystalline form but also in solution, and contain water only as individual molecules acting as integral parts of the structure. In contrast, denatured proteins are loose, unfolded, random coils. Accordingly, the water contents of emulsifier layers made up of globular proteins differ greatly from those of layers of adsorbed random coils. Adsorbed coils form layers of about 5 nm in thickness, which corresponds to $1–3$ mg/m^2. It has been estimated that this is equivalent to a water concentration of 50% in the adsorbed layer. Layers of adsorbate made up of folded, globular proteins, on the other hand, contain almost no water.

Myofibrillar proteins act as the principal stabilizers in meat products, although milk proteins and non-animal proteins are usually added too. The disentanglement of the fibrils, the breakdown of quaternary and tertiary structures, the emulsification of fats, and the formation of gel and membrane structures all occur simultaneously, for example on roasting. The processes by which proteins are denatured (together with the partial coalescence of oil droplets) on roasting should not be overlooked.

The individual classes of proteins are summarized in the table below, Table 13.2. Most of these proteins, however, are relevant as raw materials for processing of meats, for example in sausages, or are used in industry as sources of protein.

Table 13.2. Classification of proteins (globular, fibrillar, and composites with non-protein segments).

Molecular shape		Composites
Globular proteins	Fibrillar proteins	
albumins	collagen	glycoproteins
globulins	elastin	nucleoproteins
histones	keratin	chromoproteins
protamines	fibrinogen	phosphoproteins
prolamins	myosin	lipoproteins

The quaternary structure of proteins is an important factor in their emulsifiying properties. Casein in milk, for instance, exists in the form of aggregates of over 500 protein molecules. This aggregation number is changed by homogenization and the action of the protein as an emulsion stabilizer is improved. Table 13.3 shows some proteins that are employed as emulsifiers.

Flexible proteins like casein are adsorbed like heteroplymers because of their flexibility, unlike phosphoproteins and lipoproteins, which are adsorbed as granular particles at the o/w interface, for example in mayonnaise.

Table 13.3. Proteins as emulsifiers.

Source	Protein
milk	α_{s1}-, α_{s2}-, β-, κ-casein
wheat	α-lactalbumin, β-lactoglobulin

An important point to be remembered in the case of macromolecular emulsifiers concerns the time course of adsorption. Technical processes often have to be altered to fit this aspect. Not only diffusion but also the refolding of adsorbed random coils plays a role. During emulsification the large protein molecules diffuse only slowly to the newly formed droplet surfaces, over a period on the order of milliseconds. In high-pressure homogenizers, in which the droplets form very quickly, the many droplets which are not completely surrounded by polymer will floc and become linked by bridges, so that the emulsion is unsatisfactory. Emulsification with proteins is therefore better performed in slower apparatus such as colloid mills or mixing turbines.

The formation of bridges is also involved in the manufacture of butter. Proteins adsorb at the interfaces of the air bubbles incorporated on churning the cream in preference to the interfaces of the dispersed fat droplets, so that there is too little stabilizer in total for the fat/water + air/water interfaces; coalescence results.

Another aspect is the differing adhesion strengths of adsorbates. Hydrophilic proteins such as gelatins can be displaced from the droplet surface by suitable low molecular mass emulsifiers or more hydrophobic proteins (see Table 13.4 below). After emulsification the droplets have flocculated, and on addition of a small amount of a second tenside they coalesce into larger drops.

Table 13.4. Displacement of gelatin from droplets of soya oil.
Emulsion: 40 vol% soya oil in 0.4% aqueous gelatin solution. Displacement by 0.04%
second tenside [5].

Second tenside	gelatin in solution [%]	gelatin on surface of drops [mg/m²]	Drop size [µm]
none	0.10	1.9	2.5
Tween 60	0.36	0.4	3.6
glycerin monoester	0.30	0.4	3.6
sodium caseinate	0.28	1.0	3.4
lecithin	0.16	2.0	3.3

13.4 Lipids

The emulsifiers most commonly used in the food industry are not proteins but monoglycerides; their market share is 70%. These amphiphilic lipid emulsifiers are often crystalline or partly crystalline at room temperature. In water, above the Krafft point (the temperature at which monomer solubility is sufficient for the formation of micelles) they can form lamellar mesophases (cf. Section 1.8), which turn into gels on cooling (Figure 13.7).

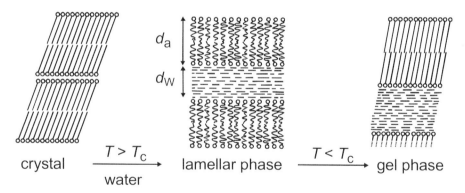

Figure 13.7. Phase transitions at the Krafft point T_c (from reference [3]):
palmitic acid monoglyceride/water   7:3 $d_a = 35.3$ Å; $d_W = 15.2$ Å
elaidic acid monoglyceride/water   7:3 $d_a = 37.9$ Å; $d_W = 14.8$ Å.

   In addition to the micellar phase, cubic, hexagonal (hexagonal I, oil inside the cylinder), and inverse hexagonal (hexagonal II, water inside the cylinder, HC chains pointing outwards) phases exist. These can sometimes create difficulties in processing, if they come into being as a result of local variations in concentration and cause blockages because of their high viscosity. The lamellar phases of monoglycerides are of great technical importance; when diluted with water, they can form vesicles, for example in conjunction with amylose. These vesicle systems with relatively low viscosity are needed, for example, in the processing of potato products.

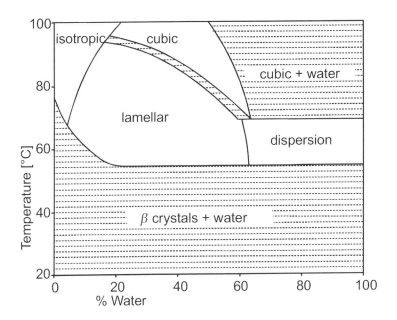

Figure 13.8. Phase diagram for palmitic acid 1-monoglyceride (according to reference [3]).

In technical processing of monoglycerides, temperature is an important factor in ensuring that the desired emulsifying properties are exploited to the full. This can be seen in Figure 13.8, the phase diagram for palmitin monoglyceride. The lamellar phase is the dominant region, but as the temperature rises it turns into the cubic phase. In common usage, the liposome region is referred to as a dispersion (for C_{14}–C_{18} monoglycerides). In the lamellar mesophase, the interlamellar water layers can swell to a width of 21 Å, corresponding to a water content of 40%. More details concerning phase diagrams for food tensides can be found in reference [4].

As has already been stated, monoglyceride lamellar phases convert into gel phases when cooled below the Krafft temperature. These gels are extremely sensitive to the presence of electrolytes. Decreased electrostatic stabilization can make the interlamellar aqueous layers shrink considerably. These gels, combined with other tensides such as propylene glycol monostearate or polysorbate 60 are used, for example, as raising agents in cake production.

Even at room temperature, the phospholipid lecithin forms a lamellar phase. For us, the dispersion region in which liposomes (more strictly, vesicles based on biological materials as colloids) are formed is the region of interest (Figure 13.9).

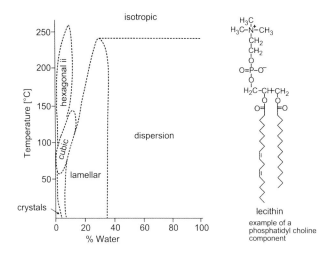

Figure 13.9. Phase diagram for soya lecithin/water (from reference [3]).

Since foods must not contain huge amounts of emulsifiers for obvious reasons, not the least of which are those of flavor and dietetics, only products which are consumed in heavily diluted form or in very small quantities can contain highly concentrated phases such as lamellar phases. Food products consist, rather, of macroemulsions or possibly liposomes. Table 13.5 lists typical concentrations, particle sizes, and surface coverages.

Table 13.5. Typical concentrations of emulsifiers in food colloids, drop size, and number of tenside layers (following reference [5]).

Product	Emulsifier type	Proportion by volume of the disperse phase	Drop size [μm]	Emulsifier concentration [g/L]	Number of mono-layers
margarine	monoglycerides	0.2	1.0	1	2
ice cream	monoglycerides	0.1	0.5	3	3
salad dressing	Tweens	0.4	0.4	10	4
mayonnaise	phospholipids	0.8	5.0	10	10

13.5 Polysaccharides

The class of products known as gums contains a wide range of long-chain poly-saccharides. Straight-chain, branched, and cross-linked members of the group are all water-soluble or at least swell in water. Chemically modified polysaccharides are included in the group as well, as can be seen in Table 13.6.

Table 13.6. Examples of polysaccharides in commercial use.

Source	Examples
seeds	maize starch, guar gum, locust bean (carob) gum
roots, tubers	potato starch, tapioca starch
seaweed extracts	alginate, carrageenan, agar
vegetable extracts	pectin
resins, exudates	gum arabic
microbial fermentation	xanthan, dextran
modified polysaccharides	methylcellulose, carboxymethylcellulose, hydroxyalkylmethylcellulose, starch acetate, starch phosphate, hydroxyethylstarch, hydroxypropylstarch, oxidized starch, hydrolysed starch

These polysaccharides have a wide range of applications, including the inhibition of the crystallization of ice and sugar, of sedimentation and creaming, of flocculation and coalescence, or of changes in the shape of gels. Further examples are gathered in Table 13.7.

Table 13.7. Functions of polysaccharides used in foods (from reference [6]).

Function	Application
swelling agent	diet products
thickener	jams, cake dough, sauces
gelling agent	instant puddings, aspic, mousses
binder	sausages
extender	sausages
adhesive	glazes, icing
coating	confectionery
emulsifier	salad dressings
protective colloid	flavor emulsions
encapsulating agent	flavor enhancers
film former	sausage skins
foaming agent	whipping cream
foam stabilizer	toppings, beer
suspension stabilizer	chocolate milk
crystallization inhibitor	ice cream, syrups
fining agent	beer, wine
clouding agent	fruit juices
flocculating agent	wine
syneresis inhibitor	cheeses, frozen foods
deforming agent	gum drops, candies

Smell, taste, color, and texture are important organoleptic properties that must be optimized so that consumers will accept the product. Amongst the characteristics that can be controlled by use of carefully selected polysaccharides are the viscosity of liquids and the tackiness and elasticity of gels.

As is also the case for other macromolecular substances, the specific viscosity η_{sp} (= η/η_s-1) of soluble polysaccharides increases sharply from a certain concentration c^*; for many polymers, the corresponding viscosity is 10 mPa·s (η_s is the viscosity of the solvent). The increase in viscosity is related to the interpenetration of the polymer coils at higher concentrations. If the intrinsic viscosity $[\eta]$, which is of course a measure of the volume of a macromolecule in solution, is known, then we can use it in the approximation $c^* \cdot [\eta] = 4$. For some polysaccharides, such as guar gum or locust bean gum, the value is lower, and is connected with a greater increase in viscosity above c^*. If gelling does not occur, the zero-shear viscosity η_0 varies above c^* in proportion to $c \cdot [\eta]^{3.3}$. This means the viscosity doubles when the concentration is raised from 1% to 1.23%. For gelling polymers like galactomannan the viscosity increase is considerably greater.

In these gels, ordered regions are linked together; they can be double helices, for example in amylose, agarose, and carrageenan, ribbons, as in locust bean gum, cellulose, and galactomannan, and "cation eggboxes" in alginate and pectin. The type and number of links can be controlled by addition of salts, as in the gelling of alginate specifically by calcium ions or of carrageenan simply by an increase in ionic strength. Dehydration caused by addition of sugar can also be used to control gelling in some cases. Gel strength and elasticity are often optimized by mixing different polysaccharides. A well-known example is the addition of locust bean gum to carrageenan, pectin, or agar.

Polysaccharides may have a linear structure, like amylose, which has a polymerization degree of 1000–16 000 and makes up some 25% of starch, or a branched structure like amylopectin, the other component of starch, which has segment lengths of 20–25 glucose units and a polymerization degree of 10^5–10^6. Alginates, with varying contents of mannuronic and guluronic acid units arranged in blocks or alternately, comprise linear polymer chains. Further details on the composition, structure, and uses of polysaccharides will be found in references [7] and [8].

References for Chapter 13:

[1] D. G. Dalgleish, Food Emulsions, in Surfactant Sci., Ser. Vol. 61, Emulsions and Emulsion Stability (J. Sjöblom, Ed.), Marcel Dekker, New York, 1996.

[2] T. E. Creighton, Proteins, Encyclopedia of Polymer Science and Engineering, Vol. 13, John Wiley & Sons, New York, 1988.

[3] N. J. Krog, T. H. Riisom, K. Larsson, in Encyclopedia of Emulsion Technology, Vol. 2 (Paul Becher, Ed.), Marcel Dekker, New York, 1985.

[4] N. Krog, J. Birk Lauridsen, in Food Emulsions (E. Friberg, Ed.), Marcel Dekker, New York, 1976.

[5] D. F. Darling, R. J. Birkett, in Food Emulsions and Foams (E. Dickinson, Ed.), Royal Soc. Chem., Cambridge, 1987.

[6]   M. Glicksman, Gum Technology in the Food Industry, Academic Press, New York, 1969.

[7]   J. M. BeMiller, Industrial Gums, in Encyclopedia of Polymer Science and Engineering, Vol. 13, John Wiley & Sons, New York, 1988.

[8]   G. G. S. Dutton, Polysaccharides, in Encyclopedia of Polymer Science and Engineering, Vol. 13, John Wiley & Sons, New York, 1988.

14 Agricultural Formulations

More and more, environmental considerations are determining the development and application of crop protection agents. Safety aspects of the handling and use of sprays and the like must be rethought. There must be no toxicological doubts about the inert substances included in formulations. Despite the limits placed on it, not least for economic reasons, research in the field of agricultural formulations is very active, including in the fields of packaging and apparatus.

Whereas active substances once had to be applied at rates of kilograms per hectare, the new pesticides that have now been developed require concentrations for application of g/ha. To facilitate correct dosage, these active substances must be appropriately diluted with auxiliary substances when they are formulated, bearing the requirement of good storage stability in mind.

Another aspect that must be considered when developing formulations is their compatibility with other formulations that might be mixed into the liquid for spraying, especially other brands.

14.1 Formulations and Targets of Active Substances

Leaving aside the direct spraying of insects or application onto the soil, the usual method of application of insecticides, herbicides, and fungicides is onto plants, either at ground level or from an aircraft. In most cases an aqueous formulation is used for the spraying, made up from emulsifiable concentrate, suspension concentrate, wettable powder, water-dispersible granules, etc. In order to improve the action of pesticides, the user often mixes other substances, known as *adjuvants*, into the spray solution, which usually influence interactions with the plants. Most often, these are special tensides, for example silicone-based tensides, pyrrolidone derivatives, polymers, modified vegetable oils, or mineral oils. The importance of adjuvants is illustrated by the fact that in some cases their addition can increase biological activity by a factor of 10 [1]. However, unsuitable or incompatible additives may cause damage; for example, they may increase phytotoxicity. Adjuvants are added to herbicides specifically.

In order to act, the active substance of herbicides applied to the surface of a plant must somehow penetrate into the living cells. The structure of the target plant at the optimal time of application is an important factor in this process. This structure differs depending on the stage of development of the plant, and determines how the components of the formulation applied are transported within the plant. Different and selective interactions will occur according to the structure of individual plants and the physicochemical properties of the formulation.

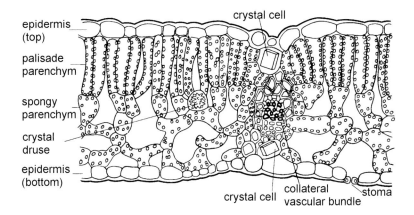

Figure 14.1. Cross-section through a broad leaf [2].

Figure 14.1 shows a cross-section of part of the leaf of a deciduous plant to illustrate the anatomy of leaves. The upper and lower epidermis surround the mesophyll, which on the upper surface of the leaf is developed into a parenchyma of palisade cells. These densely packed cells contain large numbers of chloroplasts to harvest light. Underneath this layer is an area of well-ventilated intercellular tissue, the spongy parenchyma, containing fewer chloroplasts. The intercellular system branches into fine channels between the palisade cells and communicates with the atmosphere through the stomata.

The vascular bundles running through the leaf are connected to the central vein and carry the water absorbed by the roots, nutrients, and salts (in the xylem) into the leaf and, in the opposite direction, transport the products of photosynthesis from the leaf to other regions of the plant (in the phloem, which consists of sieve tubes made of sieve cells). Xylem and phloem make up the long-distance transport system of the plant.

The epidermis, a covering, usually single-celled, layer of cells joined tightly together without gaps (Figure 14.1), is coated with a hydrophobic cuticle, which must be penetrated by the active substances before they can reach the underlying cells. The cuticle is made up chiefly of a cutin matrix of high molecular mass polyesters and varies between species. Its permeability differs depending on growing conditions and the age of the plant. Waxes are embedded in the cutin matrix and overlay it, determining the microtopography of the surface of the leaf (Figure 14.2). They form flakes, fibers, rods, or dendrites. Their chemical and physical heterogeneity is characteristic of the cuticle. Carbohydrate fibers projecting into the cuticle and sometimes the cutin matrix itself are possible routes into the leaf for hydrophilic substances, while lipophilic active substances can diffuse directly through the waxy layer of the cuticle.

Open stomata too can provide paths into the leaf for active substances in aqueous formulations. However, this is only possible when the surface tension of the formulation is less than 30 mN·m^{-1} (spreading problems; addition of adjuvants). In the empty spaces behind the stomata, depots of active substance can form that cannot be washed away by rain when the stomata are closed [3].

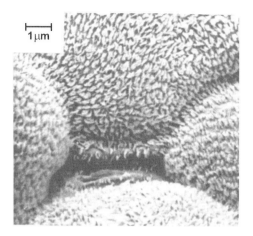

Figure 14.2. Microtopography of the cuticle on top of a leaf epidermis, showing a stoma [2].

For transport over short distances in the tissue, once the substance has penetrated the cuticle it has two possible routes (Figure 14.3). In the apoplast, the active substances diffuse through the cell walls and intercellular spaces. However, transport through the parenchyma of the symplast, the usual route for organic molecules, is faster. Active substances can diffuse through cell membranes from the symplast into the apoplast and vice versa. The transfer into the transport system of the phloem seems to begin in the apoplast, and it is this transfer that limits the movement of systemic pesticides in plants.

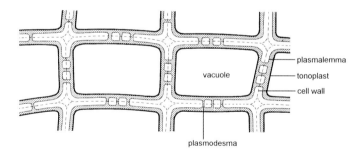

Figure 14.3. Schematic cross-section through plant tissue. Adjacent cells are connected through their cell walls by plasma channels (plasmodesmata 60 nm in diameter). The living cells all together thus form a continuum, the symplast, which is surrounded by the aqueous solution known as the apoplast (the white area of the cell walls in the figure) [2].

For a number of pesticides, transfer into the phloem must occur against a concentration gradient. This is only possible with the assistance of carrier substances. Another mechanism relies on the difference in pH between different areas of the plant. Weak acids accumulate in cells of higher pH. This is true in particular for the sieve tubes in phloem, so the transfer of weakly acidic pesticides such as those based on chlorophenoxyacetic acid, sulfonyl urea, or cyclohexanedione into the phloem is favored [4].

As described above, pesticide uptake can be enhanced by the addition of adjuvants such as tensides. The most important effect is the promotion of wetting so that even leaf surfaces rough on the microscopic scale are included (cf. Figure 14.2), enlarging the contact area. Nonionic tensides with ethylene oxide chain lengths of 5–6 units improve the absorption through the cuticle of lipophilic active substances. EO chain lengths of 15–20 are more effective for polar substances. Certain tensides are able to cross the cuticle themselves and thus, in a few cases, to affect the diffusion of the active substance.

Pesticides can enter the plant not only via the leaves but also through the root system. This is particularly important in pre- and postgermination application of herbicides. In this case, the substance is transported within the apoplast. The availability of the applied pesticide depends heavily on its absorption by the soil and its transport and degradation in the soil. Active substances that are strongly absorbed by the soil can only enter the root system with difficulty. A further problem is the contamination of groundwater, which can be tackled by the use of controlled-release formulations, for example those using cyclodextrin complex formers.

14.2 Types of Formulation

14.2.1 Emulsions, Emulsifiable and Water-Soluble Concentrates

A commonly used type of formulation is the emulsifiable concentrate (EC), based on organic solvents and emulsifiers, and emulsifiable in water. For example, an EC might contain:

30–50% active substance
40–60% solvent/s (e.g. high-boiling mineral oils)
 5–10% emulsifiers.

Active substances that are soluble in water up to at least the highest concentration used for their application can be formulated as water-soluble concentrates (soluble liquids, SL). A solvent miscible with water in which the active substance has adequate solubility is selected; tensides are seldom necessary.

Liquid active substances that are sufficiently stable to hydrolysis and are insoluble in water can be formulated as emulsions in water (EW). Since EWs contain no organic solvents, or only small amounts, they are environmentally and toxicologically more acceptable than ECs.

A new type of formulation is the gel (GL). A gel is an EC in which the viscosity has been elevated by means of thickeners or gel formers. They are sealed into water-soluble packages and are thus easily and safely handled with little risk of contamination [5].

A selection of solvents and tensides used in the formulation of ECs is given in Table 14.1.

Table 14.1. Auxiliaries used in emulsifiable concentrates.

Auxiliary	Examples
solvent	aliphatic HCs (with or without aromatic groups), C_8–C_{18} methyl esters of fatty acids, vegetable oils, alcohols, N-alkyl-pyrrolidones, tetrahydrofurfuryl ether, ketones such as cyclo-hexanone, γ-butyrolactone
tensides	*nonionic*: polyethoxylated derivatives of: C_8–C_{18} fatty alcohols, alkyl phenols, and dialkyl phenols (C_8, C_9, C_{12}–C_{16}), castor oil, alkyl amines, tristyrylphenol, sorbitan esters (cf. Section 2.1.3 too); EO/PO block copolymers and their alkyl phenyl ethers, silicone comb surfactants, alkylated polyvinylpyrrolidones *anionic*: alkyl sulfates (C_8–C_{18}, pref. C_{12}), alkyl benzenesulfonates e.g. calcium dodecylbenzenesulfonate, naphthalenesulfonate, dioctylsulfosuccinate

Toxicity, phytotoxicity, dissolving power, compatibility with emulsifiers, vapor pressure, and flashpoint are the properties that determine which solvents will be used in ECs, EWs, and GLs and how. Aromatics, for example, are more phytotoxic than aliphatic solvents [6]. Surface tension and rate of evaporation are other parameters that can affect the phytotoxicity. The type of solvent can have a considerable influence on the uptake of the active substance through foliage. Attempts have been made to optimize solvent/emulsifier/active substance systems using solubility parameters [7]. It is often possible to combine two poor solvents to obtain the same solubility as that of a good but toxic solvent (cf. Chapter 8).

Chapter 2 covers the topic of emulsifiers. Agricultural formulations make use of recently developed tensides as well as traditional ones; comb surfactants based on silicones with hydrophilic EO/PO chains, for example, are utilized as wetting agents to improve the spread of the formulation on leaves [8].

N-Alkylpyrrolidones (octyl, dodecyl), compounds with the strongly polar pyrrolidone residue, form very stable micellar systems with suitable cotensides. Besides ECs, they are also employed in microemulsions and the solubilization of pesticides [9, 10].

Alkyl polyglycosides, which have good biodegradability, behave much like ethoxylated alcohols. However, their solubility in water is higher, especially in the presence of salts, and does not decrease as the temperature rises (see under "Cloud point") [11].

Polyglyceride/sugar fatty acid esters are another type of compound with good biodegradability; they are obtained by ester exchange, and have very low toxicity [12].

14.2.2 Suspension Concentrates, Suspoemulsions, and Capsule Suspensions

Solid active substances with poor solubility in water and adequate stability to hydrolysis can be formulated as suspension concentrates (SCs or "flowables") in water. These aqueous formulations contain only small quantities of organic solvents such as glycols, which act as antifreeze. It is important that the finely dispersed particles of active substance do not grow during storage (Ostwald ripening) and that the dispersions do not form any sediment that cannot be redispersed.

Formulations combining EWs and SCs exist; these are designated "suspoemulsions" (SEs).

Agricultural active substances can be microencapsulated like pharmaceuticals and applied in the form of aqueous suspensions (CS, capsule suspension) [13]. Such formulations are safer for the user; they have much lower oral and dermal toxicities than similar EC formulations, as can be seen from Table 14.2 [14]. In addition. CS formulations have been found to be less phytotoxic and to photodegrade 2–3 times more slowly [15]. Microencapsulation also has advantages when used for volatile active substances.

Table 14.2. Comparison of the toxicities of EC and CS formulations.

Formulation	LD_{50} in rats [mg/kg]	
	oral	dermal
Furathiocarb 400 EC	81	1805
Furathiocarb 400 CS	> 3000	> 4000

The most common microencapsulation method for agrochemicals is *interfacial polymerization* [16]. In this process, the first reactant (a polyfunctional isocyanate or acid chloride) is dissolved in the liquid active substance, which is then dispersed in water and the second reactant (a polyfunctional amine) is added. The polymerization occurring at the interface between the active substance and the aqueous phase completely encloses the fine droplets of active substance with a thin membrane of polyurea or polyamide.

Compared with interfacial polymerization, the *coacervation* method has the drawback that encapsulation can only be carried out in dilute dispersions. Furthermore, aggregates tend to develop unless an additional unreactive dispersion stabilizer is mixed in. Finally, the capsules are too large for agricultural use. On the other hand, the process has the advantage that not only emulsion droplets but also solid particles can be encapsulated.

Aggregation of the encapsulated particles is the main problem with *in situ polymerization* too. In this case the polymerization or polycondensation creates nanoparticles in solution, which settle onto the surface of the dispersed solid or emulsified droplets and can form a shell there.

14.2.3 Wettable Powders, Water-Dispersible Granules, and Water-Soluble Powders

Like emulsifiable concentrates, wettable powders (WPs) are in common use. These finely ground powders containing dispersants can release a lot of dust and are therefore sometimes packaged as individual doses in water-soluble bags. An example of a WP is constituted as follows:

20–50% active substance
10–20% mixed dispersants such as ligninsulfonate + calcium alkylphenylsulfonate
30–70% fillers like kaolin

Unlike WPs, water-dispersible granules (WGs) make almost no dust [14], as is clear from Figure 14.4. They also occupy a smaller volume. The current trend in formulation is in the direction of such WGs at the expense of WPs. The products flow well and can be easily measured by volume, although they do dissolve more slowly when the solution for spraying is made up.

Very little residue (on the order of 0.01%) remains in the packaging, so it is not necessary to wash this before disposal. This is an advantage over suspension concentrates, which leave a dried crust that can be difficult to remove.

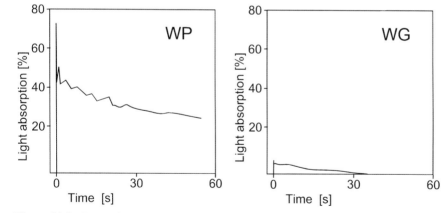

Figure 14.4. Comparison of the dusting characteristics of wettable powders (WP) and water-dispersible granules (WG), measured with Casella apparatus (taken from reference [14]).

The granulation methods are similar to those used for pharmaceutical products and described in Chapter 12:

- dish granulation
- spray drying
- fluidized-bed granulation
- extrusion
- high-shear granulation

Figure 14.5 shows the sizes and shapes of various water-dispersible granules. The substances listed below are used to formulate water-dispersible granules:

- active substances
- wetting agents, e.g. alkyl naphthalenesulfonates, alkyl sulfates
- dispersants, e.g. ligninsulfonates, naphthalene–formaldehyde condensates
- antifoaming agents, e.g. silicone oils
- fillers, e.g. kaolin
- binders, e.g. polyvinylpyrrolidone, starch derivatives
- disintegrating agents, e.g. water-soluble salts, cross-linked polyvinylpyrrolidone [17]
- adjuvants, e.g. activity enhancers

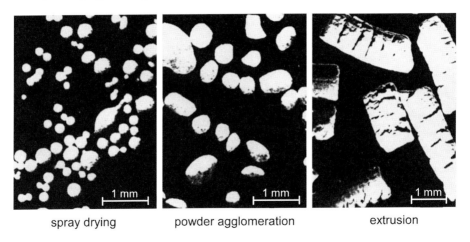

spray drying   powder agglomeration   extrusion

Figure 14.5. Water-dispersible granules (WG) manufactured by various methods (taken from reference [14]).

Wetting is only a problem if the granules are very small. Since common wetting agents tend to foam, antifoaming agents must be added to the formulation.

Dispersants not only have a stabilizing effect, but also assist the actions of binders and disintegrating agents or solution aids. More details concerning dispersants will be found in Chapter 5, and concerning binders and disintegrating agents in Sections 12.3.2 and 12.3.5 in the chapter on pharmaceutical technology.

If a solid active substance is soluble in water, it is usually mixed or ground with a soluble salt and used in the form of a soluble powder (SP).

14.2.4 Formulations Applied without Dilution

While the majority of formulations are sprayed as heavily diluted aqueous solutions, emulsions or suspensions, there are some that are applied without dilution. The most common of these are granules (GR), which are sprinkled by hand or machine. The active substance, usually in a proportion of only a few percent, is absorbed by a granulated carrier substance with the correct grain size.

Ultra-low-volume formulations (UL) are designed especially for crop-spraying from aircraft or the ground. They are sprayed as extremely fine droplets by special atomizers, at a dosage per hectare of usually no more than 0.5–2 L. In order to prevent large losses by evaporation, UL formulations must be prepared with nonvolatile solvents [18].

Dustable powders (DPs) are fine powders with a low content of active substance that can be spread directly from a variety of possible applicators. Dustable powders are no longer used in most countries.

14.3 Adjuvants

Adjuvants (auxiliary components), which enhance the activity of the active substance, for instance, can be part of the formulation of the active substance, but are usually added when the formulation is prepared for spraying in order to modify their properties.

The following list gives some examples of adjuvants: antifoaming agents, wetting agents, dispersants, spreading agents, drift controllers, evaporation inhibitors, leaching inhibitors, thickeners, buffers, adhesion improvers, activity enhancers, penetration enhancers, phytotoxicity inhibitors, marker substances, fertilizers.

Adjuvants are used with herbicides in particular. The ideal adjuvant should assist weed control while leaving the crop unaffected. Nonionic tensides and various oils are the most frequently employed adjuvants.

References for Chapter 14:

[1]   J. W. van Valkenburg, in Adjuvants for Herbicides (R. D. Hodgson, Ed.),
       Monogr. 1, Weed Science Society of America, Champaign, IL, 1982.
[2]   G. Czihak, H. Langer, H. Ziegler, Biologie, Springer-Verlag, Berlin, Heidelberg,
       New York, 1981.
[3]   R. J. Field, N. G. Bishop, Pestic. Sci. *24*, 55 (1988).
[4]   R. J. Field, F. Dastgheib, in Pesticide Formulation and Adjuvant Technology
       (C. L. Foy, D. W. Pritchard, Eds.), CRC Press, Boca Raton, New York, London,
       Tokyo, 1996.
[5]   B. Frei, P. Schmid, in Pesticide Formulation and Adjuvant Technology (C. L. Foy,
       D. W. Pritchard, Eds.), CRC Press, Boca Raton, New York, London, Tokyo, 1996.

[6]   F. A. Manthey, J. D. Nalewaja, in Adjuvants and Agrichemicals (C. L. Foy, Ed.), CRC Press, Boca Raton, FL, 1992.

[7]   K. E. Meusburger, in Advances in Pesticide Technology, ACS Symp. Ser. 254 (H. B. Scher, Ed.), American Chemical Society, Washington, D.C., 1983.

[8]   D. S. Murphy, G. A. Policello, E. D. Goddard, P. J. G. Stevens, in Pesticide Formulations and Application Systems, Vol. 12 (B. N. Devisetty, D. G. Chasin, P. D. Berger, Eds.), ASTM STP *1146*, 45, American Society for Testing and Materials, Philadelphia, 1993.

[9]   K. S. Narayanan, R. K. Chaudhuri, in Pesticide Formulations and Application Systems, Vol. 12 (B. N. Devisetty, D. G. Chasin, P. D. Berger, Eds.), ASTM STP 1146, American Society for Testing and Materials, Philadelphia, 1993.

[10] Z. H. Zhu, D. Yang, M. J. Rosen, J. Am. Oil Chem. Soc. *66*, 998 (1989).

[11] R. A. Aleksejczyk, in Pesticide Formulations and Application Systems, Vol. 12 (B. N. Devisetty, D. G. Chasin, P. D. Berger, Eds.), ASTM STP 1146, American Society for Testing and Materials, Philadelphia, 1993.

[12] J. F. Fiard, J. M. Mercier, M. L. Prevotat, in Pesticide Formulations and Application Systems, Vol. 12 (B. N. Devisetty, D. G. Chasin, P. D. Berger, Eds.), ASTM STP *1146*, 33, American Society for Testing and Materials, Philadelphia, 1993.

[13] G. J. Marrs, H. B. Scher, in Controlled Delivery of Crop-Protection Agents (R. M. Wilkins, Ed.), Taylor and Francis, Inc., Bristol, PA, 1990.

[14] E. Neuenschwander, L. Loosli, CIPAC Symp., Athens, 1989.

[15] A. J. Stern, D. Z. Becher, in Pesticide Formulation and Adjuvant Technology (C. L. Foy, D. W. Pritchard, Eds.), CRC Press, Boca Raton, New York, London, Tokyo, 1996.

[16] G. Japs, U. Nehen, H. J. Scholl, US Patent 4 847 152, 1989.

[17] L. S. Sandell, EP 501 798, 1992.

[18] A. Grubenmann, E. Neuenschwander, in Advances in Pesticide Science, Part 3, (H. Geissbühler, Ed.), Pergamon Press, Oxford and New York, 1979.

15 Pigments and Dyes

15.1 Solubility of Pigments and Dyes

The difference between pigments and dyes lies in their solubility. Pigments are practically insoluble in the medium for application (the vehicle) and are dispersed in it as solid particles, usually of size less than 1 μm. Dyes, in contrast, are absorbed by the substrate, for instance a yarn, in dissolved form and then fixed there by hydrogen bonding or chemical reaction with the substrate and the formation of poorly soluble salts or pigments so that it is difficult or impossible for the dye to diffuse out of the substrate.

Dispersion colors are a borderline case; they are used in aqueous dispersions at high temperatures, say 130°C under pressure. At this temperature they are sufficiently soluble in water to permit diffusion in the dyeing process (Table 15.1).

Table 15.1. Solubility in water of dispersion colors and pigments.

	Solubility in water at 130°C [mg/L]
dispersion colors	5–500
pigments	<0.05

Sometimes pigments are classified as either *classical* or *high-grade* pigments. High-grade pigments are often superior to classical pigments in some respects concerning their application, such as their stability to migration and recrystallization, thermal stability, lightfastness and fastness to exposure (to weathering). This is principally a result of their very low solubility.

Whether classical or high-grade pigments are used depends upon the application. Classical pigments with their higher solubility cannot be used in solvent-containing baked enamels (though they are suitable for use in air-drying alkyd resin coatings). At that high temperature, crystal growth, bleeding, and blooming can occur.

Figure 9.6 illustrates the degree to which pigment solubility can differ in different solvents (the pigment shown there is ten times more soluble than the most soluble high-grade pigment).

The rigid, planar basic structures of dye and pigment molecules, with their delocalized π bonds, determine the color range and also the solubility range of dye and pigment molecules. Substituents added onto the rigid π structure change not only the nuances of the color but also technical properties like their solubility, wettability of crystals, thermal and chemical stability.

Figure 15.1 shows the number of C, N, and O atoms in the rigid π systems for different classes of dyes and pigments (when several rigid π systems are connected by groups that can rotate in the dissolved pigment, then only one of those systems determines the size of the rigid section; for example, $Z = 11$ for isoindolinone).

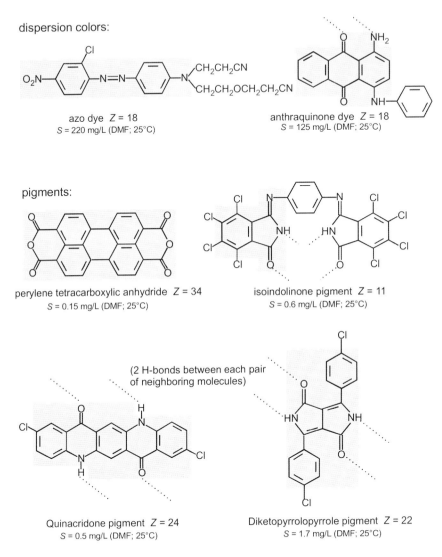

dispersion colors:

azo dye $Z = 18$
$S = 220$ mg/L (DMF; 25°C)

anthraquinone dye $Z = 18$
$S = 125$ mg/L (DMF; 25°C)

pigments:

perylene tetracarboxylic anhydride $Z = 34$
$S = 0.15$ mg/L (DMF; 25°C)

isoindolinone pigment $Z = 11$
$S = 0.6$ mg/L (DMF; 25°C)

(2 H-bonds between each pair of neighboring molecules)

Quinacridone pigment $Z = 24$
$S = 0.5$ mg/L (DMF; 25°C)

Diketopyrrolopyrrole pigment $Z = 22$
$S = 1.7$ mg/L (DMF; 25°C)

Figure 15.1. Rigid π-electron system and hydrogen bonds in disperse dyes and pigments (Z: number of atom centers in the rigid section); solubility S measured in dimethylformamide.

The degree to which the relative solubility of a dye or pigment can be influenced by substituents is shown in Table 15.2 (the parameters given only allow a rough estimate of the effect of the structure; specific solvents are not considered. See also Chapter 9 regarding calculations of solubility).

The parameters that determine the solubility of a pigment are the size of the rigid π system, the presence of intermolecular hydrogen bonds, and elevation of molecular mass by the inclusion of substituents to depress the solubility. Isoindolinone pigments, for example (Z of rigid π system = 11) are too soluble unless eight chlorine atoms at least are included. According to the figures in Table 15.2, this lowers the solubility by a factor of $1.033^{(4 \times 34.5)} \times 1.009^{(4 \times 34.5)} \approx 300$.

Prevention of the formation of intermolecular hydrogen bonds in the solid can have a dramatic effect. This is the reason that di-*o*-chlorophenyldiketopyrrolopyrrole is one hundred times more soluble than the corresponding *p*-chloro compound.

The rigid π system does not relax when the solid dissolves. However, other components of the molecule can then rotate freely or with constraints. It is this influence that has been quantified in the table. The effects of specific solvents are not considered.

Table 15.2. Influence of substituents on the solubility of dyes and pigments (compared in the same solvent; measurements by A. Grubenmann).

– Enlargement of the rigid π system by z C,N,O atoms:	diminished by $1.8^z \times$
– For n H-bonds per neighbor in the crystal lattice:	diminished by $12^n \times$
– Free $-O-$, $-N<$:	increased by $10-15 \times$
– $-CH(CH_3)_2$, $-C(CH_3)_3$:	increased by $25 \times$
– n ($-CH_2-$):	increased by $1.3^n \times$
– $-OSO_2-$, $-NHSO_2-$:	no effect
– For g grams increase in molar mass due to substituents	
halogen, $-CN$ in the rigid π system:	diminished by $1.033^g \times$
halogen, $-CN$ and other substituents in the flexible sections:	diminished by $1.009^g \times$

15.2 Pigments

15.2.1 Characterization of Pigments

The application characteristics of pigments are largely determined by parameters related to the arrangement of the molecules in the pigment crystal, crystal shape and crystallinity, specific surface area, the nature of the surface, and the chemical properties of the surface. The morphology of the powder, aggregate, and agglomerate forms (Chapter 3) are also significant. These parameters can be changed, conditioned, in such a way as to optimize the technical characteristics for specific applications.

Synthetic products can be reduced to the desired particle size by grinding, and the crystallinity thus disrupted can be restored by gentle recrystallization or thermal treatment. Sometimes the surface of the pigment must be chemically modified [1]. This treatment can improve the pigment dispersibility and the interactions between pigment and binder.

The *specific surface area* of pigment powders is in the range 10–130 m²/g for organic pigments (Figure 15.2); this is the area available for adsorption in the dispersion medium. For the manufacture of deagglomerated and unflocculated dispersions, enough stabilizer (e.g. binder) must be present to cover the entire surface. However, the important factor is not only the specific surface area, as a measure of the particle size, but also the *particle size distribution*. This affects not only the color but also the rheology of paints and varnishes, printing inks, and polymer melts.

Figure 15.2. Electron micrographs of Pigment Red 168 samples of different specific surface area: 20.8 and 35.9 m²/g (from reference [2]).

15.2.2 Absorption, Scattering, and Reflection from Pigmented Layers

Pigments are either mixed into bulk plastics or applied as a thin layer (e.g. printing ink) or a thick layer (e.g. automobile paints). The optical properties of such pigmented layers – color, transparency, covering (hiding) power, gloss – depend on both the pigment and the binder.

When light impinges on a pigmented layer, a fraction of it is reflected from the surface. This determines the gloss of the layer (Figure 15.3a); various types of surface defects diminish the gloss. The light that penetrates the interior of the sample is scattered by the pigment particles in more or less all directions away from its original path (Figure 15.3b). In addition the radiation is absorbed by the pigment particles and thus weakened (Figure 15.3c). Some of the light scattered in all directions by particles is returned to the surface, where it exits so that we can perceive it; however, this re-emission depends on the support bearing the pigment coating, which may re-emit more or less of the radiation that reaches it back to the illuminated surface. If, of all the light exiting the surface, the proportion influenced by the support is small, then the support cannot be perceived. In this case the pigmented layer has good hiding power. Otherwise, the layer appears transparent to some degree.

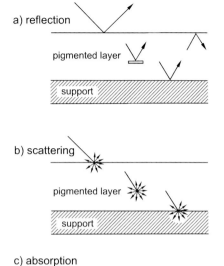

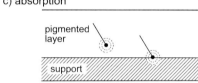

Figure 15.3. Absorption, scattering, and reflection of light in pigmented layers; from reference [2].

The effective scattering and absorption in a pigmented layer can be characterized by the scattering coefficient $S(\lambda)$ and the absorption coefficient $K(\lambda)$. The reflectance $R(\lambda)$ describes the proportion of incident light of wavelength λ that is returned by the sample. For covering colors, the Kubelka–Munk Theory [2] posits the following relationship between reflectance, scattering coefficient, and absorption coefficient (Equation 15.1):

$$\frac{K(\lambda)}{S(\lambda)} = \frac{\left(1 - R(\lambda)\right)^2}{2\,R(\lambda)} \qquad (15.1)$$

The ratio K/S can be used as a measure of the color depth, and is often applied to determine the development of color depth as a function of dispersion time.

15.2.3 Dispersibility

Pigment particles are agglomerated in the powder form. Dispersion techniques are used to try to break the agglomerates up into single particles. The aim of dispersion in a paint or an ink, for example, is the optimization of color tone (or tint) and depth. Often the maximum depth of color can be achieved only by an excessively long dispersion process, particularly when the pigment is very fine. In this case a comparison of the cost of deepening the shade by longer dispersion and the cost of deepening the shade by raising the pigment concentration is worthwhile.

Amongst other effects, dispersion:

- increases the color depth, most perceptible in white reduction
- changes the tone of the color
- increases the transparency, that is, decreases the hiding power
- enhances the gloss
- elevates the viscosity
- diminishes the critical pigment volume concentration, e.g. in loads for grinding.

When a pigment is dispersed in a vehicle, the following processes occur simultaneously:

- dispersion of the agglomerates by mechanical action (dry and wet grinding)
- wetting of the pigment surface (soaking of the liquid into the pigment powder, spreading of the liquid or binder components over the pigment surface)
- stabilization of the dispersed particles to prevent reagglomeration and flocculation (electrostatic and steric stabilization).

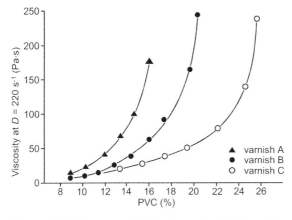

Figure 15.4. Influence of the binder on the CPVC and viscosity of Pigment Yellow 13 in offset printing ink varnishes (from reference [2]).

The volume percentage of the pigment in the nonvolatile part of the formulation is known as the pigment volume concentration (PVC); at the critical pigment volume concentration (CPVC) the pigment particles are all wetted but no excess binder is present. At this point, the viscosity approaches a limiting value asymptotically and other technical characteristics alter sharply. However, the CPVC can be steered by careful selection of the binder and other additives (Figure 15.4).

Surface properties, particle size distribution, agglomeration, and also additives such as dispersants and resins affect the dispersibility of pigment powders or press cakes. The dependence on particle size is related to the number of points of adhesion between the particles in the agglomerates, amongst other things. When the size distribution is broad, the small particles act as links between the larger ones and thus make dispersion more difficult. They reduce the number of spaces in the agglomerates and obstruct the penetration of solvent. Powders with narrower particle size distributions can therefore more easily be dispersed than similar powders with broad size distributions.

If insufficient binder or dispersant is present to surround the particles, they may *reagglomerate*.

Loose associations – *flocs* – develop owing to van der Waals attraction if the dispersion is not adequately sterically or electrostatically stabilized. These flocs are often easily redispersed by small shearing forces. Well-stabilized dispersions can settle out and, in doing so, slide into the primary energy minimum, forming a compact sediment that cannot be redispersed by agitation. In contrast, stabilized dispersions with a *secondary energy minimum* can easily be redispersed after the loosely flocculated sediments with plenty of space between the particles have settled out. The same is true of structurally viscous thixotropic samples, in which sedimentation can often be prevented.

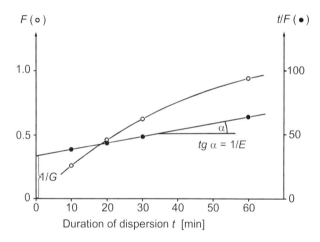

Figure 15.5. Color strength: two plots of the development of depth of color ($F = K/S$) as a function of dispersion time t; from reference [3].

On dispersion, the depth of color does not develop proportionally to the duration of the dispersion process, but rather has a parabolic course. A linear relationship appears when the reciprocal color depth $1/F = S/K$ is plotted against 1/time. However, a better linear correlation is obtained by representing t/F as a function of time (Figure 15.5), as in Equation 15.2:

$$\frac{t}{F} = \frac{t}{E} + \frac{1}{G} \tag{15.2}$$

(F: depth of color (K/S); E: depth of color for $t\to\infty$; G: rate of dispersion; t: duration of dispersion)

15.2.4 Problems in Pigmented Systems

15.2.4.1 Blooming and Bleeding

At high application temperatures, and depending on the pigment, considerable quantities of the pigment can dissolve, only to crystallize out again later, either within the vehicle or on the surface. The latter process, *blooming*, often carries on for years. It can occur in, for example, baking enamels, plasticized PVC, polyethylene, or rubber mixtures containing high-boiling-point naphthenic oils as plasticizers. Tables 15.3 and 15.4 show the relationship between solubility and blooming for a pigment in plasticized PVC. Blooming can be hindered by the incorporation of large amounts of pigment: if the pigment is present in a concentration much higher than its solubility at the processing temperature, sufficient crystals remain in the interior of the medium on which the dissolved pigment can crystallize out to prevent blooming.

Table 15.3. Solubility of Pigment Red 170 in plasticized PVC (taken from reference [2]).

Temperature [°C]	Solubility [weight %]
20	$8.0 \cdot 10^{-7}$
50	$1.2 \cdot 10^{-5}$
100	$4.0 \cdot 10^{-4}$
140	$3.7 \cdot 10^{-3}$
160	$9.7 \cdot 10^{-3}$
180	$2.3 \cdot 10^{-2}$
200	$5.1 \cdot 10^{-2}$

Table 15.4. Blooming (+) of Pigment Red 170 in plasticized PVC as a function of pigment concentration and processing temperature (taken from reference [2]).

Processing temperature [°C]	Pigment concentration [weight %]				
	0.01	0.025	0.05	0.1	0.5
140	+	−	−	−	−
160	+	+	−	−	−
180	+	+	+	−	−
200	+	+	+	+	−

Bleeding is the movement of a pigment dissolved in a vehicle out of that medium into a differently colored or uncolored medium of the same or similar composition in contact with the first. Bleeding is an especial problem in plastics and paints. In paint technology, fastness to bleeding is measured in terms of the color stability to overpainting when a new layer of paint is superimposed on the first.

15.2.4.2 Plate-Out, Chalking

Plate-out often occurs in the processing of plasticized PVC, for example calendering and rolling. Lubricants, stabilizers, and plasticizers that are not compatible with the medium and are therefore exuded carry pigment particles out to the surface with them. A colored crust forms on the surface of the processing machine. Poor wettability of the pigment by the plastics components, particularly if the specific surface area is large, aggravates the effect. Plate-out can also be a problem in powder paints.

The degradation reactions that occur in the binder when pigmented media are weathered, particularly when adequate moisture is present, uncover pigment grains and thus create a raw surface, especially in the case of systems containing titanium dioxide. This process is known as *chalking*. Photochemical processes involving oxygen are the cause, particularly with surfaces of TiO_2 and certain color pigments.

15.2.4.3 Distortion

In partly crystalline polymers such as polyolefins, certain pigments can act as nucleators for crystallization of the plastic. The change in density caused by crystallization leads to stresses, deformations, and tears, and also often accelerates degradation on weathering. Such distortions can, for example, render crates brittle and their stacking impossible. The crystallization of the plastic commences at the surface of the pigment particles and results in spheroliths, especially in the case of acicular pigment particles.

15.2.5 Applications for Pigments

15.2.5.1 Paints

The term "paint" covers liquid or powdered preparations that are thinly coated onto surfaces and turned into permanently adhering coatings by drying, baking, crosslinking, or polymerization. Careful selection of the paint to suit the surface to be coated is just as important as the optimal preparation of that surface [4].

There are numerous coating procedures, such as brushing, spraying, two-component spraying, dipping, electrostatic spraying or dipping, drum coating, centrifugal coating, pouring, rolling, and powder coating. The technique of coil coating, in which rolled steel or aluminum sheeting is coated at rates of up to 150 m/min in widths of up to about 2 m [5], has become economically important.

The huge range of paints available can be classified according to end use, e.g. automobile paints, wood varnishes, lacquers for the interior of food cans, according to application method, e.g. spray paints, dip coatings, according to their form, e.g. solvent-containing, water-thinnable, powdered, or according to their drying characteristics, e.g. baked, air-dried.

Paints contain numerous components with defined tasks in the liquid for application and in the finished coating: volatile solvents and nonvolatile components such as binders (film-formers, resins, plasticizers), additives, dyes, pigments, fillers.

Macromolecular substances such as nitrocellulose or vinyl chloride–vinyl acetate copolymers are employed as film formers, or else substances of low molecular mass which polymerize as the paint dries are used, for example unsaturated polyester resins or epoxy resins. The viscosity of a polymer solution increases with the molecular mass of the polymer, so in industry low-viscosity coatings, in which the film-forming polymers only develop as the coating hardens, are preferred. These low molecular mass components are often liquids, so such coatings require little or even no solvent.

Nevertheless, even these coatings do require some high molecular mass components. Dispersants and dispersion stabilizers ensure the deagglomeration of powdery additives like pigments and fillers during the dispersion process in manufacture, and impede their flocculation in the prepared paint and on application.

The term "*resins*" is used for a group of film-forming substances of resinous consistency, which as a rule are readily soluble and are used to increase the solid content of paints, as well as enhancing their adhesiveness and their gloss. In addition, resins increase the hardness of the film and shorten the drying time for systems that cross-link by oxidation.

Plasticizers are nonvolatile organic liquids of oily consistency, such as dioctyl phthalate.

Assorted auxiliaries like driers, antiskinning agents, hardening accelerators, running improvers, sedimentation inhibitors, matting agents, wetting agents, and antiflooding agents are added to improve the properties of the liquid paint or the finished coating.

15.2.5.1.1 Oxidatively Dried Coatings

These are principally air-drying alkyd resin coatings with medium- and long-chain drying oils [4]. Their broad spectrum of uses ranges from artists' paints to industrial coatings. The important criteria for the pigments used are fastness to light and to weathering. For systems that dry at room temperature, pigment sensitivity to the solvents used (aliphatic and aromatic hydrocarbons, oil of turpentine, higher alcohols) is seldom a problem.

15.2.5.1.2 Baked Enamels

Baked enamels are used chiefly in industry; their use requires costly plant. They are designated *industrial coatings*.

For many pigments, the solvents typically used (glycols, glycol ethers, esters, ketones, chlorohydrocarbons, nitroparaffins) are not without problems.

The binders employed vary widely, for example melamine–formaldehyde or urea resins that harden on heating by polycondensation with other resins such as epoxy resins, alkyd resins with short-chain oils, or acrylic resins. The coating is baked at 100–200°C for anything from a few minutes to more than an hour.

In two-pack coating systems, hardening commences as soon as the two components are mixed. In industry, it is often accelerated by raised temperatures of up to 120°C. This is the case for, amongst others, isocyanate cross-linked hydroxy-containing polyester or acrylic resins used, for example, in automobile repair paints, or the unsaturated polyester or acid-hardening alkyd–melamine or urea varnishes used in the furniture industry.

In order to reduce solvent emissions, research has been carried out into developing *low-solvent* or *solvent-free* coating systems for several years.

High-solids systems contain a higher proportion of solids than conventional baked enamels. The binders used have low molecular masses.

Nonaqueous dispersions (NAD systems) consist of a mixture of dissolved and dispersed binders. If the pigment particles present have large specific surface areas, signs of flocculation may become visible unless sufficient binder is dissolved to cover the surface.

Systems that can be *diluted with water* (water-thinnable paints) contain water as the principal liquid, along with organic solvents. The dispersion step may be difficult because of problems with the wetting of nonpolar pigments, but the mixture for grinding can be formulated to circumvent this [2, 6].

UV-hardening systems may not set properly if the layer is thick, since pigments absorb in the UV range. However, it is possible to harden black coatings containing up to 3% carbon black all the way through a dry layer thickness of 45–50 μm. With colored organic pigments, dry film thicknesses of 35–40 μm and pigment–binder ratios of 1:7 are possible [7].

The pigmentation of *powder paints* is particularly problematic: reaction between the pigment and the hardener can alter the color tone as a function of temperature and time. Plate-out (formation of a pigment scale on processing machines and the surface of the system) is generally a result of inadequate wetting of the solids in the paint by the binder.

15.2.5.1.3 Aqueous Dispersions of Synthetic Resins

Aqueous paints for brush application with synthetic resin bases give films similar to those of other paints and varnishes. These dispersions, based on poly(vinyl acetate) copolymerized with vinyl chloride, dibutyl maleinate, ethylene, acrylates, or polyacrylic resin copolymers, and poly(vinyl propionate) and styrene–butadiene copolymers, require emulsifiers in the inner or outer phase for their manufacture. They must also be sterically stabilized with a substance such as poly(vinyl alcohol), starch, gelatin, or cellulose derivatives. Such dispersions can only be pigmented with preparations in which the surface-active substances do not seriously disturb the equilibrium of the dispersion. Nonetheless, it is sometimes not possible to prevent fading of the color under the influence of various additives such as solvents, antifoaming agents, or plasticizers, as a result of recrystallization or flocculation during storage.

15.2.5.2 Printing Inks

In terms of volume, this sector is the most significant consumer of organic pigments. As printing techniques develop, printing ink formulations must continually be altered to match the new requirements. As printing speeds keep on increasing, good rheological properties and color depth are crucial properties of heavily pigmented printing inks, as are often gloss and transparency too. In some applications the ink must also be resistant to various external factors.

For use in *offset* and *letterpress printing*, pigments must be very well dispersed; agglomerates in the 0.8–1 μm thickness of the ink layer impair the color depth and gloss. The high temperatures at which dispersion is carried out in high-performance agitated ball mills, sometimes over 90°C, demand good stability to recrystallization from pigments. Surface-active additives can cause problems in blending; resins are used for steric stabilization of the pigments. Inks must also be stable to water and not run.

In *full-color printing*, there are special requirements for fastness to overprinting, to calendering, and to light, heat stability, migration resistance, transparency, hiding power, and rheology.

For *tin printing*, a special variety of offset printing, the criteria are good heat stability (setting temperatures of 140°C or more), fastness to sterilization processes (in water at 120°C, 2 bar), and fastness to overprinting. UV-drying inks are also used in tin printing.

Rotogravure is split into two forms, for illustrations and for packaging or special purposes.

In illustration rotogravure, used for magazines, illustrated papers and so on, the inks need particularly good rheological properties. The halftone dots, 40 μm deep, must be able to fill with ink and transfer it immediately to paper moving at speeds of up to 12 m/s. This calls for soft pigments with grains that do not cause abrasion. The inks used contain up to 60% toluene and sometimes gasoline as well, and have a pigment content of 4–15%. Therefore, only pigments fast to solvents and stable to recrystallization can be used. Yellow inks contain pigments treated with amines, which have low viscosities even when they have large specific surface areas.

Inks for *packaging rotogravure* contain a mixture of at least two solvents like alcohols, glycol ethers, esters, ketones, and aromatic hydrocarbons. The pigments must be unaffected by these solvents, and also unaffected by the plasticizers in the substrates to be printed upon, such as dibutyl phthalate, dioctyl phthalate, or epoxides. Transparent inks in particular must contain pigments with no tendency to recrystallize. Other requirements include resistance to acids and alkalis, soaps, detergents, fats, spices, paraffin, wax, etc. The rheology of the ink is also important: the pigment surface should be treated such that the ink flows well.

Decor printing is used for molded laminate boards. For polyester laminates, the ink must be fast to certain solvents, for melamine laminates the transfer properties of the ink are important. In all cases, excellent lightfastness is necessary.

The demands placed on pigments by *flexography*, *screen printing*, and other printing techniques are concerned with stability, for example lightfastness and fastness to solvents.

Letterpress, offset, and screen-printing inks are thick pastes containing *binders* [8] that consist of stand oils (made by heating linseed oil), phenol-modified colophony, mineral oils, linseed oil and/or alkyd resins (combination coatings), or modern UV-crosslinking systems made up of radical-polymerizing prepolymers and monomers together with photoinitiators. Binders for flexographic and rotogravure inks comprise principally resins such as collodion cotton, polyamide resins, vinyl polymers, maleinate, phenol, amine, or acrylic resins.

15.2.5.3 Pigments for Plastics

The selection of a pigment should take into account not only the plastic to be colored and the technical properties of the pigment but also the conditions of processing. A huge number of different additives are mixed into plastics to improve their processing properties: antioxidants, photostabilizers, heat stabilizers, fire retardants, antistatic agents, plasticizers, lubricants, wetting agents, blowing agents, diluents, fungicides, fillers, and colorants.

A wide variety of apparatus exists for the premixing of plastics and additives; different machines can give different color results. Unless the mixtures are liquid or powdered, thermoplastics are granulated after mixing.

Pigments are usually dispersed in plastics by shearing. For thermoplastics like PVC and polyolefins, the shearing forces required to deagglomerate the pigments can only be achieved at levels of plasticization so low that they are almost never encountered in normal processing. Problems can occur in coloring extruded threads, for example, as insufficiently dispersed agglomerates can cause the spun thread to break, or in thin self-supporting films, where inadequate dispersion can result in the formation of pinholes.

To avoid these and other problems, masterbatches, concentrated pigment preparations containing well-dispersed pigment, are increasingly being used to color plastics. Pigment masterbatches are available as granules or as pastes, for example with plasticizer.

Polyolefins, the most common plastics, can be divided into three principal groups that differ mostly in terms of processing temperature:

- low-density polyethylene (LDPE)   160–260°C
- high-density polyethylene (HDPE)   180–300°C
- polypropylene (PP)   220–300°C

Polyolefins are partially crystalline in form and therefore scatter light; this effect makes pigmented polyolefins appear brighter. Ethylene–propylene copolymers have elastomeric properties and resemble natural rubber when the propylene content is 20% or more.

Some pigments tend to migrate, given their solubility in polyolefins and the low glass transition temperatures of the polymers, especially in LDPE and under the influence of various additives. Migration is analogous to blooming and bleeding in pigmented paints. Distortions due to the presence of pigment, on the other hand, occur chiefly in HDPE.

Pigments can affect the stability of polyolefins to light. A range of pigments can limit the effectiveness of HALS-type antioxidants (hindered amine light stabilizers) in polyolefins, for instance.

At room temperature, *polystyrene* (PS) is far beneath its glass transition temperature. The migration of dissolved pigments is therefore of little importance; in fact, it is even possible to use soluble dyes, which, when combined with pigments with good hiding power, yield brilliant tints. As polystyrene is processed at high temperatures (up to 300°C), the pigments used must be very stable to heat.

PS yellows in air under the influence of UV radiation and must therefore be stabilized with UV absorbers and antioxidants. However, these additives can cause problems in impact-resistant copolymers like those with acrylonitrile and butadiene (ABS), in which the individual phases have different dissolving powers for additives.

Poly(vinyl chloride) (PVC) is manufactured by several different processes. The product is classified as batch, block, suspension, solution, or emulsion PVC. The method of polymerization of the PVC to be colored has a clearly visible effect on the resulting color for many pigments. Recrystallization and migration of pigments is particularly pronounced in plasticized PVC because of the presence of the plasticizer. The dispersibility of pigments in PVC is a crucial property affecting its use, as it is in other plastics. The high processing temperatures chosen for economic reasons (up to 200°C) and the resulting low viscosities render the dispersion of powdered pigment difficult, so that masterbatches must be used. Figure 15.6 shows the influence of processing temperature.

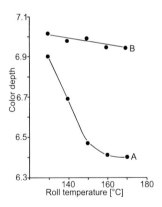

Figure 15.6. Effect of processing temperature on depth of color: Pigment Violet 19, γ modification, in plasticized PVC (33% dioctyl phthalate), pigment content 0.1%, TiO_2 0.5% (taken from reference [2]). Curve A: powdered pigment; curve B: masterbatch.

Polyurethane (PU), one of the most versatile plastics, behaves like plasticized PVC when it comes to pigmentation, and can be processed with plasticizers in the same way.

The factors described above apply by and large to the mixing of pigments into other plastics such as *polyamides* (PA), *polycarbonates* (PC), *poly(ethylene terephthalate)* (PET), and *cellulose derivatives*. Selection of a pigment must take account of the processing conditions, the additives present, the intended application, and economic criteria.

15.3 Dyes

Dyes [9] diffuse in solution, most often aqueous, into the substrate to be dyed and must be fixed there, either by chemical reaction with the substrate or by formation of insoluble salts and pigments.

Direct or *substantive dyes* are able to dye native or regenerated cellulose directly from neutral, aqueous solution, without previous treatment with a mordant. They are usually azo or sometimes anthraquinone dyes containing sulfonic acid groups. However, their fastness to liquids (water, sweat, laundering) becomes so poor as the depth of color increases that the dyes have to be converted into insoluble compounds by a subsequent treatment with cationic substances or complex formation with copper.

Acid dyes likewise possess one or more sulfonic acid groups which form salts with basic groups. They are used for the dyeing and printing of wool and polyamides, and also of silk, leather, base-modified polyacrylonitrile, and for foods in the form of the soluble sodium salts.

Dispersion dyes, which are relatively poorly soluble in water, are used to dye semi-synthetic or synthetic fibers such as polyester, polyamide, and polyacrylonitrile. The dyestuff is dispersed in the liquor but, as shown by the dispersion coefficient, has a greater affinity for the hydrophobic fibers than for water and is almost entirely absorbed by the fibers. Since it diffuses only into the amorphous parts of the fibers, the dyeing should be carried out at a temperature *above the polymer's glass transition temperature*.

Cationic or *basic dyes* contain positively charged nitrogen atoms. They are employed as water-soluble salts in the dyeing of polyacrylonitrile and acid-modified polyester fibers, and for paper and leather.

Vat dyes are pigments that are absorbed by the substrate in reduced, water-soluble form (leuco form, for example reduced with sodium dithionite) and there converted back into the insoluble colored form (the pigment) by oxidation, for example by atmospheric oxygen. Leuco vat dyes are also used as esters, which must be saponified after absorption and before oxidation.

Like vat dyes, *developing dyes* are only transformed into their final colored form once they have been absorbed by the substrate. Amongst the members of this class are naphthol AS azo dyes and phthalocyanine developing dyes.

Certain suitable azo dyes too react within the substrate. They are converted into *metal complex dyes* by chromium, cobalt, or copper ions. These complexes can also be used as pigments. However, their significance is declining for environmental reasons.

Reactive dyes contain reactive groups that form covalent bonds with corresponding groups in the substrate. Suitable residues include the hydroxy groups of cellulose, the amino and thiol groups of wool and silk, and the amine and carbonyl groups of polyamides.

15.3.1 Dye Technology

In synthesis, dyes form as 1–10% suspensions and are turned into an easily filtered form by salting out, pH adjustment, or temperature change. A number of intermediate steps are necessary to convert the dye into a commercial product such as a powder, granules, an aqueous fluid paste, or a concentrated solution [10, 11].

After filtration, the dye has the form of a briquette with a solids content of 15–60%. The addition of hydrotropic substances such as urea or glycols permits direct conversion to highly concentrated solutions. The dry dyes, with a residual moisture content of 0.5–5%, are usually ground down to grain sizes of 1–50 μm in impact mills before further processing.

Since individual batches may differ slightly in tint and shade owing to different by-products and salt content, other dyes must be mixed in to adjust the tint (toning) and salts must be added to correct the depth of color (cutting).

This addition of standardizing agents has the further advantage of improving the properties relevant to the application of the dye. Depending on the form of the commercial product, different agents are used; common ones are:

– neutral inorganic salts: sodium sulfate, sodium chloride
– alkaline inorganic salts: sodium carbonate, sodium bicarbonate, trisodium phosphate
– acidic standardizing agents: sodium bisulfate, amidosulfonic acid, oxalic acid
– buffers: mono- and disodium phosphate
– complex formers: ethylenediaminetetraacetic acid, polyphosphates such as sodium hexametaphosphate
– nonelectrolytes: dextrin, sugar, urea, benzamide
– antidusting agents: mineral oil (combined with emulsifiers), phthalate esters, triacetin
– dispersants, antifoaming agents, wetting agents, antigelling agents.

When dyeing with dispersion dyes, the rate of dissolution depends on the particle size and the crystal modification. Therefore, it is not enough to set the particle size of such dyes by grinding (0.5–2 μm); the dyestuff must be converted into the most stable crystal modification by *forming* by addition of emulsifiers and solubilizers during grinding, or by heating the aqueous suspension. If this is not done, even small deviations in the process parameters could lead to faulty dyeing.

Lignin sulfonates or the condensation products of naphthalenesulfonic acid and formaldehyde are used as dispersants. Long-chain alkyl sulfonates or alkyl naphthalene-sulfonates are added to act as wetting agents, and various polyethoxylated products as emulsifiers (cf. Figure 5.17 for dispersants and Figure 5.18 for stabilizers).

The formulation of aqueous suspensions demands the addition of preservatives and of humectants to retard drying out (glycols, glycerin). If dispersion dyes are made commercially available in dried form, the very fine dispersion must be stable; a large excess of dispersant is therefore added to such products to prevent aggregation, in proportions of 1:1 or, better yet, 2:1 to the dye.

Concentrated solutions are gaining more and more popularity as commercial forms, not only because the dosing is easier but also because the problems caused by dust can thus be avoided; inhalation of dye dust can result in allergies and illness for sensitive people. The solvent is chosen according to the intended substrate; the choice must take guidelines for environmental protection into consideration. Polyols (e.g. ethylene glycol), ether alcohols (e.g. diethylene glycol monoethyl ether), and, depending on the class of dye, carboxylic acids (e.g. acetic acid), acid amides (dimethylformamide, urea) are used as solvents or solubilizers for addition to water.

15.3.2 Textile Dyeing

The dissolved dye is present in the liquor in the form of individual molecules or ions and aggregates. The dye should be absorbed by the substrate through diffusion to the fiber surface and into the interior of the fiber. Subsequent fixing, by steaming, dry heat, or chemical conversion, stabilizes the absorbed dye [12].

Aggregates can collect on the surface of the fiber and molecules of the dye can then diffuse from there through the fiber pores to amorphous regions. The rate of diffusion depends, amongst other factors, on the morphology of the substrate, the size of the dye molecule, and the temperature. When the dye is absorbed an equilibrium is established, which can be represented by a *Freundlich adsorption isotherm* (Equation 15.3; C = dye concentration):

$$K = \frac{C_{\text{fiber}}}{C_{\text{liquor}}^n} \qquad (15.3)$$

The constant K is a measure of the affinity of the dye for the fiber. The exponent n varies according to the fiber and is 0.6 for cotton, for example.

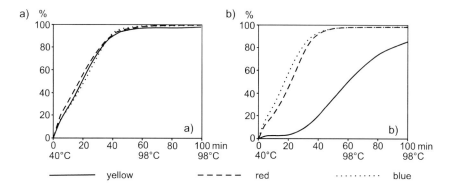

Figure 15.7. Absorption behavior of dye mixtures (reactive dyes for wool): a) ideal combination; b) unsuitable combination; the yellow component is absorbed too slowly. pH 5.5, heating rate 1°C/min (data kindly provided by U. Strahm, Ciba SC, Basel).

In order to avoid unevenness when dyeing with a mixture of dyes, the absorption characteristics of the individual components must not be too dissimilar. Figure 15.7a depicts the dyeing behavior of a mixture of red, blue, and yellow dyes that are absorbed at the same rate and yield an even tone, while in example b the yellow component is absorbed more slowly. With such a mixture, it is impossible to guarantee even coloring.

The temperature is very important. The higher the temperature, the more the equilibrium shifts towards the liquor, but the faster diffusion occurs. If the temperature is raised over the glass transition temperature of the polymeric substrate, crystalline regions can be dyed as well as amorphous ones.

Fibers can accumulate a surface potential in water, which has an effect similar to that in electrostatically stabilized dispersions. *Cellulose*, for example, develops a negative surface potential on immersion in water, thus repelling negatively charged ions. For dyeing, salts such as sodium sulfate or sodium chloride must therefore be added to reduce the thickness of the electrical double layer and thus enable negatively charged sulfonate-containing dye molecules to penetrate it. The added salts also favor the aggregation of dye ions in and on the fiber (salting out).

Figure 15.8 shows fibers of modified cellulose dyed with reactive dyes for cotton. Depending on the way the dyeing was carried out, the yarns may be colored only at the perimeter (annular coloring) or completely.

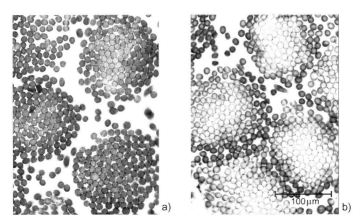

Figure 15.8. Dyeing of Lyocell fabric with reactive dyes for cotton: a) complete coloring; b) annular coloring (kindly provided by U. Strahm, Ciba SC, Basel).

Hydrophobic fibers cannot be dyed with water-soluble dyestuffs in an aqueous liquor without further treatment. More hydrophobic dyes are required for this task, and these are, of course, less soluble, like dispersion dyes. Hydrophobic substrates include polyester (PES), cellulose acetate (CA), and cellulose triacetate (CT). The rate of absorption can be controlled by means of *carriers*, or dye accelerators. Carriers are aromatic compounds such as *o*-phenol with an affinity for the fibers and a solubilizing effect on dispersion dyes. Alternatively, the fibers can be dyed in a high-temperature (HT) process (120–130°C).

Wool is a polypeptide, made up of various amino acids. Since it contains both basic (3.3% lysine, 9.8% arginine, 1.2% histidine) and acidic (6.8% aspartic acid, 14.5% glutamic acid) groups, it can be dyed by both cationic and anionic dyes, according to the pH of the dye liquor. However, cationic dyes for wool no longer have any economic significance, chiefly because of their unsatisfactory lightfastness. Diffusion of the dyes through the cuticular layer normally does not occur until the temperature reaches 70–90°C, unless the layer is mechanically damaged. Chemical change too, for instance chlorination, can alter the permeability. *Assistants* with an affinity for the dye form adducts with the dye in the liquor, retarding absorption and helping to level the dye deposition. More aid to even coloring is provided by assistants with an affinity for the fibers, which block sites on the wool fiber with particular affinity for the dye and are then gradually displaced by the dye.

Polyamide fibers, such as those made of polyamide 6 (polycaprolactam) and polyamide 66 (from hexamethylenediamine and adipic acid) are used principally in carpets and floor coverings. They can be dyed with anionic dyestuffs and, as such, behave much like the natural polyamide fibers of wool and silk. The dyes form salts with the basic amino end groups (in the $-NH_3^+$ form) of the polymer molecules, thus becoming bound. The ability of the fiber to absorb dye correlates with the number of ammonium groups present. In order to ensure dyeing is even, the dyeing process is initiated in a weakly basic pH range, in which only about 20–30% of the dye is absorbed by the fiber, and the pH is then lowered continuously to convert the amino groups into salt-forming ammonium groups.

Polyacrylonitrile (PAC) fibers resemble wool in their characteristics. However, it is difficult to dye the homopolymeric fibers; only when comonomers containing acidic groups were built into the polymer did it become possible to use cationic dyes, a breakthrough which permitted the use of PAC fibers in the textile sector. The saturation value is given by the concentration of acidic groups that form salts with the cationic dyes. The PAC fiber only interacts with dyes and assistants above its glass transition temperature of 70–80°C.

Dispersion dyes are favored for *PVC* fibers. The fibers have the advantage of being nonflammable and are used, for example, for seat covers in aircraft. The lightfastness of dyes on PVC is often worse than it is when they are applied to PES, and the color yield is also lower.

15.3.3 Textile Printing

To print onto textiles [13], a high concentration of dye is mixed into a medium that obstructs its diffusion, the thickener, and applied to the textile as a paste. The thickener must prevent capillary flow between the threads of the fabric. After a drying stage to avoid soiling of the guide rollers, the dye diffuses out of the dried paste into the fiber interior, where it is fixed by saturated or moist steam. As with dyeing from a liquor, an equilibrium is established between the dye remaining in the paste and the dye absorbed. The thickener and the remaining dye is then washed out.

The various printing methods, *direct*, *discharge*, and *resist printing*, differ in the mode of action of the auxiliaries and the dyes, but can all be performed by hand or automated.

In *direct printing*, the paste is applied locally in a sharply defined pattern. If the color of a dyed fabric is destroyed locally by the printing paste applied, the process is known as *discharge printing* (*white discharge* if there is no dye in the paste, *color discharge* if there is). In *resist printing*, local application of printing paste before the cloth is dyed in a bath prevents absorption of the dye in the covered areas. *Color resists* also contain a dye that diffuses into the fibers on steam treatment.

Printing pastes usually contain thickeners in concentrations of 2–10% [14]; these include:

Starch products, like torrefied starch, which is only soluble in hot water, or torrefied swellable starch, soluble in cold, can be used as the primary thickener in quantities of 30–40%. They have excellent resistance to alkali and are used, for example, as components in the mixture for printing acrylic fabrics to deepen the colors obtained.

Degraded *vegetable gum resins* are available as the cold-water-soluble products "crystalline gum" or "industrial gum"; their rheology is almost Newtonian and they have good film-forming properties, and are used for high-quality combination thickeners.

Methyl, ethyl, hydroxyethyl, and hydroxypropyl starch derivatives are widely used as stabilizers in emulsion pastes containing large amounts of gasoline.

The advantages of *alginates* (sodium, ammonium, and magnesium salts) are their good water-solubility and the speed at which they can be washed out after the dye has been fixed, even at high temperatures. Since the introduction of reactive dyes they have become very important as thickeners.

Carboxymethylated polysaccharides, such as carboxymethylcellulose, being good film-formers, and anionic ethers of high-grade guar flour are often used in mixed thickeners; guar derivatives are used alone to make up thickener solutions.

Xanthan gum, obtained from the controlled fermentation of glucose and having a relative molecular mass of around two million, has an unusual plastic rheology in aqueous solution. After the initial yield value has been exceeded, the viscosity declines far faster than it does for other polysaccharide thickeners. Xanthan is used in the printing of pile carpets.

Of the synthetic thickeners, only *polyacrylic acid* has a significant role nowadays.

Emulsions can also be used in thickening systems; they consist of emulsifiers, water, and heavy gasoline. When these emulsions are added to soaked and swollen thickeners, a so-called "*half-emulsion*" results.

Similar requirements apply to the dyes used for printing as to those used for dyeing from a liquor. However, there are additional criteria to be taken into account according to the printing method.

Substantive and acid dyes, both of which contain acid groups that favor solubility in water, differ in their affinities for different fibers. They are used together to print cellulose, for example, because of their mutually complementary shades. In direct printing, they are stirred into the paste with urea and solvents as additives, for instance ethylene glycol, thioethylene glycol, or glycerin. The optimum color yield and color fastness is only achieved after fixing with saturated steam for 30–60 min. For the best fastness, further treatment with quaternary polyammonium compounds is necessary in the subsequent washing stage to improve fastness to wetting. For discharge printing, dyes from the group of direct dyes, easily reductively cleaved at the azo bond, are the obvious choice. Sodium hydroxymethanesulfinate is used as the reducing agent, alone or together with discharge printing auxiliaries like anthraquinone (catalyst for reduction) or quaternary ammonium compounds.

Vat dyes are amongst the most stable types of dyes. The dye is applied in a paste containing the pigment, then reduced to the leuco form on the fabric by means of alkali and a reducing agent in an atmosphere of saturated steam, and can thus be absorbed by the fibers. The insoluble pigments are then recreated by oxidation. In the subsequent boiling stage, not only is pigment adhering to the surface removed, but the vat pigment is converted into the optimum crystal size and shape by recrystallization, so that the optimum fastness, brilliance, and tint are achieved. Vat dyes are seldom used as the ground color in discharge printing, but are added to the discharge paste. The same reducing agent that cleaves the ground dye (direct dye, insoluble azo dye, or reactive dye) into products that can be removed by washing is also responsible for the conversion of the vat dye into its leuco form in the paste.

Before *coupling dyes* are used, the fabric is coated with 2-hydroxy-3-naphthoic acid anilide (naphthol AS) or its derivatives. Alongside the thickener, the printing paste contains diazotizing solution, sodium acetate, and acetic acid. The most commonly used diazotizing agents are complex double salts of the diazonium salts, for example with zinc chloride, 1,5-naphthalenedisulfonic acid, or boron tetrafluoride (true salts).

Phthalocyanine developing dyes form in the fiber as insoluble pigments from an iso-indolenine ("phthalogen") and a heavy metal compound. These pigments are important in resist printing, as many fast dye colors can be used neither for discharge nor for resist printing. The resist-printed article is padded with the dye liquor containing the phthalogen, the metal salt, and ammonia (that is, the textile is soaked with the liquor and the excess is pressed out between rollers), then dried and steamed. In direct printing, the phthalogen, the metal salt, ammonia, and thickener (e.g. starch ether and alginate) are all present in the paste. The fixing step involves, for example, steaming at 100–102°C for 3–5 min.

Reactive dyes react with the hydroxy groups of fibers such as cellulose. Polymers containing OH groups (e.g. starch derivatives, locust bean gum derivatives, etc.) are therefore not suitable for use as thickeners; alginates are used instead. Alginates are also utilized as emulsifiers in emulsion pastes. As well as the reactive dye and the alginate thickener, the printing paste contains 10–20% urea to bind water and dissolve dye in the fibers, sodium hydrogencarbonate as an alkali for the reaction with the fiber, and a weak oxidizing agent, for example 3-nitrobenzenesulfonic acid, to prevent any reduction of the reactive dye. The color is fixed either by steaming or by dry heat. In discharge printing, the dye color is destroyed, as usual, by the reducing agent sodium hydroxymethanesulfinate; in resist printing the paste contains nonvolatile organic acids, which neutralize locally the alkali necessary for the reaction of the reactive dyes.

References for Chapter 15:

[1]   P. Bugnon, Prog. Org. Coat. *29*, 39 (1996).
[2]   W. Herbst, K. Hunger, Industrial Organic Pigments, VCH Verlagsgesellschaft mbH, Weinheim, 1983.
[3]   U. Kaluza, Physical/chemical fundamentals of pigment processing for paints and printing inks, Edition Lacke und Chemie Elvira Moeller GmbH, Filderstadt, 1981.
[4]   U. Biethan et al., Lacke und Lösemittel, Verlag Chemie, Weinheim, New York, 1979.
[5]   M. Schmitthenner, Farbe + Lack *104(5)*, 50 (1998).
[6]   W. Herbst, O. Hafner, Farbe + Lack *82*, 393 (1976).
[7]   A. Valet, T. Jung, M. Köhler, Farbe + Lack *104*(2), 42 (1998).
[8]   Karsten, Lackrohstofftabellen (various editions), Curt R. Vincentz Verlag, Hanover.
[9]   Colour Index, 3rd ed., Vol. 1–5, Additions and Amendments, Soc. of Dyers and Colorists, Bradford, and Amer. Assoc. of Textile Chemists and Colorists, Research Triangle Park, 1971.
[10] Technical Manual and Yearbook, Amer. Assoc. of Textile Chemists and Colorists, Research Triangle Park, annually since 1923.
[11] Ullmanns Encyklopädie der technischen Chemie, 4th ed., Verlag Chemie, Weinheim, 1976, Vol. 11, p. 135 (Farbstoffe, synthetische/Dyes, synthetic).
[12] Ullmanns Encyklopädie der technischen Chemie, 4th ed., Verlag Chemie, Weinheim, 1982, Vol. 22, p. 635 (Textilfärberei/Textile dyeing).
[13] Ullmanns Encyklopädie der technischen Chemie, 4th ed., Verlag Chemie, Weinheim, 1982, Vol. 22, p. 565 (Textildruck/Textile printing).
[14] H. Dahm, Bayer Farben Revue, "Sonderdruck Verdickungsmittel und Kleber", 6th ed., 1981.

Index